Theoretical Models of Chemical Bonding

Atomic Hypothesis and the Concept of Molecular Structure

Theoretical Models of Chemical Bonding

Part 1: Atomic Hypothesis and the Concept of Molecular Structure

Part 2: The Concept of the Chemical Bond

Part 3: Molecular Spectroscopy, Electronic Structure and Intramolecular Interactions

Part 4: Theoretical Treatment of Large Molecules and their Interactions

Theoretical Models of Chemical Bonding

Part 1:

Atomic Hypothesis and the Concept of Molecular Structure

Editor:
Prof. Dr. Zvonimir B. Maksić
Zagreb

AKADEMIE-VERLAG BERLIN

Die Originalausgabe „Theoretical Models of Chemical Bonding" (Part 1—4)
erscheint im Springer-Verlag
Berlin · Heidelberg · New York · London · Paris · Tokyo · Hongkong

Vertriebsrechte für die sozialistischen Länder:
Akademie-Verlag Berlin

Vertriebsrechte für alle Staaten mit Ausnahme der sozialistischen Länder:
Springer-Verlag Berlin · Heidelberg · New York · London · Paris · Tokyo · Hongkong

ISBN 3-540-51578-X Springer-Verlag Berlin · Heidelberg · New York

Maksić: Theoretical Models of Chemical Bonding (Part 1—4)
ISBN 3-05-500807-3 Akademie-Verlag Berlin
Part 1 ISBN 3-05-500808-1

Erschienen im Akademie-Verlag Berlin, Leipziger Straße 3—4, Berlin, DDR-1086
Printed in the German Democratic Republic
Gesamtherstellung: Druckhaus „Thomas Müntzer" GmbH, DDR-5820 Bad Langensalza
LSV 1215
Bestellnummer: 764 115 5 (9296/1)

To the memory of my parents
Olivera and Branko Maksić

Series Preface

"Imagination and shrewd guesswork are powerful instruments for acquiring scientific knowledge . . ."

J. H. van't Hoff

The last decades have witnessed a rapid growth of quantum chemistry and a tremendous increase in the number of very accurate ab initio calculations of the electronic structure of molecules yielding results of admirable accuracy. This dramatic progress has opened a new stage in the quantum mechanical description of matter at the molecular level. In the first place, highly accurate results provide severe tests of the quantum mechanics. Secondly, modern quantitative computational ab initio methods can be synergetically combined with various experimental techniques thus enabling precise numerical characterization of molecular properties better than ever anticipated earlier. However, the role of theory is not exhausted in disclosing the fundamental laws of Nature and production of ever increasing sets of data of high accuracy. It has to provide additionally a means of systematization, recognition of regularities, and rationalization of the myriads of established facts avoiding in this way complete chaos. Additional problems are represented by molecular wavefunctions provided by the modern high-level computational quantum chemistry methods. They involve, in principle, all the information on molecular system, but they are so immensely complex that can not be immediately understood in simple and physically meaningful terms. Both of these aspects, categorization and interpretation, call for conceptual models which should be preferably pictorial, transparent, intuitively appealing and well-founded, being sometimes useful for semi-quantitative purposes. They should reduce *a posteriori* the overwhelming information embodied in intricate wavefunctions to manageable and memorable qualitative results and concepts. Physically sound models have to provide an interpretation of experimental and accurate ab initio results and serve as a tool in unearthing common roots between seemingly unrelated data. It is noteworthy that chemistry operating on the molecular level uses its own scales provided by characteristic, standard or gauge molecules. Hence, an adequate model should offer a rationale for the trends of changes of properties within families of related compounds. Last but not least, qualitative models should try to bridge a gap between rigorous quantum mechanics and the empirical concepts of the phenomenological chemistry, or to provide their reconciliation if possible.

It follows that models, both quantitative and qualitative, are inevitable ingredients of the scientific method representing a link in the cognitive chain between theory and experiment. A harmonious development of molecular sciences requires a rather uniform progress in experimental research and quantum, computational and interpretive, chemistry. For that purpose it is of utmost importance to establish a common language — a sort of scientific Esperanto — thus avoiding a scientific tower of Babel. Important words in the scientific vocabulary are qualitative concepts.

It is the aim of the present series of books "Theoretical Models of Chemical Bonding" to provide a conceptual basis of modern interpretive quantum chemistry. It will be shown that there are qualitative models which meet all the attributes mentioned above. A general theme of the series is the most important building block of the macroworld — a molecule and its properties. The problems addressed cover a wide spectrum of topics ranging from the basic postulates of classical chemistry to a fine description of subtle spectroscopic properties, electron correlation and relativistic effects, intramolecular and intermolecular interactions, chemical reactivity and biochemical activity etc. Chapters are produced by a number of leading experts in the fields and they reflect much of the current thinking. No attempt has been made to avoid an overlap between the related articles, because some overlapping is necessary if continuity and coherent presentation of the topics considered is desired. A general level of theory is intermediate and excessive mathematical formalism is avoided wherever permitted. In keeping with the general philosophy of the Series an emphasis has been laid on chemical phenomena and experimental facts and their interpretation in terms of qualitative models close to chemical intution. The simplest possible description of molecular properties is striven for, carefully avoiding the Scylla of complexity and the Charybdis of oversimplification. It was impossible to give a full account of quantitative models like the modern VB approach and high-performance MO procedures because of space limitations. Fortunately, they are well covered in other books and review articles.

The introductory book of the Series is devoted to the notion of atoms immersed in chemical environments and to the concept of the molecular structure. A necessity for modelling in chemistry and its striking features are expounded in the Prologue. In the second book, which has been prepared in parallel, the third most important postulate of classical chemistry is tackled: the phenomenon of covalent bonding and a wide variety of its manifestations.

It is common wisdom today that molecular structure is the central and most fruitful theme of modern chemistry, biology, and to some extent of medicine. Knowledge of molecular geometric structural parameters gives important clues for understanding electronic behaviour in molecules and it is a starting point for more detailed studies. It is, therefore, rather disappointing that it is extremely difficult, if not impossible, to derive the concept of molecular structure from the first principles of quantum mechanics (Chapter 1). It indicates at the same time that a deductive method is not always the best way for acquiring scientific information. Chapter 2 written by the late Professor O. E. Polansky shows that some molecular properties are influenced by topology alone. A brief discussion of the graph theoretical indices is given in the Prologue too.

Two subsequent Chapters deal with molecular symmetry which, revealing some of the most fundamental laws of Nature, gives a penetrating insight into the structure and properties of matter. Interestingly enough, there is a kinship between molecules and elementary particles. Symmetry has a prominent role in chemical and biochemical reactions to mention only chirality for example.

The following two chapters describe how good theory blended with experiment can provide valuable information about molecular architecture. It illustrates rather nicely how the approximate but shrewed guesswork of Born, Oppenheimer and Huang yielded a model of molecular structure which proved enormously useful in rationalizing spatial arrangements of atoms and chemical bonds in nearly rigid molecules. It should be strongly pointed out in this connection that clamped nuclei approximation actually makes quantum mechanics a suitable theory for chemistry. Chapter 6 dwells on the simple ball-and-elastic spring model of molecules which underlies the molecular mechanics methods. Various applications to stereo-chemistry are thoroughly discussed.

The last two chapters elaborate on another pivotal topic: a concept of atoms embedded in intramolecular potentials. It appears, namely, that molecules have a "good memory" and "remember" atoms which took part in their formation. In other words, atoms are not scrambled by chemical bonding and retain some individuality within molecular systems. Hence, the modified atoms are natural constituents and basic units of molecules, which is by no means obvious at first glance starting from the Schrödinger equation. Bader's elegant "zero-flux" theory of modified atoms is discussed in Chapter 7. It turns out that a number of empirical notions of classical chemistry can be translated into quantum mechanical language by using the rigorously defined concept of topological atoms. The concluding chapter describes modified atoms in molecules at the qualitative model level illustrating a power and limitations of the conceptual approach. Naturally, the last two chapters could have been placed at the very beginning of this volume, for it is impossible to discuss molecular structure without adopting the atomic hypothesis. Hoewever, they both lean heavily on the use of the clamped nuclei approximation and the molecular geometry concept. Therefore, the latter had to be discussed first.

The present Series will hopefully strengthen interplay and increase creative interaction between the rigorous quantum theory, conceptual modelling of the electronic structure of molecules, and chemical phenomenology. Rapid concerted progress could be expected by increasing the overlap between these areas, because a fine balance between the very detailed results and development of unifying ideas is its *conditio sine qua non.* If this goal is at least partially fulfilled, our efforts will be greatly rewarded. It is also hoped that the interested reader will share some of the fascination and enthusiasm which played a great part in the preparation of these volumes.

Finally, I would like to express my sincere thanks to all the authors for their fine contributions and support which made this endevour possible. Special thanks go to Professors E. Clementi, T. Cvitaš, R. Gleiter, W. Kutzelnigg, W. H. E. Schwarz, J. Tomasi and M. C. Zerner for fruitful discussions and useful suggestions. A good deal of editing was done during my multiple stays at the Organic Chemistry Institute of the University of Heidelberg and financial support from the Alexander von Humboldt-Foundation is greatefully acknowledged.

Z. B. Maksić

Table of Contents

Prologue

If you know a thing, it is simple,
If it is not simple, you don't know it.

Oriental proverb

Modelling — A Search for Simplicity

Z. B. Maksić

Ruđer Bošković Institute, POB 1016, 41001 Zagreb, Croatia, Yugoslavia and Faculty of Natural Sciences and Mathematics, University of Zagreb, Marulićev trg 19, 41000 Zagreb, Croatia, Yugoslavia

1 Introduction

Modelling is an important ingredient of the scientific method. It is present in all sciences including humanities as an indispensable tool for tackling otherwise intractable complex problems and phenomena. We shall briefly comment here on modelling in chemistry with particular emphasis on theoretical models. It should be noted in passing that the present reflections are general and that they might be of some relevance in other natural sciences too.

There is no precise and unique definition of a model and consequently different people have different opinion about its notion, meaning and relevance, which is somtimes a cause for serious misconceptions and heated disussion. The reason is that models vary considerably in their nature and serve quite different purposes. Another reason is that the art of modelling is not always mastered leading to artificial results and consequently to a low reputation of models in general. To put it succinetly, the definition, classification and taxonomy of models is subjective and perhaps they should be modelled themselves. Hence, what follows is necessarily a personal view. The interested reader may also consult several scholarly written reviews [1–6] and books [7–9] or special subject issues of some journals [10–12]. Needless to say, he is also encouraged to read the contributions collected in this series of book written by experts in the field, who discuss in depth some specific models and their applications in interpreting molecular phenomena.

A salient feature of all models that they represent an approximate description of the corresponding complex systems — the so called objects, originals or prototypes. In modelling we sacrify perfect truthfulness because the object is distorted by simplification. This is, however, more a strength than a weakness because the model gains greatly in transparency and conceptual clarity. It deliberately neglects the less relevant details and emphasizes the most important facets. Therefore, a metaphor of the known physicist Y. I. Frenkel that a good model is a good caricature is well taken. It should always be kept in mind that a model is different, and sometimes quite different, from its object in very many ways. It is also useful to remember that a reliable model exhibits the essence of the property or a phenomenon under study, since attention is focused only on selected and dominant features (effects). They are manageable substitutes for their intractable prototypes.

Models can be broadly classified as: (a) *iconic*, (b) *analog*, and (c) *abstract*. (a) *Iconic models* are usually material and differ from the original in scale i.e. in size. For instance, a globe is an iconic model of the planet Earth. Ball-and-stick models of molecules represent a grossly enlarged picture of the "frozen" originals. In particular, Corey-Pauling-Koltun (CPK) molecular models involving coloured interlocking balls that give a close approximation of bond lengths and angles were of paramount importance in revealing complex structures of large biological molecules. A fact that the ball-and-stick model was an essential tool in one of the greatest discoveries of this century — the double helix structure of DNA — is remarkable indeed. Sometimes the iconic model has reduced dimension. The 3D electron density distribution in atoms and molecules can be represented by 2D pictures. (b) *Analog models* can be quite dissimilar to their prototypes, but they mimic a function, mechanism or a process. Analog models share a similarity in action. An example is provided by the ball-and-spring model of molecules which is partly iconic but at the same time analog because is simulates molecular vibrations. It gives a basis for the molecular mechanics method which is quite successful in yielding useful data on structure, vibrational spectra and thermodynamic properties within its field of validity [13]. (c) *Abstract, symbolic or conceptual* models give a description of complex situations and systems which usually lack any resemblance in form to the examined object. They are mental constructions which may range from the simply descriptive and sometimes pictorial models to rigorously defined expressions or equations. The former is exemplified by a picture of atomic orbitals and their squares which give an idea about the probability density distribution of an electron. An algebraic expression pV = nRT holding for an "ideal" gas, where p, V and T denote the pressure, the volume, and the temperature, respectively, is a very approximate description of the behaviour of gases at low pressure. Abstract models serve as an intellectual eye opener. A language itself is a highly abstract model. Theoretical chemistry of course offers a number of abstract models which will be considered at a later stage.

2 The Necessity of Modelling in Chemistry

After these general remarks we can focus our attention on chemistry — a part of natural science which systematically examines the properties of substances and their changes and transformations at the molecular level*. One can, however, argue that molecules do not exist because they are quantum objects. Quantum mechanics in turn is a holistic theory meaning that coupled systems form an inseparable unit. The state of the whole system cannot be reconstructed from the states of the partial systems or in other words: in macroscopic bulk matter consisting of many molecules described by a global state a single molecule can no longer exist in a pure state. Concomitantly, the usual notion of a "molecule" disappears [14]. Hence, a concept of a molecule in a strict sense is a model itself. Nobody can deny, however, that the molecule is a well-founded concept which is an elementary and essential building block of the magnificent edifice of molecular physics and chemistry. It is common wisdom that the behaviour of macroscopic substances can be reduced to the structure, properties and interactions of relatively independent subunits called molecules. Their collective effects on experimental devices explains measured data and in this connection they are real as long as a better interpretation at the microscopic level is not available. We shall turn back to this point in discussing the meaning of models.

Chemical reactions take place in the liquid, gas, and solid phases. In all cases, myriads of molecules are involved. Description of such complex systems necessarily invokes modelling. Furthermore, one is tempted to say that (bio)chemistry is and always will be, in ultima linea, an empirical or at least semiempirical science (see later, however, an alternative view pertaining to the Clementi ab initio computational model of chemistry).

* In view of Bader's definition of "zero-flux" modified atoms in chemical environments, this level can be further reduced to the atomic level

Hence we shall briefly consider the phenomenological models, but some preliminaries are appropriate first.

A model description in chemistry (and physics) includes, in general, the identification of a complex system to be studied and the specification of its environment if required. Then the problem (question to be asked) is defined. Next, the fundamental subunits, whose internal structure is ignored, are identified within the system. Then appropriate features are ascribed to these fundamental subunits and their mutual interactions are determined. Finally, properties of the system are estimated and interpreted in terms of subunits and interactions. Dependence of properties of the system in question is frequently given as a function of parameters of the model. The latter absorb properties of subunits and their interactions. Now we can discuss some specific examples.

2.1 Empirical Models

The most important model of classical chemistry is the notion of molecular composition and structure, that is to say, a molecule consists of atoms which exhibit certain valencies and have a definite spatial arrangement within the molecular domain. Hence the most prominent empirical model is the structural formula. It is a pictorial (iconic), partly analog and has a highly abstract content, illustrating that the taxonomy of models is always somewhat abitrary. The structural formula implies namely the clamped nuclei (or Born-Oppenheimer) approximation, a particular bonding pattern etc. The fundamental subunits are atoms and their interactions could be loosely defined in classical terms as valency forces. The structural formula has a surprisingly high information content based on experience and analogy. Theoretical discussion of the basic postulates of classical structural theory of molecules is presented in vols. 1 and 2 of this series.

Another outstanding classical model is the Mendeleev periodic system of elements. It is an empirical finding which is purely symbolic and abstract. The subunits are atoms but explicit dependence and periodicity of their properties is difficult to express in a mathematical form. In spite of its qualitative nature, the periodic system had a predictive power and anticipated existence and properties of several missing elements which were subsequently discovered. It is common knowledge that periodicity of atomic properties and the Mendeleev system itself can be interpreted by the shell model of the quantum electronic structure theory of atoms.

Empirical models play an extraordinary important role in everyday chemistry. They ensure economical thinking and classify similar compounds into families exhibiting characteristic properties (albeit with some variation). Chemistry with its myriads of molecules would otherwise be governed by chaos. The concept of the functional group will serve an illustrative purpose here. Organic chemistry in particular is structured around the empirical recognition that compounds containing the same reactive group of atoms (or a fragment), e.g. R-NH_2, R-COOH etc., where R stands for the rest of a molecule, have similar behaviour and closely related properties [7]. Gould writes: "One of the most important corner stones in the framework of contemporary chemical thought is the knowledge that reactions of compounds are largely determined by their functional groups" [15]. Indeed, the model of the functional group enables a vast number of individual reactions to be categorized thus serving as a mnemotechnic device and as an aid in their rationalization.

As a special case we mention briefly Pearson's model of chemical reactivity based on an idea of hard and soft acids and bases (HSAB) [16]. The functional groups and other reacting species are divided into two classes: acids and bases. Each class is then dissected into two groups consisting of hard and soft members. It is stated that a hard acid reacts more readily with a hard base in agreement with experience. An analogous conclusion holds for soft acid and bases. The HSAB principle was originally a completely empirical concept. Its theoretical rationalization within the frame of the density functional formalism is presented in vol. 2 of this series [17].

2.2 Theoretical Models

It is pertinent to start here with a widely quoted (and sometimes misquoted) classical statement of Dirac in 1929 on quantum mechanics [18]: ... "in the consideration of atomic and molecular structure and ordinary chemical reactions it is, indeed, sufficiently accurate if one neglects the relativity variation of mass with velocity and assumes only Coulomb forces between various electrons and atomic nuclei. The underlying physical laws necessary for the mathematical theory of a large part of physics and the whole of chemistry are thus completely known, and the difficulty is only that the exact application of these laws leads to equations much too complicated to be soluble. It therefore becomes desirable that approximate practical methods of applying quantum mechanics should be developed, which can lead to an explanation of the main features of complex atomic systems without too much computation." Of course we know today that the relativistic effects are of importance in molecules involving heavy atoms [19], but the rest of the statement holds completely. Quantum mechanics coupled perhaps with some statistical methods (e.g.

Monte Carlo [20]) suffices for chemical purposes. Needless to say, semi-classical approaches are extremely important in modelling [4, 20] as we shall see in what follows. It is also true that computational intricacies and obstacles in treating complex atomic and molecular systems involving a large number of particles were so pronounced that many brilliant theoreticians turned in despair to other problems of theoretical physics. Others like L. Pauling, J. H. Van Vleck, R. S. Mulliken, and C. A. Coulson concentrated on development of simple conceptual models of chemical bonding. The latter had a profound effect on the chemical thinking in spite of their qualitative nature (or perhaps because of that). Thus Dirac's prophecy about the importance of qualitative models quickly came true. The development of quantitative models, in contrast, took more than 4 years. Namely, only the tremendous progress in computer technology and numerical techniques in the last two decades has dramatically changed the situation. It has led to theoretical methods possessing precision which is comparable to the experimental accuracy in small molecules. Spectacular success of computational procedures announced the so called "Third age in quantum chemistry", which is characterized by a combined use of theoretical and experimental methods thus opening new vistas and possibilities for investigating molecular systems [21]. In particular, accurate ab initio methods are the only means of studying molecules not easily amenable to experiment — short lived species, interstellar molecules etc. There is still, however, a persisting problem of interpretation of data provided by rigorous ab initio theory and/or experiments. Wavefunctions consisting of 10 or 100 thousand of CI determinants employing large basis sets are not easy to analyze in physically meaningful terms. Simple models are evidently badly needed for that purpose.

Therefore one of the main tasks of conceptual theoretical models is to extract chemically relevant information from highly sophisticated wavefunctions. Equally important is the rationalization of experimentally established trends of changes of molecular properties in families of related molecules. Finally, it is impossible to overestimate the significance of qualitative models in building bridges between rigorous quantum mechanics and intuitive notions of phenomenological chemistry.

There is a wide spectrum of models in quantum chemistry varying in their origin, aims and degree of complexity. One can classify them by using different criteria. For instance, according to the level of sophistication and performance one can distinguish three categories: (a) quantitative; (b) semi-quantitative and (c) qualitative models. It should be stressed, however, that their boundaries are not sharp. As an example of the quantitative model we mention the Hartree-Fock (HF) + Møller-Plesset (MP) procedure of Pople et al. [5]. It employs HF wavefunctions as a convenient starting point which are supplemented by the pertubation theory as originally proposed by Møller and Plesset [22]. This approach provides well-defined results for energies of electronic states for any arrangement of fixed nuclei leading to reliable ground state geometries and a set of continuous potential surfaces. It has the desired size-consistency property meaning that an application to an ensemble of isolated molecules gives additive results. Finally, the calculated electronic energies for the ground states are upper bounds to the exact solution of the Schrödinger equation (i.e. the procedure is variational). Performance of this model is remarkable and well documented [23], but limited to relatively small molecules.

Another important ab initio computational model for chemistry was put forward by Clementi [20]. This is a global approach to simulation of complex chemical systems which can mimic liquid water structure, hydration networks in a crystal, water and ion structures for DNA and transport of ions through membranes. It is also capable of predicting the structure of proteins. The strategy of computations is of some interest. In the first step HF calculations are performed on individual molecules and the correlation energy is taken into account to some extent. In the second phase intermolecular interactions are treated by atom-atom potentials carefully calibrated vs some accurate ab initio results. Three and higher many-body effects are added if necessary. Finally, a transition from physics to chemistry is made by statistical Monte Carlo and/or molecular dynamics methods. It is important to note that nontrivial chemical problems are attacked completely free of empirical parametrization.

In addition, the chemically relevant (electrostatic) model of Tomasi [4] is worth mentioning. It is a useful vehicle in exploring solvent effects. To summarize, the models of Clementi and Tomasi represent an important step toward computer chemistry and biochemistry.

(b) The class of semi-quantitative models is reserved in principle for semi-empirical methods. It is rather dissapointing and unsatifactory that they leave so much to be desired [3]. Broadly speaking they can be split into two families: (I) Those which employ some kind of neglect of diatomic differential overlap and (II) schemes which fully appreciate overlapping of AOs. The first group of semiempirical methods is obtained by approximate solution of HF-SCF equations treating inner-shell electrons as inactive, impenetrable and unpolarizable cores. Hence the units of the model are cores and valence electrons. Their Coulomb interactions are given in approximate or empirically parametrized form by fitting some experimental observables like heats of formation, atomic and molecular spectral data and the like. The number of these adjustable parameters is very large as a rule and one cannot help a feeling that many of semiempirical methods are stretched on the bed of Procrustes. Some of them do not satisfy even minimal requirements of quality of the produced wavefunctions which can be traced down to the use of inadequate basis sets [24]. The choice of the latter is crucial and should be compatible with the set of approximations employed [25, 26].

In the group (II) molecular wavefunctions are generated by solving the secular equation employing an effective hamiltonian. A representative example is provided by the Iterative Extended Hückel Model (IEHT), which incidentally gives charge distribution and one-electron properties substantially more accurate than NDDO schemes like CNDO, MINDO and MNDO. It is remarkable that an extremely simple maximum overlap hybridization model has a surprisingly good performance within the limits of applicability. Obviously, orbital overlapping is of utmost importance.

Semiempirical methods are useful if used with due care and caution. They should always represent a good compromise between the true and simple, but Trindle [3] is probably right in stating that many of them are not true enough to excuse their lack of simplicity.

(c) Qualitative quantum chemistry models are designed to answer the question what is going on and are relatively little concerned with the quantitative aspect of bonding. The simplest concepts are perhaps hybridization [12] and atomic charges (monopoles) which have a surprisingly high information content and even some semblance to the truth.

Modern Valence Bond (VB) and contemporary ab initio Molecular Orbital (MO) methods have reached the stage of powerful quantitative models. Here we are interested in their original qualitative form as put forward by the pioneers of quantum chemistry: Heitler, London, Pauling, Slater, Hund, and Mulliken. Application of VB and MO methods to the H_2 molecule deserves some more comments. It is not only a homage to the founding fathers but serves also as an illustration of their conceptual difference and complementarity. In the Heitler-London treatment of the H_2 molecule, two hydrogen atoms retain their identity. The wavefunction is constructed in such a way that the two electrons can never be found simultaneously on the same atom. Their motion is correlated. An alternative concept forms the basis of the MO model. Neglecting the existence of atoms at the outset, the system is treated as an ensemble of two protons and two electrons. Amusingly enough the proscribed notion of atoms in a molecule immediately entered into the model via the LCAO approximation. The resulting MO, however, embraces the whole of the molecule being populated by two electrons of opposite spins implying that they are totally uncorrelated. Therefore, it is not surprising that at this elementary level the VB function is superior. Concomitant with their conceptual difference VB and MO models describe different aspects of electronic properties of molecules. MOs being delocalized belong to irreducible representations of the point groups. Their orbital energies equal the ionization potentials via Koopman's approximation and provide in general a useful tool in rationalizing molecular spectra. It provides qualitative rules for some aspects of molecular structure (Walsh) and reactivity (Woodward-Hoffmann). VB wavefunctions are better suited for description of local bond properties. Local symmetry is built in by hybrid AOs and the whole picture is close to the structural formula and chemical idea (Lewis) that covalent bonds are realized by localized pairs of electrons. It follows that elementary VB and MO wavefunctions are two different but complementary models of the same phenomenon. This is just one more illustrative example of a dichotomy which is so characteristic for the quantum microworld, or as N. Bohr put it in a seemingly paradoxical lapidary sentence: "*Contraria sunt complementa*" in the invisible wonderland of molecules, atoms and subatomic particles. A well-known fact that VB and MO models are just different starting points and that their consecutive refinements would eventually lead to the same wavefunction reveals the very nature of modelling. For modelling the choice of the starting point is decisive. It should be selected so that the studied feature is adequately described in the simplest and most economic way.

It should be mentioned once again that each classification is arbitrary and the one presented above is not an exception. Namely, essentially qualitative models may give semi-quantitative information for some specific properties as exemplified by the Independent Atom Model (IAM) [27].

Other classification schemes are of course possible as well. One can divide models into theoretically analytical (or deductive) and inductively synthetical depending on the way they are constructed. The former are obtained by a simplified solution of the correct Schrödinger equation of the problem under study. On the other hand the latter starts from the intuitively simple physical idea which is subsequently refined with more subtle details. Hybridization — one of the brilliant inductive models of Pauling — will serve an illustrative purpose. Consider for example the hybridization in the HFC = CClH molecule. The ideal sp^2 hybridization will give an essentially correct bur very crude picture. Then one can take into account that all bonds are different and that bond angles deviate from 120°. Introduction of non-integer hybridization will do more justice here. Then one can generalize the concept further by optimizing the non-linear parameters ζ of hybrids for each direction separately, introduce more a flexible basis set and perhaps relax slightly the orthogonality condition. Finally, one can account for the ionic character by involving Coulson-Fisher two-center orbitals [28]. An alternative deductive way is provided by starting from the HF-MO wavefunction which is then localized by the appropriate unitary transformation [29]. Localized molecular orbitals can be further broken down to bonding HAOs. It is also possible to obtain directly, e.g. natural hybrid orbitals by analyzing the first order density matrix [30].

It is noteworthy that there are other theoretical approaches which are close to modelling but the resulting constructions are not models in a strict sense of the word although they are called that way. We shall mention two examples of some interest. The first is provided by hypotheses. In general they are tentative explanations of phenomena which might be true but this is by no means certain. Very old examples are given by models of the universe by Ptolemy and that of Copernicus, which was subsequently refined by Kepler. The former gave admirable agreement with observations. It might have been true but was not. On the contrary, the latter hypothesis proved essentially correct. Another important hypothesis was the 19th century idea of atoms (Dalton, Boltzmann), which was fiercely disputed by contemporaries. Their existence was finally and unequivocally proved in this century.

Mathematical fitting of data without the underlying physical picture is not a model because of the lack of interpretive ability. There is an abundance of graph-theoretical indices which are purely mathematical constructions and yet they are correlated by some researchers with various molecular properties implicitly suggesting a deeper meaning of these indices than is justified. In our opinion they belong to the realm of hypotheses. They might be true, but they are not. In graph theory as applied to chemical bonding the level of abstraction is exaggerated so that the whole physics disappeared. One should just recall that hundreds of diatomics have the same molecular graph — two vertices connected by an edge. The same holds for larger molecules too. Generally speaking, ad hoc assumptions that certain mathematical structures should have a semblance to molecular properties is arbitrary. This could have been an excellent theoretical framework in the time of Couper and Kekulé in the last century, but today it is obvious to almost everyone that the notion of orbitals, their overlapping, their charge populations and hybridization are pivotal. They represent irreducible subunits of models in quantum chemistry. This conjecture does not imply that graph theory is completely outdated. It has been shown that graph-theory is equivalent to the Hückel MO model in its simplest form.

There is something which is called a topological effect in molecular orbitals, topological charge stabilization etc*. [31]. However, one has to be very careful to bypass the traps lying on the path of the application of graph theory to chemical bonding problems, because as Lord Kelvin nicely put it: "It is as dangerous to let mathematics take charge of physics (and chemistry!) as to let an army run a government."

3 The Art of Modelling

It is clear that the design of quantitative models requires high mathematical skill and computer expertise. On the other hand qualitative models necessitate a deep understanding of chemical phenomena and intuition. Since we are interested here in interpretive aspects of quantum chemistry, some general characteristics of successful qualitative and semiquantitative models are worth mentioning. They are:

(a) Self-consistency
A model should be non-contradictory and logically consistent. It must obey the fundamental physical laws and basic principles. In particular, the conservation laws (symmetry) known to apply to the prototype should hold for the model.

(b) Simplicity
A good model must be simple and economical. It should neglect all unnecessary details and ad hoc asumptions. Obviously, a proper model is attained only by a delicate balance between simplicity and adequacy. The simplest possible description of the studied property which gives reasonably close estimates of the true values are the best in the sense of the Occam razor principle. It can be also said that these models reached Pauling's point of perfection regarding simplicity. Once the dominant factors are identified they may be empirically adjusted through parameters to increase the performance of the model. At this stage extreme care has to be exercised because of the overparametrization threat. There is a constant danger that the right answer might be obtained for the wrong reasons. The geometric system of Ptolemy is a warning which conclusively shows that a wrong model could be in perfect accordance with (experimental) observations.

A distinct advantage of models is that sometimes a small effect can be isolated and, *ceteris paribus*, parameterized to

---

* Effect is probably not the right word. Topological component or influence would serve the purpose better.

yield very useful results which in turn can be obtained by rigorous computational methods only with extreme difficulty and/or cost.

Simple models are easy to memorize which is an additional advantage.

(c) Stability and Flexibility

A model should be stable to small errors. A good model should be apt to inductive generalization thus producing models of increasing complexity yielding more and more accurate descriptions of the object. The Hückel π-MO scheme for hydrocarbons provides an illuminative example. It can be generalized by including overlap integrals and by recognizing heteroatoms. The next decisive step is the inclusion of all valence electrons leading to the Extended Hückel Theory (EHT). This approach gives plenty of good qualitative results but grossly overestimates the intramolecular charge drift. This situation is remedied by the inclusion of explicit dependence of the diagonal matrix elements (of the effective hamiltonian) on the charge density. Additional refinements of the off-diagonal elements yield the Iterative Extended Hückel scheme (IEHT), where charge distribution is obtained in iterative cycles until consistency is achieved. Finally, relativistic effects may also be introduced [32, 33]. This kind of gradual and controlled building of more sophisticated models should hopefully approach the accurate solutions in an asymptotic fashion.

(d) Utility

A model should provide quick and inexpensive information on large molecular systems of chemical interest. It should possess predictive power and serve as guidance in executing accurate ab initio calculations and/or experimental measurements by suggesting target experiments and crown computational cases. It should provide a diagnostic tool yielding a crude test for the rigorous computations and observed results (by estimating at least an order of magnitude of the studied entity). In addition to predictive power it should have high explanatory power. It must — above all — offer interpretation of properties and phenomena under study and rationalization of trends of changes along a series of related compounds. A good model should be capable of revealing hidden similarity between widely different objects by providing a unifying principle. By the same token it should explain empirical relations between various observables. Hybridization is a good illustrative example indicating at the same time that it is a hidden phenomenon itself [28]. Visuality is a distinct advantage of models. They should provide a picture of a forest neglecting individual trees and leaves (details).

(e) Ability to fail

This rather jovial requirement is essential. A model that is never wrong would represent the truth itself. Flaws are inevitable and important. It is of major significance to establish limits of applicability of a model. The latter should be used only within its range of validity. For example, a point charge model does a good job in alkali halides but it is rather irrelevant in hydrocarbons. In contrast, hybridization plays a minor role in alkali halides. Some models have mastered the failing ability to perfection.

It is now perfectly clear that good model-building is a complicated and demanding scientific procedure, where enthusiasm should never be ahead of sound and sober reasoning. One can call model-building an art because it involves a lot of imagination, but it should be always kept in mind that proper modelling is a scientific method with all the attributes of the latter. If a comparison is necessary then the diamond cutting skill is perhaps appropriate.

4 Discussion and Conclusion

As pointed out earlier, modern ab initio computational quantum chemistry gives for the selected classes of molecules and for interaction phenomena a very good description at a level of accuracy comparable with that of sophisticated experimental measurements. The value of these quantitative models is undeniable because they answer the question how much of something is happening. It is inevitable, however, that so intricate algorithms and unwieldy wavefunctions blur the picture and that the essential features are difficult to recognize. Very sophisticated wavefunctions usually give us much more than we need to know. Here qualitative models enter stage answering the question what is going on. Analogously, simple pictorial models are useful in deciphering coded messages of Nature provided by experiments. Conceptual models are thus a necessary link in the cognitive chain, just like quantitative models, theories and experiments. We would like to suggest a square with four equivalent vertices (Fig. 1) as a schematic representation of the scientific method. The square is rooted in observations and experiments as it should be. A natural transmission between experiments and theories placed on the top is provided along the edges via qualitative and quantitative models. However, all routes of creative interactions are possible and opened as indicated by arrows. Of course, one has to solve also the implicit semantic problems. Whereas the meaning of

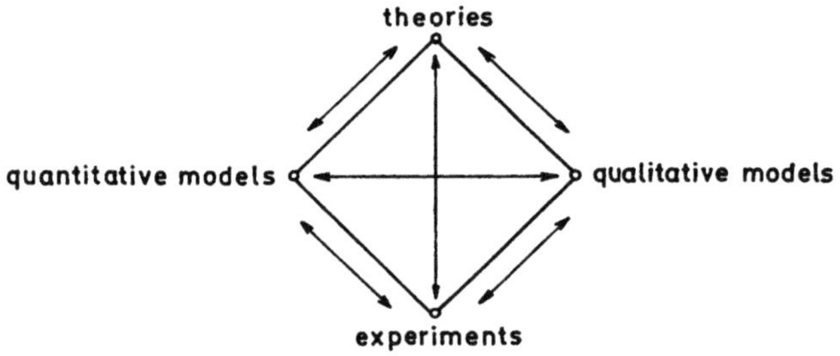

Fig. 1. Schematic illustration of the scientific method

experiments and models thoroughly discussed here are clear, the notion of theory seems to be vague. For example, Extended Hückel Theory (EHT) is not a theory at all, but a simple qualitative model. Chemical Graph Theory is not a chemical theory, as the name implies, but a mathematical discipline applied to some chemical problems. Theory, succintly embodied in a set of axioms, provides a logical abstract framework for the whole or a large part of a particular natural science (since we consider here only the latter). Relations between experiments, models and theories presented in Fig. 1 are highly schematic, but they give an insight into the research in natural sciences thus being a model itself.

Having established models as inevitable components of the gnoseological process an obvious question arises: what is the relationship between the models in general and reality? In our opinion there are many misconceptions in this respect. A widespread fallacy is that models cannot be right or wrong implying that they are to be used but not trusted. In other words, models are by definition never correct or incorrect [14] and their possible utility is determined solely by agreement with observations. This is false statement. A reliable model has a grain of truth. It is true within the limits of the approximation(s) involved and within carefully determined limits of applicability — no more but at the same time no less. In this sense molecules are real, although they are inseparable subunits of bulk matter (*vide supra*).

It is a game of fate that models are sometimes poorly understood. The very famous mathematician N. Wiener argued that the best model of a cat is another cat or, better, the cat itself. The last part of the statement is in direct contradiction with the very definition of a model.

There is also a danger of overestimating the role of mathematics in quantitative models. This has a long and respectable tradition. I. Kant (1724—1804) said: "Ich behaupte, dass in jeder besonderen Naturlehre nur so viel Wissenschaft angetroffen werden könne, als darin Mathematik anzutreffen ist.". This was repeated later in some more words by Quetelet [34]: "The more progress physical (natural!) sciences make, the more they tend to enter the domain of mathematics, which is a kind of center to which they all converge. We may even judge the degree of perfection to which a science has arrived by the facility with which it may be submitted to calculation." Mathematical algorithms are essential parts of models in carrying out all their consequences, but their perfection does not replace the model itself. The value of the model lies in its physical content and chemical performance. One can say that mathematics is in natural sciences a good servant but a poor master. It is so comforting that one of the greatest physicists of all time A. Einstein did not care much about mathematics. Actually, he reportedly said that mathematics alone is the best way of self-deceiving.

Quantitative and qualitative modelling should be well balanced if concerted progress in science is desired. There is a tendency that more weight is given to quantitative aspects because of the omnipresent and psychologically addictive (super)computers. It should be stressed that many, and perhaps the most of great issues of science are qualitative and not quantitative. One should just recall that the most important relations in quantum mechanics — Heisenberg's relations of uncertainty — are not equations but inequalities instead. It is therefore amusing that some researches deny any value of models and yet they work with quantitative models even without knowing it. One should strongly point out that it is modelling which gives a characteristic flavour and authenticity to quantum mechanics of molecules alias quantum chemistry.

To summarize, well-established models provide a useful vocabulary, constitute the frame of mind and organize our patterns of thoughts. They lend themselves to classification purposes. Models give a pervasive physical insight and extract the key features of very complex phenomena thus revealing their essence, simplicity and beauty. They provide ways of *knowing things* as indicated in the motto. Metaphorically speaking, models extend the range of our senses. They make it possible to "see" mentally what cannot be seen.

References

1. Pierls R (1980) Contemp. Phys. 21: 3
2. Del Re G (1974) Adv. Quant. Chem. 8: 95
3. Trindle C (1984) Croat. Chem. Acta 57: 1231
4. Tomasi J (1988) J. Mol. Structure (Theochem) 179: 273; Alagona G, Bonaccorsi R, Ghio C, Montagnani R, Tomasi J (1988) Pure and Appl. Chem. 60: 231

5. Pople JA, Binkley JS, Seeger R (1976) Int. J. Quant. Chem. Symp. No. 10: 1
6. Durand P, Malrieu JP (1987) Adv. Chem. Phys. 67: 155
7. Suckling CJ, Suckling KE, Suckling CW (1978) Chemistry Through Models, Cambridge University Press, Cambridge
8. Hammond GS, Osteryoung J, Crawford TH, Gray HB (1971) Models in Chemical Science —An Introduction to General Chemistry, W. A. Benjamin, Menlo Park, California
9. Maksić ZB (ed) (1987) Modelling of Structure and Properties of Molecules, Ellis Horwood, Chichester
10. Maksić ZB (ed) (1984) Conceptual Quantum Chemistry. Models and Applications, Vol. 1 and 2, Croat. Chem. Acta 57: 765, 1231
11. G. Del Re (ed) (1984) Proceedings of the 1983 Colloquium on Mechanisms of Elementary Physicochemical Processes, Int. J. Quant. Chem. 26: 563
12. Maksić ZB, Orville-Thomas WJ (1988) Six Decades of the Hybridization Model. A Tribute to Linus Pauling, J. Mol. Structure (Theochem) 169: 1
13. Meyer AI (1990) chapter 7 of this volume
14. Amann A, Gans W (1989) Angew. Chem. Int. Ed. Engl. 28: 268
15. Gould ES (1959) Mechanism and Structure in Organic Chemistry, Holt Reinhart and Winston
16. Pearson RG (1967) Chem. Brit. 3: 103
17. Pearson RG (1990) In: Maksić ZB (ed) Theoretical Models of Chemical Bonding, Vol. 2, Springer, Berlin Heidelberg New York, p 45
18. Dirac PAM (1929) Proc. Roy. Soc. A 123: 714
19. Schwarz WHE (1990) In: Maksić ZB (ed) Theoretical Models of Chemical Bonding, Vol. 2, Springer, Berlin Heidelberg New York, p 593
20. Clementi E (1985) J. Phys. Chem. 89: 4426 and the references cited therein
21. Richards G (1979) Nature 278: 507
22. Møller C, Plesset MS (1934) Phys. Rev. 46: 618
23. Hehre WJ, Radom L, Schleyer PvR, Pople JA (1986) Ab Initio Molecular Orbital Theory, Wiley, New York
24. Maksić ZB, Supek S (1988) Theoret. Chim. Acta 74: 275
25. Cook DB, Hollis PC, McWeeny R (1967) Mol. Phys. 13: 553
26. Cook DB (1978) Structures and Approximations for Electrons in Molecules, Ellis Horwood, Chicester
27. Maksić ZB (1990) chapter 10 of this book
28. Maksić ZB (1990) In: Maksić ZB (ed) Theoretical Models of chemical bonding, Vol. 2, Springer, Berlin Heidelberg, New York, p 137
29. Edmiston C (1990) In: Maksić ZB (ed) Theoretical Models of Chemical Bonding, Vol. 2, Springer, Berlin Heidelberg New York, p 257
30. Foster JP, Weinhold F (1990) J. Am. Chem. Soc. 102: 7211
31. Polansky OE (1990) chapter 2 of this book
32. Lohr Jr. LL, Pyykkö P (1979), Chem. Phys. Lett. 62: 333; Gleghorn JT, Hammond NDA (1984) Chem. Phys. Lett. 105: 621
33. Pyykkö P (1988) Chem. Rev. 88: 563
34. Quetelet A (1828) Instructions populaires sur le calcul des probabilité, Tarlier, Brussels

The Concept of Molecular Structure

Brian T. Sutcliffe

Department of Chemistry, University of York, York, Y01 5DD, England

After a brief discussion of the classical origins of the idea of molecular structure it is shown what steps are necessary to exhibit molecular structure in solutions of the Schrödinger equation for a system that might intuitively be believed to be an isolated molecule. It is demonstrated that there is no straightforward or independent way in which molecular structure, as exemplified by a definite molecular geometry and associated properties, can be derived from such solutions. It is argued that the idea of a molecular structure does not fit easily into any feasible computational scheme for obtaining accurate solutions and it is suggested that as calculations develop in complexity and sophistication it is likely that quantum chemists will abandon the idea of molecular structure even as a starting point. An attempt is made to pur these observations in the broader context of the current debate initiated by such workers as Woolley and Claverie.

1 Introduction

Few chemists would dissent from Ballhausen's remark, made in 1979, that "today we realise that the whole of chemistry is one huge manifestation of quantum phenomena" [1]. Although this view appears innocuous enough, chemists usually make an unarticulated assumption in assenting to it. They usually assume that quantum theory accounts for chemistry by accounting for its traditional theoretical structures, namely molecules. It is certainly the case that quantum theory claims the phenomena of chemistry as lying within its domain of validity but it is clear that (even assuming the claim to be valid) there is no *logical* necessity that the account of chemical phenomena given by quantum theory should be cast in terms of entities like molecules, whose provenance is determined in an altogether different theory.

However that may be, the view generally held by chemists is that the basic connection between traditional chemical theory and quantum theory was adumbrated by Lewis in 1911 in his electronic theory of valency (see, for example [2] or [3]). This adumbration was then clearly realised by Heitler and London in 1927 when they treated the electronic structure of the hydrogen molecule quantum mechanically. By this treatment, the molecule, which is undoubtedly the basic unit of standard chemical theory, was seen as being properly placed in quantum theory. All that has happened since then is seen as confirming that fact.

Since the late 1960's however, the perception that quantum theory did in fact contain the traditional theory has been subject to some fairly rigorous questioning. This has centred on the validity of the idea of molecular structure in a quantum theoretical context. Unsurprisingly this questioning has led to much bitter argument in the literature and it is the object of this article to expound, in as neutral a way as is possible, the background to the problem and to provide something of a guide to the literature on it.

Although the origin of this questioning lies within the chemical community and has been stimulated by experimental advances (particularly in high-resolution molecular spectroscopy) which seem to yield results difficult to explain in molecular terms, it has been taking place against the background of more general discussions about the meaning of quantum theory. In these circumstances it is perhaps not surprising that the absence of logical necessity referred to above, which for so long seemed an irrelevant banality to chemists, should seem to present a possible way out of difficulties.

Quantum theory has always presented interpretational difficulties and the very greatest minds in the field have taken up what seem to be strongly opposing positions on how the theory should be interpreted. The most celebrated of these interpretational conflicts arose between Einstein and Bohr and is presented in its most stark form by their attempts to understand the likely outcome of a thought-experiment proposed by Einstein, Podolsky and Rosen (EPR) in 1935. Since it was not possible actually to perform the EPR experiment it was not absolutely necessary for ordinary working quantum mechanicians to take up a position on it and most did not. However in 1964 JS Bell was able to re-formulate the theory relevant to the EPR experiment in such a way as to suggest how an actual experiment might be performed. Bell's result was in the form of an inequality. If this inequality was violated then it was extremely difficult to continue to hold Einstein's view. If Bohr's view prevailed, however, then quantum theory had to be interpreted in a very odd way. It had to be an holistic theory in which it was impermissible in principle to talk about isolated, localised objects. In 1981 Aspect was able to perform an experiment to test Bell's inequality and almost everyone agrees that the results show conclusively that Bohr was correct and that quantum theory is a very odd theory.

It was against this background that the questioning of the idea of molecular structure took place. It was natural therefore that from time to time features of the more general discussions got incorporated into the particular ones. This was particularly so because the Bohr interpretation seemed to strike at the heart of local realism and for chemists a molecule is nothing if not real and localised in space. It is not at all clear however that it would be helpful to consider those wider problematical issues in quantum theory in order simply to *understand* why there is a problem with molecular structure. It may, of course, be the case that such consideration will be necessary to *solve* the problem, but for the purpose of this article these more general considerations will be supressed as far as possible. Some idea of the general background can be found in Jammer's two volumes [4, 5]. More recent and more popular accounts of some aspects of the problem can be found in Gribbin [6] and Rae [7] and also in the articles by de Witt [8] and Mermin [9]. A serious and learned atempt to place the theory of molecular structure, indeed the whole of quantum chemistry, in the context of the general discussion can be found in the book of Primas [10]. A reader wishing an account of this book might care to consult the author's review of it [11] and the contrasting review by Jørgensen [12].

It seems appropriate to begin the exposition with a section that outlines informally, the development of the traditional idea of molecular structure up to its high point in the middle of the Nineteenth Century. It is then useful to try to see the idea from the point of view of Schrödinger's equation and its solutions. This discussion is inevitably somewhat technical and though every attempt will be made to avoid the more sordid details it does seem important to present sufficient mathematical results to enable the reader to see that what is involved is not just linguistic nonsense.

In the light of these two sections some attempt will be made to summarise what the problems really are. As for their resolution, that must remain a matter for the reader.

2 The Traditional View

It is generally believed that Chemistry began to go in the right direction with the publication by Robert Boyle in 1661 of his book "The Sceptical Chemist". In this book he enunciated the axiom that only what can be demonstrated to be the undecomposible constituents of bodies are to be regarded as elements. He also offered a distinction between chemical compounds and mixtures and characterised a compound as resulting from the combination of two constituents so as to form a third that has properties completely different from either of its constituents alone. Of course Boyle's theoretical position begs a number of questions but associated at it was, with a firm grasp of the need for careful experiment, it did provide the basis for an operational program that was pursued very fruitfully throughout Europe in the 18th Century. It was the carrying out of this program (and particularly in the work of Lavoisier) that led to the discovery of the law of constant combining proportions and later the law of multiple proportions in the context of which Dalton in 1808 proposed his atomic theory.

Dalton's theory was centered on the ideas that every element is made up of homogeneous indestructible atoms whose weight is constant and that chemical compounds are formed by the union of the atoms of different elements, in the simplest numerical proportions. It is difficult now to appreciate the great imaginative leap made by Dalton in asserting that compounds were to be understood in terms of combinations of the atoms of elements, for there were at that time no known physical forces that seemed capable of accounting for this combination. In making this assertion therefore Dalton in some ways began the separation of chemistry from physics. The idea that atoms had some latent power (first called atomicity and later valency) of combining with other atoms in a manner that seemed peculiar to chemical change, had no counterpart in the physicist's view of atoms.

The historical development of views of chemical combination through to the development of a theory of molecular structure in the 1860's is extremely complex with many events in it still a matter of dispute among scholars. It would therefore be inappropriate to go too deeply into these matters here. However, it does appear to be the case (see e.g. von Meyer [13]) that Laurent in 1846 was the first person to use the world molecule in a way that is recognisable today. Laurent supposed that an atom was the smallest quantity of an element which can be present in a compound. He further supposed that a molecule was the smallest quantity that can be employed in order to produce a compound. He understood the molecular weight of an element or compound as meaning the quantity which, under like conditions, occupies the same volume as two atoms of hydrogen; the latter quantity he looked upon as a molecule of hydrogen. He was thus able to arrive at what would now be considered correct empirical formulae such as Cl_2, O_2, N_2, HCl and so on, and correspondingly correct molecular weights.

The extent to which Laurent's ideas actually influenced chemical thinking is a matter of some uncertainty. It is usual to regard these ideas as becoming really influential only after Cannizzaro in 1857 reinterpreted in a convincing way Avogardo's work of 1811.

However that may be, by 1852, Frankland was clearly thinking along molecular structure lines for in a paper in that year he put forward the idea of the *saturation capacity* of an element, an idea that was central to the development of the notion of valency and hence of molecular structure.

Frankland's early work was on inorganic compounds and he seems to have understood that what would now be called variable valency was a possibility, but he observed that the variability was strictly limited. He said that "no matter what the character of the uniting atoms may be, the combining power of the attracting element, if I may be allowed the term, is always satisfied by the same number of atoms".

In later work on organo-metallic compounds Frankland recognised the tetravalency of carbon and in a little book "Lecture notes for chemical students" published in 1866 he developed the graphical formulae introduced by Crum Brown in 1861, identifying the lines joining the atom symbols as *bonds*. This seems to be the first use of this word. These developments led rapidly to the writing of structural formulae for molecules that were to all intents and purposes, by 1870, the modern ones.

Frankland's ideas (as developed by Kekulé in particular) led to the theory of structural isomerism. Eventually at the hands of Van't Hoff and Le Bel the structural formulae took on three-dimensional form to explain optical activity and what came to be called geometrical isomerism. Thus by the late 1870's the modern molecule had emerged as a chemical concept, *sui generis*.

If one reads the nineteenth century literature or looks at a history of chemistry (for example, Russell's modern book on the history of valency [14], or a more nearly contemporary history such as that of von Meyer [13] which was first published in 1888), it becomes very clear that there was much argument about whether molecules really existed "out there" or whether they should be regarded simply as theoretical constructs with no "objective" existence. Very distinguished names can be cited on both sides of the argument. Those who espoused the "really out there" view were seen as making it very difficult to relate chemistry to physics because there was no suitable physical law of force to describe bonding. Those who espoused the "purely theoretical" view were seen as actually attempting to separate physics from chemistry and certainly to many, this point of view seemed sterile and unsatisfying. The qualities of the dispute had much in common with the more contemporary dispute between those who adopt the so-called "Copenhagen interpretation" of quantum mechanics and those who adopt a more "realist" position.

However this may be, developments both in physics and in chemistry from about 1880 up to 1916 seemed to most people to settle the question firmly in favour of the reality of molecules as objects existing "out there". Among the experiments that seemed to compel the realist view were those of Perrin on Brownian motion and of the Braggs on the X-ray diffraction patterns of the alkali halides. The results of both these sets of work seemed to speak for the reality of atoms and hence, by implication, that of molecules. Less well remembered now perhaps, was the work of Bjerrum who in 1913 was able to get a somewhat imperfect but still very plausible interpretation of the infrared spectrum of CO_2 by treating the molecular model as a semi-rigid rotor.

Also during this period there were theoretical developments following the "discovery" or "invention" (the term used depended on ones philosophical position) by Thomson of the electron. By 1904 Thomson himself had developed a rather primitive theory of molecular structure involving electrons and Lewis had begun work (not to be published until 1916) on his theory. Lewis' theory, which was to become known as the electronic theory of valency, is still widely taught to chemists and so need not be considered in detail here. The history of its development can be found in Lewis' monograph [2] and an account of it at its high point can be found in Sidgwick's 1927 book [3].

The theory was much admired by chemists with a theoretical turn of mind for it seemed to account for molecular structure in a way that provided a potential link to physics. Suitable pairs of dots and crosses taken to stand for electrons in the standard Lewis symbolism, could be interpreted as chemical bonds and it was hoped, that given suitable equations of motion for the electrons, the bond could be understood in physical terms.

As Woolley [15] has pointed out, much work was done within the old quantum theory to try and get the equations of motion for electrons in molecules, but with relatively little success. However, this work did have an influence on the way that the "new" quantum mechanics was used after 1927 in attempts to describe molecules. In particular in all the old quantum theory work it had been assumed that the nuclear motion could be treated classically and that to all intents and purposes the chemist's molecular model could be used to describe the "equilibrium geometry" of a molecule. Indeed these notions were at the heart of early attempts to account for molecular spectra in the old quantum theory.

Thus when Heitler and London applied the new quantum mechanics to the H_2 molecule in 1927 they did so against this background. They did not worry at all about nuclear motion but simply concentrated on applying quantum mechanics to the electrons at some fixed nuclear separation to see if the dissociation energy could be calculated reasonably well. In fact the Heitler and London work, though pioneering, is somewhat imperfect at a technical level (they estimated one of the integrals and used perturbation theory). The first full account of H_2 in quantum mechanical terms is by Wang [16]. His paper was published early in 1928 and reports a variational approach, with all the integrals calculated *ab initio*. The equilibrium internuclear distance was determined by plotting the potential energy curve as a sum of the quantum mechanical electronic energy and the classical nuclear repulsion energy and extensive comparisons were made between the calculated and experimental results.

The interesting thing about all these efforts to tie the molecule into quantum mechanics is how much they all leaned on the classical chemist's picture of the molecule. It is often supposed that the work of Born and Oppenheimer on nuclear motion, published in 1927 [17], was central to the thinking about the quantum mechanics of molecules. A survey of the literature up to about 1935 shows, however, that their paper was hardly if ever mentioned, and when it was mentioned, its arguments were used as *a posteriori* justification for what was being done anyway. The way that molecules got treated in quantum mechanics arose from the way that chemists looked at molecules and was not strictly compelled by the quantum mechanics.

Of course this is no criticism of the approach used. It can hardly be thought of as "wrong" to import ideas from one theory into another if they help to solve what seems otherwise to be a very difficult problem. But it was both good luck and bad luck that in the case of H_2, the imported ideas (classical treatment of the nuclei and the consequent idea of an equilibrium bond length) seemed so natural and obvious as not to excite comment. Once the realm of diatomic molecules is left, however, things become much less clear and the next section is devoted to an account on what can be learned from quantum mechanics alone, trying to avoid chemical preconceptions, in an attempt to show the nature of the difficulties.

3 Molecules and Quantum Mechanics

In Chemistry, quantum mechanics is nearly always thought of as synonymous with the solution of Schrödinger's equation for stationary states. To describe an isolated molecule it is presumed sufficient to write down the Schrödinger form of the Hamiltonian operator for the appropriate number of nuclei and electrons and to attempt to find its eigenfunctions. The dominant interaction between the particles is taken to be the coulomb interaction and, generally speaking, the spin is considered only in so far as it determines the statistics of the particles.

In what follows this minimum position will be the one assumed as describing the least complicated set-up in which there is a likelihood of identifying a molecule.

To make the discussion concrete let us imagine an attempt to describe quantum mechanically the ammonia molecule, NH_3. Here it might be thought that the ground is reasonably firm. It might be expected that the chemical molecule would be the eigenfunction of lowest energy and that the wave functions corresponding to that energy could be processed to yield the characteristic pyramidal geometry and a definite dipole moment.

The eigenfunctions lying close to this lowest one, it might be hoped, could be interpreted in terms of the rotation of the molecule as a whole and would be described pretty well in terms of the levels of a symmetric-top appropriate for the determined geometry. One might also hope to be able to identify somewhat higher energy levels as arising from the umbrella motion thought characteristic of high vibrational states of the system.

If all these features, or at least a substantial fraction of them, could be picked out of the solutions to the problem, without actually putting them in surreptitiously, then it would seem reasonable to consider the classical molecular structure notions for ammonia as pretty much confirmed by quantum mechanics.

Ammonia is a ten-electron four-nucleus problem so it is sensible to start off by writing down Schrödinger's equation in a laboratory-fixed cartesian reference frame for such a system. Denoting the particle coordinates by $\mathbf{x}_i$ ($\equiv x_i, y_i, z_i$) and assuming each has a charge $Z_i e$ and mass m_i ($Z_i = -1$ for an electron and $m_i = m$ for all electrons) then the Hamiltonian is

$$H(1, 2, \ldots, 14) = \frac{-\hbar^2}{2} \sum_{i=1}^{14} \frac{1}{m_i} \nabla^2(\mathbf{x}_i) + \frac{e^2}{4\pi\varepsilon_0} \sum_{i \geqq j=1}^{14} \frac{Z_i Z_j}{r_{ij}} \tag{3.1}$$

where

$$r_{ij} = ((x_j - x_i)^2 + (y_j - y_i)^2 + (z_j - z_i)^2)^{1/2}$$

and

$$\nabla^2(\mathbf{x}_i) = \frac{\partial^2}{\partial x_i^2} + \frac{\partial^2}{\partial y_i^2} + \frac{\partial^2}{\partial z_i^2}$$

The NH_3 molecule would then seem to be associated with the lowest lying solutions of the eigenvalue problem.

$$\hat{H}\Psi(\mathbf{x}_1, \dots, \mathbf{x}_{14}) = E\Psi(\mathbf{x}_1, \dots, \mathbf{x}_{14}) \quad (3.2)$$

In order to see what kinds of solution are possible the sensible first step is to see what the symmetries of the problem are. The three sets of operations that are immediately apparent as leaving $\hat{H}$ invariant are:

1) All uniform translation $\mathbf{x}_i \to \mathbf{x}_i + \mathbf{a}$
2) All orthogonal transformation $\mathbf{x}_i \to \mathbf{R}\mathbf{x}_i$, $\mathbf{R}^T\mathbf{R} = \mathbf{E}_3$ where $\mathbf{R}$ is an orthogonal matrix and $\mathbf{E}_3$ a unit matrix of order three. $|\mathbf{R}|$ can be $+1$ or -1.
3) All permutations of identical particles, $\mathbf{x}_i \to \mathbf{x}_j$, if $m_i = m_j$ and $Z_i = Z_j$.

The uniform translations constitute the translation group in three dimensions T(3). The orthogonal transformations constitute the orthogonal group in three dimensions O(3). Both of these groups are continuous groups. The permutations of a set of N identical particles constitute the symmetric group S_N of degree N; and if there are a number of sets of identical particles then the full permutation group for the problem is the direct product of the symmetric group for each set. This group is finite.

The group T(3) is a non-compact group and thus has no finite dimensional irreducible representations. What this means is that the energy of Eq. (3.2) is actually completely continuous and that none of the eigenfunctions is square integrable in the usual sense. This is clearly bad news if one is looking for molecules but fortunately it is possible to avoid the problem. Just as in classical mechanics, it is possible to separate the centre-of-mass motion from the full Hamiltonian, Eq. (3.1), and for that motion to carry the continuous part of the spectrum.

To be precise if there are N particles (fourteen in the present case) then it is always possible by means of a linear transformation to factorise the variable space, R^{3N} as $R^3 \times R^{3N-3}$ where the variables in R^{3N-3} are translationally invariant (or translation free). This process is sometimes called setting up a space-fixed frame. Such a frame is parallel to the lab-fixed frame but it moves with the centre-of-mass.

The required linear transformation may be written as:

$$(\mathbf{X}\mathbf{t}) = (\mathbf{x})\,\bar{\mathbf{V}} \quad \text{where} \quad \bar{\mathbf{V}} = (\mathbf{a} \mid \mathbf{V}) \quad (3.3)$$

with $a_i = M_T^{-1} m_i$, $M_T = \sum_{i=1}^{N} m_i$ and $\sum_{j=1}^{N} V_{ji} = 0$.

The transformation must have an inverse

$$\mathbf{x} = (\mathbf{X}\mathbf{t})\,\bar{\mathbf{V}}^{-1} \quad \text{where} \quad \bar{\mathbf{V}}^{-1} = \begin{pmatrix} \mathbf{b} \\ \hline \tilde{\mathbf{V}} \end{pmatrix} \quad (3.4)$$

with $b_i = 1$ and $\sum_{j=1}^{N} \tilde{V}_{ij} m_j = 0$.

The matrices $(\mathbf{x})$ and $(\mathbf{X}\mathbf{t})$ are three-by-N matrices of all the variables. Simple algebra then confirms that the $\mathbf{t}_i$, $i = 1, 2 \dots N - 1$ are invariant under the translation $\mathbf{x}_i \to \mathbf{x}_i + \mathbf{a}$ and that the centre-of-mass coordinate $\mathbf{X}$ carries all the translations.

Rather more tedious algebra shows that the kinetic energy operator in Eq. (3.1) can then be re-written as

$$\frac{-\hbar^2}{2M_T}\nabla^2(\mathbf{X}) - \frac{\hbar^2}{2}\sum_{i,j=1}^{13} G_{ij}\,\vec{\nabla}(\mathbf{t}_i)\cdot\vec{\nabla}(\mathbf{t}_j) \tag{3.5}$$

where $\mathbf{G} = \mathbf{V}^T\mathbf{m}^{-1}\mathbf{V}$with $\mathbf{m}$ a diagonal matrix of particle masses.

The potential energy then becomes rather a nasty mess, but it is easy to see from Eq. (3.4) and from r_{ij} that it depends only on the $\mathbf{t}_i$ so that the Hamiltonian, Eq. (3.1), can be separated in the standard way to yield solutions of the form

$$\Psi(\mathbf{x}_1, \mathbf{x}_2 \dots \mathbf{x}_{14}) = P(\mathbf{X})\,Q(\mathbf{t}_1, \mathbf{t}_2 \dots \mathbf{t}_{13})$$

where

$$\frac{-\hbar^2}{2M_T}\nabla^2(\mathbf{X})\,P(\mathbf{X}) = E_T P(\mathbf{X})$$

with $P(\mathbf{X}) = \exp(i\mathbf{kX})$, $E_T = |\mathbf{k}|^2/2M_T$, $\mathbf{k} = (k_x, k_y, k_z)$.

Because there are no restrictions on the k_α; E_T is continuous in $(0, \infty)$ and obviously $P(\mathbf{X})$ is not square integrable. However, it is possible to ignore it and to concentrate on $Q(\mathbf{t}_1, \mathbf{t}_2 \dots \mathbf{t}_{13})$ in the search for molecular structure. It is the space-fixed Hamiltonian expressed in terms of the space-fixed variables $\mathbf{t}_i$, and comprising the second therm in Eq. (3.5) together with the potential energy term, which needs to be investigated for the possible molecule-like solutions $\mathbf{Q}$.

It is very easy to see that this space-fixed Hamiltonian is still invariant under orthogonal transformations $\mathbf{t}_i \rightarrow \mathbf{Rt}_i$ in the space-fixed frame. The permutational symmetry is however rather more obscure.

This is because in the transformation to a space fixed frame, a variable has been "lost". In the present case there are only thirteen $\mathbf{t}_i$ and what these $\mathbf{t}_i$ represent depends upon the choice of $\mathbf{V}$ in Eq. (3.3). Under a permutation $\hat{\mathbf{P}}$ the laboratory-fixed variable set changes

$$\mathbf{x} \rightarrow \mathbf{xP} \tag{3.7}$$

where $\mathbf{P}$ is a standard orthogonal permutation matrix with columns which are zero except for one unit entry. From Eq. (3.3) and (3.4) it follows that change induced among the $\mathbf{t}_i$ by this permutation is just

$$\mathbf{t} \rightarrow \mathbf{tH}\,, \qquad \mathbf{H} = \tilde{\mathbf{V}}\mathbf{PV} \tag{3.7}$$

Although $\mathbf{H}$ is, by construction, a non-singular matrix it is not, in general, orthogonal neither is it of standard permutation form. It is therefore a somewhat tedious process to show that if $\hat{\mathbf{P}}$ is a permutation among identical particles then the space-fixed Hamiltonian is invariant, but it can be shown to be so.

The restrictions placed on the choice of the $\mathbf{t}_i$ by Eqs. (3.3) and (3.4) are really minimal but the permutational invariance requirements do suggest that it is sensible, when possible, to choose the $\mathbf{t}_i$ so that **H** becomes a standard orthogonal permutational matrix.

In the case of ammonia this can be done by choosing the $\mathbf{t}_i$ as variables relative to the nitrogen nucleus. Thus using $\mathbf{x}_{14}$ to denote the laboratory-fixed nitrogen variables on could choose

$$\mathbf{t}_i = \mathbf{x}_i - \mathbf{x}_{14}\,, \qquad i = 11, 12, 13 \quad \text{for the protons}$$

and similarly for the ten electrons. It is then easy to see that under any permutation of identical particles the $\mathbf{t}_i$ simply go into each other in the ordinary way and that **H** is of standard permutational form.

The choice made in terms of the electronic variables does however have a disadvantage. If it is made as above then it is found that coupling terms arise in the kinetic energy between the operators dependent on the "electronic" $\mathbf{t}_i$ and the "nuclear" $\mathbf{t}_i$. This may be avoided by means of a Jacobi coordinate choice, which in this case means referring the electronic variables to the centre-of-nuclear mass, so that

$$\mathbf{t}_i = \mathbf{x}_i - \mathbf{X}_n \qquad i = 1, 2, \ldots, 10 \tag{3.8}$$

with

$$\mathbf{X}_n = M^{-1} \sum_{i=11}^{14} m_i \mathbf{x}_i\,, \qquad M = \sum_{i=11}^{14} m_i$$

the $\mathbf{t}_i$, as so defined, go into one another under electron permutations and remain unchanged under nuclear permutations.

The inverse relations are:

$$\begin{aligned} \mathbf{x}_i &= \mathbf{X}_n + \mathbf{t}_i - \mathbf{T} && i = 11, 12, 13 \\ \mathbf{x}_{14} &= \mathbf{X}_n - \mathbf{T} && \\ \mathbf{x}_i &= \mathbf{X}_n + \mathbf{t}_i && i = 1, 2, \ldots, 10 \end{aligned} \tag{3.9}$$

with

$$\mathbf{X}_n = \mathbf{X} - M^{-1} m \sum_{i=1}^{10} \mathbf{t}_i$$

$$\mathbf{T} = M^{-1} \sum_{i=11}^{13} m_i \mathbf{t}_i$$

and where m denotes the electron mass.

It is tempting to think of the $\mathbf{t}_i$, $i = 1, 2 \ldots 10$ as the electronic variables and the $\mathbf{t}_i$, $i = 11, 12, 13$ as the nuclear variables. However such an identification is purely conventional. If the origin for the whole problem had been the nitrogen nucleus, one would still have thought of ten of the $\mathbf{t}_i$ as electronic coordinates but they would have been quite different ones. On the whole it is probably better to think of the $\mathbf{t}_i$ as pseudo-particle coordinates since they do not actually correspond to particles. Neither is it always possible to choose them in such a way that they transform simply into one another under a permutation. This is most easily seen by imagining how a homonuclear triatomic could be dealt with. In this case it is easy to see that one can define at most two $\mathbf{t}_i$ in terms of the nuclear variables and thus under a permutation, one variable must go into a linear combination of the two.

However in the case of ammonia there is no trouble. The kinetic energy operator from Eq. (3.5) becomes

$$\frac{-\hbar^2}{2}\sum_{i=11}^{13}\frac{1}{\mu_i}\nabla^2(\mathbf{t}_i)-\frac{\hbar^2}{2}\sum_{\substack{i,j=1\\ 1\neq j}}^{13}\frac{1}{m_{14}}\vec{\nabla}(\mathbf{t}_i).\vec{\nabla}(\mathbf{t}_j)$$
$$\frac{-\hbar^2}{2\mu}\sum_{i=1}^{10}\nabla^2(\mathbf{t}_i)-\frac{\hbar^2}{2M}\sum_{\substack{i,j=1\\ i\neq j}}^{10}\vec{\nabla}(\mathbf{t}_i).\vec{\nabla}(\mathbf{t}_j) \qquad (3.10)$$

where $\mu_i^{-1} = m_i^{-1} + m_{14}^{-1}, \mu^{-1} = m^{-1} + M^{-1}$
and the permutational invariance of this is now obvious.

From the expression for r_{ij} and the form of the $\mathbf{x}_i$ in terms of the $\mathbf{t}_i$ it follows that the electron-electron and nucleus-nucleus repulsion terms in the potential energy take pretty standard forms. Had the nitrogen nucleus been taken as the origin for the electronic variables, the forms for the electron-nucleus attraction terms would also have been standard. With the centre-of-nuclear mass choice however one gets complications. Thus

$$r_{14,1} = |\mathbf{T} + \mathbf{t}_i|$$

and clearly integrals over such an operator must involve highly complicated multipolar expansions. Such integrals could be avoided in the present case by using the nitrogen nucleus as the general origin but as has been remarked, that would lead to coupling terms in the kinetic energy operator. However this may be, such potential energy operator forms cannot in general be avoided, even by relaxing the separation requirement on the kinetic energy operator.

Rather delicate mathematical considerations (see e.g. [18]) can be used to show, at least for neutral systems and positive ions, that Schrödinger Hamiltonians arising from Eq. (3.5) and hence from Eq. (3.10) do have square integrable eigenfunctions in their spectrum. It might therefore be sensible to start looking for a molecule among the lower lying of such solutions, Q(**t**). Before doing so, it is appropriate to say something about the particle statistics.

The coordinate transformations that have so far been considered have, of course, no effect on the spin-space. Thus there are ten-particle spin-eigenfunctions describing the electrons in ammonia, a three-particle spin eigenfunction describing the protons and a one-particle spin eigenfunction describing the nitrogen. These spin-eigenfunctions must themselves be basis functions for irreducible representations of the relevant symmetric group. As is well known (see e.g. [19]) fermions, such as the protons and electrons, can provide a spin-function basis only for irreducible representations that correspond to two-rowed Young diagrams. In order that the overall wave function for a set of fermions be antisymmetric then, the only space functions allowed must provide a basis for irreducible representations that correspond to Young diagrams with two columns. There are equivalent restrictions for bosons. Thus not every function which is a solution of the space-fixed problem actually corresponds to a physical state of the system. Care must be taken then to pick from among the solutions, only the allowed ones.

A possible way of looking for allowed functions would be to think how approximate solutions to the space-fixed problem for ammonia might be constructed. First one would decide upon the spin-state that was required. This might be expected to be a singlet for the electrons and a doublet for the protons. A very simple trial function that satisfied all the permutational symmetry requirements for electrons would than be a determinant of doubly-occupied orbitals for the $\mathbf{t}_i$, $i = 1, 2 \ldots 10$ since the electron spins here can be associated with the $\mathbf{t}_i$. In the present case this sort of thing can also be done for the $\mathbf{t}_i$. $i = 11$, 12 and 13, by associating each orbital with a proton spin variable, leaving the nitrogen spin "bare". Usually things would be a bit more complicated for the nuclei and the general theory of the symmetric group would have to be deployed to construct approximate functions. It should be noted too, that distinctly different functional forms would emerge for the problem if one or more protons were replaced by deuterons, as deuterons are bosons.

Given that one could construct an orbital function in this kind of way, then one could by standard arguments, assuming orthogonal orbitals, set up an SCF procedure to determine the best orbitals. One could realise it in a linear combination-of-basis-functions manner if that seemed desirable. Of course, given the possible complexities of the potential and the kinetic energy operator, it would be a very tricky task in practice, but it is possible to see how such an approach could be carried through and indeed it has been attempted for ammonia [20] (see also [21, 22]).

Even without going into detail, it is pretty clear that an approximate solution of the kind suggested, is going to have much more in common with the standard atomic solution than it has with what is usually thought of as a molecular wave function. Indeed when it is remembered that the space-fixed Hamiltonian is invariant under the operations of O(3), just as is the conventional atomic Hamiltonian, then atomic-like behaviour, as far as symmetry is concerned, is forced on the solutions.

Even in the case of exact solutions, the levels will be characterised by an angular momentum quantum number, J, for the space part of the problem and each will be associated with $2J + 1$ degenerate states. They will be further sub-specified by spin quantum numbers S_i, for the groups of identical particles and particular "terms" will occur at particular energies according to the allowed permutational symmetries. Furthermore, levels of different parity will in general have different energies.

This last observation leads to rather a puzzling result. It is easily seen that the dipole-length operator in the laboratory-fixed coordinates, $e \sum_{i=1}^{N} Z_i \mathbf{x}_i$, is of odd parity. It follows then from Eq. (3.3) that the dipole operator when expressed in space-fixed coordinates is also of odd parity. Expectation values of the dipole-length operator between eigenfunctions of the space-fixed Hamiltonian therefore vanish, because the effect of the inversion operator on an eigenfunction is, at most, to change its sign and hence the product of a pair of functions has unchanged sign.

This means that on the face of it, quantum mechanics predicts that no molecule can have a permanent dipole moment or exhibit optical activity. Of course this might not quite be as it seems. It is well known, for instance, that excited states of the hydrogen atom appear to have dipole moments as a consequence of the applied field mixing degenerate states. It is perfectly possible that this sort of thing could happen here. In particular it could be that, for some reason, the eigenfunctions occurred in accidentally degenerate parity doublets. However this may be, it is clear that if the problem was looked at from the present perspective, it would be a rather surprising thing to find a molecule with a permanent dipole moment and the existence of one would call for particular investigation.

Now if quantum mechanics is correct and the space-fixed Hamiltonian as specfied is the appropriate one, then all the "molecular" solutions must lie among its eigenfunctions. If however, there were no preconceptions about what a molecule ought to look like there is nothing in the solutions of the space-fixed problem that would force or even guide one towards looking at them in the traditional molecular way. The most likely analysis of the results would be much more akin to the traditional methods for atoms.

In this context it should be noticed however, that some recent work on *atomic* structure (see e.g. [23, 24]), comes to the conclusion that at least certain aspects, particularly of excited states structure, can be viewed helpfully in terms of *molecular* structure ideas of symmetry.

It could be argued that what was causing confusion at the space-fixed level of description was the ubiquitous rotational symmetry. If that could be factored away then the molecule might be observed in the remaining "internal" part of the problem. It is indeed possible to factor the space R^{3N-3} into the direct product of $R^{3N-6} \times S^3$. The space S^3 is the space of motions on a sphere and is usually characterised by three Euler angles. The space R^{3N-6} then contains all the variables, usually called internal coordinates, that are invariant under orthogonal transformations of the $\mathbf{t}_i$.

Actually to perform such a factorization involves choosing an axis set somehow embedded in the system, in order that the three angles may be specified and in relation to which, the internal coordinates are defined. The process is sometimes called constructing (or embedding) a body-fixed frame. The actual coordinate transformation is a non-linear one to and from the set $\mathbf{t}_i$ and the set of three angles θ_m and $3N - 6$ internal coordinates q_k.

It is clear that a suitable set of internal coordinates could be chosen in terms of scalar products of the $\mathbf{t}_i$, for the scalar product is, by construction, invariant under all orthogonal transformations of the $\mathbf{t}_i$. Precisely how the body-fixed axes should be chosen is considerably more problematic.

In general terms the process involves specifying an orthogonal matrix, **C**, that relates the space-fixed unit axis vectors $\hat{e}_\alpha$ (α = x, y or z) to a set of body-fixed unit axis vectors $\hat{\varepsilon}_\alpha$ according to

$$\hat{\varepsilon}_\alpha = \sum_\beta \hat{e}_\beta C_{\beta\alpha}$$

The transformation specified by $\mathbf{C}$ is then such that

$$(\mathbf{C}^T \mathbf{t}_i)_\alpha = f_{\alpha i}(\mathbf{q}) \tag{3.11}$$

where the 3N — 3 functions $f_{\alpha i}$ are functions of only the 3N — 6 internal coordinates q_k. (It is possible that the transformations results in some, but no more than three $f_{\alpha i}(\mathbf{q})$ that are identically zero). Once the matrix $\mathbf{C}$ has been specified the sign of $|\mathbf{C}|$ is known so the handedness of the body-fixed frame is known and the required three angles are then implicitly specified by $\mathbf{C}$.

Using just these general considerations it is possible to write down formally a Hamiltonian in terms of the θ_m and q_k which describes the motion of the system. (See for example [25, 26]).

The algebra involved in the process is truly horrible and it would be inappropriate to attempt to detail it here but the symbolic form of the Hamiltonian is

$$\hat{H} = \hat{K}_J + \hat{K}_{q\theta} + \hat{K}_q + V(\mathbf{q}) \tag{3.12}$$

The term $\hat{K}_J$ contains the angular momentum operators $\hat{L}_\alpha$ which are operators involving the θ_m only. It is of the form

$$\hat{K}_J = -\frac{1}{2} \sum_{\alpha,\beta} M_{\alpha\beta}(\mathbf{q})\, \hat{L}_\alpha \hat{L}_\beta \tag{3.13}$$

The term $\hat{K}_q$ contains operators and functions that depend only on the q_k, while the term $\hat{K}_{q\theta}$ couples the angular momentum operators with operators that depend on the q_k. $V(\mathbf{q})$ arises from the electrostatic interactions and, since these are invariant under the operations of O(3), it depends only on the q_k.

Quite general group theoretical arguments can then be used to show that the formal solutions of the problem specified by the body-fixed Hamiltonian, Eq. (3.12), are of the type.

$$\sum_{k=-J}^{+J} {}^J\Phi_k(\mathbf{q})\,|JMk\rangle \tag{3.14}$$

Here $|JMk\rangle$ is an angular momentum (symmetric-top) eigenfunction and Φ is a function of the internal coordinates alone. The angular momentum eigenfunction is a function of the Euler angles alone. J denotes the total angular momentum, M its component along the space-fixed $\hat{e}_z$-axis and k its component along the body-fixed $\hat{\varepsilon}_z$-axis. It can be shown that the energy eigenvalue does not depend on M, so that the internal motion function need be labelled only by J and k. The solution, Eq. (3.14), is strictly one representing only SO(3) symmetry and the angular momentum eigenfunctions should be extended to $|JMkp\rangle$ to allow for parity arising from O(3) symmetry. It should also be stressed that the solutions are purely formal. The transformation from space-fixed to body-fixed coordinates is always a non-linear one so that its jacobian matrix is always a function of the internal coordinates and the Euler angles. It can therefore be singular for particular values of the coordinates and the transformation does not exist where it is singular. It can in fact be shown that any transformation that factors the space to include an S^3 part *must* give a jacobian matrix which is singular somewhere. This can be appreciated at an informal level by observing that in a non-rigid system it is not always possible to define *three* Euler angles for there is always some configuration of particles at which the definition of at least one of the angles fails. Singularities in the jacobian manifest themselves as singular terms in the Hamiltonian so that the space of eigenfunctions must be limited to avoid such behaviour.

These matters will be considered later but for the time being let it be assumed that functions like Eq. (3.14) can be found that lie in the valid domain of the Hamiltonian for some body-fixed coordinate choice. It is perhaps worth pointing out here that expected values of the dipole operator calculated between functions like Eq. (3.14) vanish in J = 0 states, even without considering parity. It is easy to show this using the properties of 3-j symbols. Thus moving to a body-fixed frame does nothing to elucidate the problems of permanent dipole moments and optical activity.

If **M** in Eq. (3.13) were not a function of **q** then $\hat{K}_J$ could be seen as describing the rotation of a rigid body whose inertia tensor, **I**, was $\mathbf{M}^{-1}$. It might be hoped therefore that molecular structure could be picked up by examining **M** in a particular case. Indeed Casimir was able to show in 1931 [27] that if one *assumed* that the molecule was a nearly-rigid body with an inertia tensor determined from the classical molecular geometry then the operator like Eq. (3.13) was capable of describing much molecular rotation spectra.

From this standpoint then the natural path to follow is the classical path, that is, to construct **I** in the space-fixed coordinates and then to diagonalise it by means of an orthogonal matrix so that its principal axes are the body-fixed axes. The orthogonal matrix that puts **I** into diagonal form is then to be used as **C**. The only variation that might be contemplated is to construct **I** from the nuclear variables only because the electron mass is negligible compared with any nuclear mass. If all goes according to plan the the full **M** should be closely related to the inverse of the diagonal form of **I**, and so Eq. (3.13) should take the diagonal form characteristic generally of an asymmetric top, but in the case of ammonia, a symmetric top.

This approach was first tried by Eckart [28] in 1934 and, almost simultaneously and independently by Hirschfelder and Wigner [29]. To cut a long story short the results they obtained bore no resemblance at all to the hoped-for ones. The moment of inertia terms were what are now called fluid moments of inertia (see e.g. Buck et al. [30]) and consist of quotients of sums and differences of powers of the principal moments of inertia, rather than individual principal moments. In particular if two of the principal moments become the same (as would happen for a symmetric top) the jacobian will vanish and the Hamiltonian contain divergent terms. Thus in this approach, the hoped-for solutions to the ammonia problem would not lie within the domain of such a Hamiltonian. There were also other difficulties in this approach, associated with the use of normal coordinates (familiar in the classical mechanics of vibrations) as internal coordinates.

In response to these difficulties Eckart made another attempt at describing a set of body-fixed coordinates for what he called normal molecules (those which would now be described as nearly-rigid).

Central to Eckart's approach [31] was the idea that the nuclei of a molecule moved in a potential that was itself invariant under translations and rotations and that for "normal" molecules this had a deep minimum at some framework or equilibrium geometry for the nuclei. "Anomalous" molecules were ones where such a minimum could not be found. Although Eckart did not actually identify the origin of the potential, it is reasonable to suppose that he thought of it as arising from the sum of the electronic energy calculated in the clamped nucleus approximation and the classical nuclear repulsion energy. However this cannot be completely certain because Eckart worked using classical mechanics and had no occasion to refer to the quantum mechanics of the problem. What is clear is that for Eckart the idea of a potential for nuclear motion was as uncontroversial and as obvious as it was to the earlier workers on diatomic molecules.

In terms of the present exposition the idea of a potential for nuclear motion has not been considered yet and it is convenient to defer its consideration for a time and simply to take the idea as given and to consider the nuclear motion. What is important is that Eckart took the equilibrium geometry to be the classical molecular geometry and argued in such a way as to show that Casimir's results could be justified from this standpoint. Eckart chose the definition of his body-fixed frame so that the coordinates which specify the equilibrium nuclear positions were, when expressed in the body-fixed frame, constants. He expressed all internal motions in terms of displacement coordintes from these positions. The internal coordinates that he chose were linear combinations of these displacements. When these displacements are zero the inertia tensor is just the constant inertia tensor defined at the equilibrium geometry. Eckart was able to show that the classical Hamiltonian determined in this embedding was such that, for sufficiently small displacements, **M** was exactly the inverse of the equilibrium inertia tensor and that $\hat{K}_{q\theta}$ vanished. He was further able to show that the internal coordinates could be chosen to be normal coordinates and that $\hat{K}_q$ and $V(\mathbf{q})$ together approximate the equations for a collection of simple-harmonic oscillations if $V(\mathbf{q})$ is expanded to second order in a Taylor series about the minimum. It follows then that in the Eckart approach the internal motions and the rotation motions can, in first approximation, be separated so that ${}^J\Phi_k(\mathbf{q})$ does not depend on k and so can be removed outside the sum Eq. (3.14) to be replaced inside the sum by constants, c_k, determined by the equilibrium geometry of the problem. These coefficients determine the standard top characteristics of a rigidly rotating body. In the case of ammonia for example a symmetric-top sum would arise, but in general the sum would yield an asymmetric top function. The energy of the system would then be a simple sum of vibration and rotation energies.

Thus even before a fully quantum mechanical form of the Hamiltonian was derived (and this was not completed in a fully satisfactory manner until the work of Watson in 1968 [32]) the Eckart Hamiltonian was widely seen to be the correct one with which to describe the molecular spectra of nearly rigid molecules.

An accessible account of the derivation of the Eckart Hamiltonian by classical methods and an early attempt to put it into quantum mechanical form can be found in Chapter 11 of Wilson, Decius, and Cross [33]. A direct quantum mechanical derivation, together with some account of the background and history of the problem since Eckart's time can be found in Chap. 7 of Biedenharn and Louck [34].

A more modern account of the classical derivation of the Eckart Hamiltonian can be found in the monograph by Ezra [35]. In that monograph he also discusses the problem of constructing the angular parity eigenstates, |JMkp⟩, appropriate to a body-fixed frame. He argues most convincingly within the Eckart approach that, provided that the system has neither a linear nor a planar equilibrium geometry, then the parity eigenstates will, in general, be accidentally degenerate.

Much of Ezra's monograph is devoted to surveying the literature on and describing a method for the extension of the Eckart approach to non-rigid (anomalous) molecules. It is clear that the standard Eckart approach will break down if the internal motions are very large, because at some energy every system becomes "anomalous". Thus if ammonia is calculated about a pyramidal equilibrium geometry, it is easy to show that the jacobian matrix becomes singular if the internal coordinates are chosen to force a linear geometry. This fact is signalled by the elements of the matrix **M** increasing without limit. However this is not a defect of the Eckart approach *per se*, for as explained above this sort of thing must happen somewhere in any body fixed approach. Another difficulty that occurs in the Eckart approach, but is not peculiar to that approach, arises from the requirements of permutational invariance and might perhaps have been anticipated from the previous discussion of spaced-fixed coordinates. It can be shown (see e.g. [35]) that in the Eckart approach only permutations that correspond to point group operations on the framework geometry of the problem can be realised without upsetting the rotation-vibration separation that makes the Eckart approach so attractive. This point will be considered in more detail later but it is clear that this implies that there would be no trouble in describing ammonia where the pyramidal group C_{3v} is isomorphic with S_3. However a molecule like ethene would cause trouble, because the full permutation group is $(S_2 \times S_4)$ and is not isomorphic with D_{2h}, the standard framework point group. This kind of difficulty must arise in any approach in which the nuclei are not treated symmetrically in defining the body fixed frame.

Although the idea of a potential for nuclear motion was vital in the way Eckart thought about his embedding, having seen how it is to be done, there is, formally at least, no reason why the same sort of embedding with respect to the nuclear variables should not be made in the full problem.

A standard geometry could simply be *assumed* for the nuclei, the appropriate transformations defined on the whole space and the complete problem put into body fixed form. As will be seen this would lead to an Eckart-like part for the nuclear motion and a part for the electronic motions expressed in the body-fixed frame with coupling terms between them which might be hoped to be small. There would therefore, in principle, be no difficulties in attacking the full problem in the Eckart approach.

However in this context it is possible to see quite clearly that it is the *assumption* of a molecular geometry that makes the approach possible rather than the reverse. Thus it is sometimes said that the Eckart approach does not so much justify the idea of molecular structure as the idea of molecular structure justifies the Eckart approach. If however the idea of a potential for nuclear motion could be accommodated properly then the approach would seem to be much more firmly founded. It could be asserted that molecular structure, at least for normal molecules, could be justified in terms of a suitable minimum in the potential. It is therefore the idea of a potential that must be considered next.

The idea of a potential for nuclear motion is usually approached through the arguments of Born and Huang [36] (see also Slater [37]) and this will be done here, for if the approach can be justified then it includes the older Born-Oppenheimer results. Born and Huang argue quite formally in terms of a laboratory-fixed Hamiltonian where it is possible unambiguously to divide the coordinates into a set of H coordinates $\mathbf{R}_i$ describing the (heavy) nuclei and L coordinates $\mathbf{r}_i$, describing the (light) electrons. They assume a total wave function for the system of the form

$$\Psi(\mathbf{r}, \mathbf{R}) = \sum_{n} \Psi_n(\mathbf{R}) \, \varphi_n(\mathbf{r}, \mathbf{R}) \tag{3.15}$$

and by quite formal argument appear to show that if $\varphi_n(\mathbf{r}, \mathbf{R})$ is determined as a solution of an electronic problem with the nuclear variables treated as parameters then the electronic energy considered as a function of these parameters, when combined with the classical nuclear repulsion energy acts, to a first approximation, as a potential for the nuclear motion problem.

From what has been said before it is clear that this argument must be at best formal. It is not possible to carry the expansion argument through in the laboratory-fixed frame because of the ubiquitous continuous spectrum arising from translations. This difficulty could be avoided by going to a space-fixed frame but then the resulting electronic energy will still have explicit spherical symmetry since it is defined in R^{3N-3}. This can be seen by recognizing that the solutions of the space-fixed problem like (3.10) must be angular momentum eigenfunctions and if the electronic part of the solution is to be obtained separately from the nuclear part, as implied by the form (3.15), then the solutions of the electronic part must be chosen to be eigenfunctions of the electronic angular momentum, as must be nuclear motion part be of the nuclear angular momentum. The resulting solutions must then be coupled to give the required total angular momentum eigenfunctions. Of course neither the electronic nor the nuclear angular momentum operators separately commute with the Hamiltonian but the electronic energy will be dependent upon the electronic angular momentum. (Some discussion of a closely related problem to this may be found in the paper of Czub and Wolneiwicz [38] which describes work arising from a proposal by Hunter [39] whose aim was to refine the Born and Huang approach).

The usual picture of a potential for nuclear motion however is one in which the potential is expressed in R^{3N-6} in terms of the rotationally invariant internal coordinates and this can be achieved only by expressing the Hamiltonian in body-fixed coordinates. The natural starting point for a discussion of the potential is therefore a body-fixed form of the Hamiltonian with an appropriately modified form of the Born-Huang expansion.

Consider the Born-Huang division of the laboratory-fixed coordinates and define a set of H-1 spaced fixed coordinates $\mathbf{t}_i$ entirely in terms of the nuclear laboratory-fixed coordinates. Define also a set of L space-fixed coordinates by expressing the L electronic laboratory-fixed coordinates with respect to the centre-of-nuclear mass. The space-fixed coordinates chosen earlier for ammonia correspond to these definitions. Now choose the matrix $\mathbf{C}$ that defines the transformation from space-fixed to body-fixed coordinates entirely in terms of the H-1 "nuclear" $\mathbf{t}_i$. This restriction means that diatomic systems must be excluded from discussion, for in such systems there is only one nuclear $\mathbf{t}_i$ so that only two angles, rather than the required three, are available to define $\mathbf{C}$. This is not a serious defect however since it is easy to treat diatomic molecules as a special case (see e.g. [40]).

If this restriction is made then the internal coordinates may be chosen as 3H-6 coordinates, q_k, that arise entirely from the original nuclear variables and L variables $\mathbf{z}_i$ that arise from the electronic $\mathbf{t}_i$, according to

$$\mathbf{z}_J = \mathbf{C}^T \mathbf{t}_i \tag{3.16}$$

With an internal coordinate and embedding choice of this kind, the Hamiltonian (3.12) can be shown to factor (again in symbolic form) as

$$H = \hat{K}_J + \hat{K}_{z\theta} + \hat{K}_{q\theta} + \hat{K}_{qz} + \hat{K}_q + \hat{K}_z + V_1(\mathbf{z}) + V_2(\mathbf{q}) + V_3(\mathbf{q}, \mathbf{z}) \tag{3.17}$$

The operator $\hat{K}_J$ has exactly the same form as (3.12) but $\mathbf{M}$ depends only on the q_k, it does not have any $\mathbf{z}_i$ dependence. Coupling operators $\hat{K}_{z\theta}$ and $\hat{K}_{q\theta}$ now couple the "electronic" and "nuclear" motions separately to the angular motion and the coupling operator $\hat{K}_{qz}$ couples the "electronic" and "nuclear" motions via $\mathbf{M}$, $\hat{K}_q$ is the kinetic energy operator for the q_k alone and $\hat{K}_z$ is the kinetic energy operator for the $\mathbf{z}_i$ alone. $\hat{K}_z$ is in fact

$$-\frac{\hbar^2}{2\mu}\sum_{i=1}^{L}\nabla^2(\mathbf{z}_i) - \frac{\hbar^2}{2M}\sum_{\substack{i,j=1\\ i\neq j}}^{L}\vec{\nabla}(\mathbf{z}_i).\vec{\nabla}(\mathbf{z}_j) \tag{3.18}$$

with $\mu = m^{-1} + M^{-1}$.

It is obviously very similar to the usual fixed nucleus kinetic energy operator. The provenance of the potential terms V_i is quite apparent. The solution form given in Eq. (3.14) continues to hold but for present purposes it is convenient to distinguish the internal coordinates by writing the internal coordinate function as ${}^J\Phi_k(\mathbf{q}, \mathbf{z})$.

In this context the standard Born-Huang argument would suggest that a potential for electronic motion could be defined if, in good approximation, one could write the internal motion wave functions in Eq. (3.14) as the simple product

$${}^J\Phi_k(\mathbf{q}, \mathbf{z}) = {}^J\chi_k(\mathbf{q})\,\psi(\mathbf{q}, \mathbf{z}) \tag{3.19}$$

The function ψ is assumed constructed as a solution of the problem

$$(\hat{K}_z + V_1(\mathbf{z}) + V_3(\mathbf{q}, \mathbf{z}))\,\psi(\mathbf{z}, \mathbf{q}) = E(\mathbf{q})\,\psi(\mathbf{z}, \mathbf{q}) \tag{3.20}$$

where $\mathbf{q}$ are to be treated as parameters, since they do not occur in the kinematics of the problem. Since all the angular momentum is carried by the $\hat{L}_\alpha$ the solutions of Eq. (3.20) do not depend on J.

To make the discussion definite let it be supposed that ammonia is being considered and that the body-fixed frame had been obtained by an Eckart-like process with respect to a pyramidal equilibrium nuclear geometry. Particular choices of the six q_k corresponding to the nuclear positions could be made and Eq. (3.20) could in principle be solved to yield a particular value of E($\mathbf{q}$) and the electronic wave function at that configuration. In practice this solution would only be approximate and could be imagined as being obtained using the standard clamped-nucleus technique of calculation, with the nuclear geometry being mapped into the q_k.

Clearly it is perfectly possible to perform a calculation using Eq. (3.20) where the nuclear geometry is linear and to obtain a value of E($\mathbf{q}$) and the wave function $\psi(\mathbf{z}, \mathbf{q})$ at that configuration. However what such a solution might mean is clearly problematic because at such a configuration the transformation to body-fixed coordinates fails, because the angular definitions fail. However Eq. (3.20) contains no reference to the angular part of the problem so it is perfectly possible to generate formally part of a seemingly impossible solution. To put these observations in context, let it be supposed that an attempt was made to solve the *whole* problem for ammonia in the Eckart pyramidal embedding, without further separating the electronic motions but with the traditional internal coordinates for the nuclei. Assuming that there are such solutions they must be of a form in which the wave function vanishes extremely strongly where the jacobian is singular for no others could possible be in the domain of the body-fixed Hamiltonian. Alternatively if approximate solutions are constructed for the problem then approximate wave functions must have that vanishing property too so that expectation values exist. From this it follows that even if $\psi(\mathbf{z}, \mathbf{q})$ does not vanish strongly where the jacobian is singular, Eq. (3.19) could still be an approximate solution for the full problem if ${}^J\chi_k(\mathbf{q})$ alone vanished sufficiently strongly.

So at this level of approximation it seems likely for ammonia that if the potential arising from the sum of E($\mathbf{q}$) and the classical nuclear repulsion has a deep minimum at a pyramidal nuclear geometry then an Eckart prescription for embedding about a pyramid and the traditional choice of internal coordinates will be adequate and effective. It should be possible to find a set of nuclear motion functions that describe all vibrations of not too large an amplitude without getting into trouble. Indeed numerical experiments seem to indicate that this is the case.

Of course this is not to avoid a certain logical puzzle for at one level it can obviously be maintained that the solutions obtained are those and only those that might be expected in terms of the information used in posing the problem. Nevertheless, it would seem that the Eckart approach can offer some justification of the molecular structure idea at this level of approximation within the electronic potential approach. Furthermore providing that the potential idea is secure it seems easy to see how to deal with non-rigid systems by appropriate design of the nuclear-motion wave functions for the large amplitude motions so that divergences are avoided.

It remains to be seen however how secure the potential idea is under refinement of the present simple product wave function and when permutational symmetry is taken into account.

The general definition of a potential energy surface depends on the ability to define such a surface without explicit reference to the angular embedding of the system. It has already been assumed that this is possible when it was agreed that solutions to Eq. (3.20) could be obtained, at least approximately, from clamped nucleus calculations and indeed it is generally true. Once the space fixed form of the Hamiltonian, Eq. (3.5), has been decided upon then it is sufficient to assume that there exist body-fixing transformations **C**, **C**′ etc., that are invertable into overlapping regions of R^{3H-3} to establish that in principle any chosen set of internal coordinates can be mapped into any other set through the scalar products in R^{3H-3}. This may be established using the theory of differentiable manifolds (see e.g. Schutz [41]), and is simply the process of specifying a suitable atlas to cover R^{3H-3} by means of local charts of the form $R^{3H-6} \times S^3$. It follows therefore that the **q** in Eq. (3.20) can be chosen in any way that is consistent with them being independent and expressible in terms of scalar products of the space-fixed variables. Thus the form of Eq. (3.20) is, up to a point, independent of any assumptions about the embedding. (Its form does of course depend on the choice of the space-fixed variables and on the choice of **C** to be a function of the nuclear $\mathbf{t}_i$ only). This observation can be regarded as the basis for topologically motivated approaches to the problem of nuclear motion like those of, for example, Mezey [42]. At the simplest level, the observation also justifies the use of the usual clamped nucleus Hamiltonian to approximate the full electronic Hamiltonian.

While there is no doubt then that a problem like Eq. (3.20) can be solved for arbitary electronic states and in terms of arbitary q_k, the association of these solutions with solutions of the full problem will depend on being able to do a full transformation and since such full transformations can only be valid locally a full description of the problem at the body-fixed level will involve many different body-fixings. This seems to have been explicitly noticed first by Duchinsky [43] in some work in which he attempted to analyse vibration-rotation transitions in which changes of electronic state were involved. He was attempting an Eckart-like approach in both states and, as the equilibrium geometry was different, there was, naturally enough, a different body fixing transformation in each of the states and it was necessary to establish the relationship between them in the common domain.

It follows therefore that while an expansion of the Born Huang type, Eq. (3.15), in terms of solutions to the electronic Hamiltonian is formally possible in body-fixed coordinates, it cannot be accomplished in terms of a single body-fixed set. It is therefore not possible to say to what extent the molecular structure that might have been chosen at the simple product level, Eq. (3.19), with respect to the lowest electronic state, will persist on refinement. It is perhaps most easy to appreciate the difficulties that can arise here by thinking of a potential energy surface, generated by some solution to Eq. (3.20), that has no minimum in it and hence no "natural" embedding choice. It will be appreciated too that acute problems will arise if, for some choice of **q**, the electronic states in the expansion become degenerate.

This kind of problem is made no easier when permutational symmetry is considered. As has been seen permutation symmetry causes no problems at the electronic level but that it can at the nuclear level. In the context of the Eq. (3.20) it is clear that $V_3(\mathbf{z}, \mathbf{q})$ must be totally symmetric under any permutations of like nuclei and that $E(\mathbf{q})$ and $\psi(\mathbf{z}, \mathbf{q})$ must be adapted to form a basis for the appropriate symmetric group of the nuclei. It is clear also that $V_2(\mathbf{q})$ is totally symmetric and that to form a valid potential for the nuclear motion problem it follows that $E(\mathbf{q})$ must also be totally symmetric and so therefore must $\psi(\mathbf{z}, \mathbf{q})$ be.

What that means in general should now be perfectly clear; in doing a calculation in which the nuclei are treated classically, not only is it necessary to consider the **q** values that generate the "classical" structure and its near neighbours but it is also necessary to consider all the isomers of that structure which are possible, for at the mathematical level all possible isomers will be generated by the permutations of like nuclei. It must be the case that E(**q**) be invariant under the nuclear permutation that has the particular property of mapping one "classical" isomer into another. Of course this does not mean that E(**q**) must have the same "classical" value at one isomer as it has at another, all that it means is that E(**q**) must provide the same mapping rule, independently of any particle names that may have been used for the nuclei in the part of the calculations generated by Eq. (3.20) in which they are identified and moved as classical particles. It is perfectly possible too that $\psi(\mathbf{z}, \mathbf{q})$ if evaluated at some isomeric geometry, is, in some well defined sense, negligibly small. The formal possibilities here are many but the point to be made is that for E(**q**), which is derived in terms of the nuclei in a classical fashion, to be used in a quantum mechanical calculation of nuclear motion, it must be properly symmetrised in terms of nuclear permutations. This process involves, in principle, investigating all classical isomers.

When working practically on simple systems in their electronic ground states often no trouble is caused by this requirement. For example the potential in ammonia calculated in the usual way is automatically invariant. It is similarly very easy, at least locally, to deal with "symmetry-related minima" such as those that arise in calculations on O_3. These and similar problems are discussed in standard texts like [44] on potential energy functions. Real difficulties arise however when there are minima that can be thought of as isomers and hence attract a different local embedding for the full problem in the Eckart approach. To appreciate why consider what happens to a particular embedding under a permutation.

It has been seen that an embedding generates a relation, Eq. (3.11)

$$\mathbf{t}_i = \mathbf{C}\mathbf{f}_i(\mathbf{q})$$

and it can be shown that if the $\mathbf{t}_i$ are subject a permutation so that as as in Eq. (3.7)

$$\mathbf{t}'_i = \sum_{j=1}^{H} \mathbf{t}_j H_{ij} \tag{3.21}$$

then the embedding prescription is, in the permuted system,

$$\mathbf{t}'_i = \mathbf{C}\mathbf{U}^T\mathbf{U} \sum_{j=1}^{H} \mathbf{f}_j H_{ji} \tag{3.21}$$

where **U** is an orthogonal matrix composed of the q_k alone. The transformations of **C** arising from the permutations of the variables $\mathbf{t}_i$ used to define **C** can be shown to induce the change.

$$|JMk\rangle \rightarrow \sum_{n=-J}^{+J} D^J_{kn}(\mathbf{U})\, |JMn.\rangle \tag{3.22}$$

in the angular functions. Here $D^J_{kn}(U)$ is an element of the standard Wigner matrix defined in terms of $\mathbf{U}$ (see e.g. Chap. 6 of ref [34]). The transformation of the internal coordinates will be such as to send ${}^J\chi_k(\mathbf{q})$ of (3.19) into some other function ${}^J\chi'_k(\mathbf{q})$ but will leave $\psi(z, \mathbf{q})$ invariant. (A more detailed account of such transformations in the Eckart approach can be found in [35] but the results are quite general).

This behaviour is highly inconvenient because it is in general extremely difficult to find $\mathbf{U}$ by analytical means and also to get closed expressions for the transformed internal coordinate functions. It is not in principal however disastrous from the point of view of approximate calculations as it is always possible to accommodate such behaviour by increasing the basis set in the expansion of the wave function and an example of such a calculation will be cited later.

The behaviour can however have extremely perplexing consequences in the Eckart approach. If the vibration rotation separation spoken of earlier as characteristic of the Eckart approximation is realised explicitly in the present context then the wave function from Eqs. (3.19) and (3.14) will be in first approximation

$$ {}^J\chi(\mathbf{q})\, \psi(\mathbf{z}, \mathbf{q}_0) \sum_{k=-J}^{+J} c_k \,|JMk\rangle \qquad (3.23)$$

Here $\psi(\mathbf{z}, \mathbf{q}_0)$ is the electronic wave function evaluated at the equilibrium framework geometry $\mathbf{q}_0$ and the c_k are the coefficients that determine the angular functions appropriate to the equilibrium geometrical structure of the molecule (what kind of top it is).

It can now be seen however that the separation, Eq. (3.23), must be spoiled by virtue of the transformations, Eq. (3.22), induced by a permutation unless $\mathbf{U}$ (and hence $D^J_{kn}(U)$) has constant elements. As mentioned earlier this is only the case if the permutation is isomorphic with an element in the point group of the framework.

Thus in the Eckart approach the idea of an equilibrium molecular structure as leading to a good separation would be perfectly secure in the case of ammonia but would be rather problematic in the case of ethene as there the separation would not persist. Of course it may be that in ethene and in like cases, the internal-coordinate-dependent terms generated in the sum, Eq. (3.22), by non-point group permutations are negligibly small in regions where the, it is to be hoped, dominant ${}^J\chi(\mathbf{q})$ are large. However this may be, they must in principle be present for otherwise the functions could not provide a basis for the symmetric group of the problem. Hence the process of body fixing would have broken a fundamental symmetry and how this could have happened is not at all apparent. Indeed at the practical level when it comes to assigning statistical weights in the Eckart approach, that is to say determining the allowed nuclear spin eigenfunctions, their presence is assumed. It will be remembered (see e.g. [45]) that in cases like ethene in the Eckart approach it is assumed that the point group (D_{2h} for ethene) is isomorphic with a subgroup of the full permutation group. Hence it is possible to put the operators into the classes in which they belong in the full group. From their characters in the classes and the *full* group irreducible characters, the appropriate irreducible representation for the nuclear motions is determined, hence the associated spin-eigenfunction forms and thus the statistical weight.

What emerges from a study of the potential surface idea, therefore, does not much help to clarify the puzzles that it was hoped that it might. At the level of abstract differential geometry the idea of a potential as a solution to a clamped nucleus electronic problem seems secure enough so that there can be no principled objection to clamped nucleus electronic structure calculations. However in the construction of such a surface the nuclei are naturally identified and treated as giving rise to classical distinguishable variables in the problem but to use the surface for quantum mechanical calculations, permutational invariance must be required to obliterate the distinguishability. Furthermore, even granted that one can get a suitable potential energy surface, any full embedding that is made in terms of it can at most be of local validity so that in general there will be have to be a number of different choices to be made to cover any surface and many surfaces must be employed in refining the Born and Huang approach. With any given choice of embedding however, permutational symmetry requires a certain behaviour for the nuclear motion wave functions which can further obscure any geometrical structure choces made to achieve a particular embedding. In the absence of systematic numerical investigation of these matters it simply is not clear what effect they will have on the persistence of the idea of molecular structure.

What should be clear however from this section as a whole is that the idea of a molecular structure in terms of a three-dimensional geometry does not seem to emerge in any natural way from a quantum mechanical discussion of the isolated molecule. Rather it appears in certain circumstances to be a helpful approximation to require a molecular structure but whether this is so or not seems to depend in any particular case on detailed numerical considerations. As yet very little such numerical work has been attempted, and this is particularly the case when considering what happens in going beyond the first term in the Born and Huang approach.

4 Discussion

It has been the aim of this article to show that the idea of molecular structure, as exemplified by a definite geometry and associated properties, is not without its difficulties even from the point of view of the most down-to-earth sort of quantum mechanical considerations.

Thus is has been seen that it is not at all easy to say what could be meant by the statement that a molecule has a dipole moment or is optically active. But it has also been seen that such properties could be attributed if there was an accidental degeneracy of the parity eigenstates associated intuitively with the molecule. Whether or not this happens in any particular case is a contingent matter and cannot be decided *a priori*.

It has further been seen that the notion of a molecular geometry is equally contingent and must depend to a large extent on the domain over which a particular embedding persists in practice. Furthermore it is clear that sometimes assigning a molecular structure to determine an embedding, is seemingly to violate the permutational invariance of the problem.

Thus it is perhaps not unfair to say that the idea of a molecule with structure does not fit easily into a quantum mechanical treatment of the isolated system neither does it seem to arise naturally out of such a treatment.

Whether or not this lack of easiness in matching the traditional description of a molecule to its quantum mechanical counterpart is a matter with profound consequences for the way in which the world is to be viewed, is a question that has exercised many scientists over the past ten years or so. If it were fundamental then it would certainly support particular philosophical views opposed to reductionism and also support certain kinds of holistic ideas. It counts too, in the arguments about realism and localisation which are such a feature of discussions of quantum mechanics in the aftermath of Aspect's work (see e.g. [6, 7]). Relevant discussions of this kind of thing can be found in the work of Primas [10] and of Woolley [15] mentioned earlier and indeed Wooley (see e.g., [46, 47, 48, 49]) has been among the most active participants in these debates and has made many subtle and perceptive contributions to them as has Claverie (see e.g., [50, 51]).

It is not at present clear whether the outcome of these debates will have any fundamental effect on the way that chemists actually look at molecules, nor indeed is it clear that it ought to have any effect no matter what the outcome. However the lack of easiness of fit in practice between the traditional and the quantum mechanical picture must affect the way in which approximate quantum mechanical calculations are done. If it turns out to be easier to get decent agreement between experiment and calculation by abandoning the idea of a molecular geometry at the body-fixed level or by abandoning the idea of a potential surface and the Born and Huang separation, then those involved in quantum mechanical calculations will almost certainly abandon them. What that will do to the relations between theoretical chemists and general chemical community remains to be seen. It could well be however that it will lead to the same kind of tensions that developed when electronic structure theorists abandoned the Valence Bond view in favour of the Molecular Orbital view because of the latter's computational simplicity.

At the moment however there is little numerical work available in which geometrical ideas at the body-fixed level have been abandoned and none in which the idea of a potential has been abandoned. What work there is, is mostly on triatomic systems. The results of some work in which the author has been involved [52, 53, 54] do seem to indicate that it is perfectly possible to get good results without any geometrical information being fed into the embedding conditions and, without choosing internal coordinates that are based on a classical molecular model. This said, however, it must be admitted that it was sometimes found that more quickly convergent approximations could be obtained by working in a coordinate system in which the internal coordinates could be regarded as bond lengths and a bond angle. It was also found to be helpful to parametrize the basis functions in terms of quantities that could be thought of as equilibrium bond lengths. Thus it cannot be claimed that molecular intuition was put completely aside for the purposes of such calculations and how they should therefore count in any discussion of molecular structure must remain uncertain.

The work mentioned above was undertaken in terms of just two space-fixed coordinates and among the systems considered was H_3^+. As explained earlier it is inevitable that in this sort of case a permutation induces a general linear transformation among the space-fixed coordinates and in the chosen body-fixing, this meant that for most of the permutations in S_3 (the symmetric group of H_3^+), **U** in (3.21) was not a constant matrix. It is further the case that in the chosen body-fixing, the internal coordinates go into very awkward combinations of one another. It was not possible to determine **U** explicitly neither was it possible to find a basis that naturally accommodated the internal coordinate change. In practice these challenges were met by using a very extensive basis set and what happened in the calculations was that the degeneracies expected as a consequence of S_3 symmetry, developed in the required manner as sufficient basis functions were included. It therefore seems pretty certain that there is no real symmetry breaking here. It is difficult to know however whether to count this observation for or against geometrical ideas. Some pioneering calculations [55, 56] on H_3^+ were done within the Eckart approach assuming an equilaterial triangle equilibrium geometry. In this case the **U** in Eq. (3.21) arising from the Eckart embedding *are* all constant matrices, so that vibrational-rotational separation is not spoiled. However in these calculations the internal coordinates transformed awkwardly and the trial functions were not accomodated to their change so that it was similarly necessary to attempt to converge the degeneracies by the use of an extensive basis. Such convergence was achieved in practice at least for $J = 0$ states. Thus in this case the calculations made with and without geometry assumptions seem to be equivalent.

Although there have as yet been few calculations attempted even on triatomic systems in which more than one potential surface is considered, what seems to be emerging from the work that has been done is that in practice, an Eckart-like assumption of equilibrium geometry is a hinderance to effective calculation. It seems to be preferable to choose a common set of coordinates to describe the molecule on each surface but, for the reasons outlined earlier, the extent to which this is sensibly possible is clearly highly problem-dependent.

It does seem likely therefore that for detailed *ab-initio* calculations on small systems, molecular geometry ideas will be abandoned by computational quantum chemists. The idea of a potential seems unlikely to be abandoned, however for the sort of results that can be obtained in the potential approach seem to be highly accurate and in good agreement with experiment.

Among the approaches that have been suggested which might enable the abandonment of the idea of a potential as specified in the Born and Huang approach are the the method of Hunter mentioned earlier [39] and the generator coordinate approach (see e.g. [57]). There appear to be serious difficulties in the way of pursuing the Hunter approach (see e.g. [38]) and even though the generator coordinate approach seems very attractive, there is not yet enough work on polyatomic systems using it to be sure if it is really feasible.

There is no reason to suppose that those involved in calculations are about to make the radical departure that would be involved in working in a space-fixed frame. Such a course of action would make molecular geometry ideas completely redundant, and avoid all the problems of body-fixing. But as must be clear from what has been already said about the approach, it would be fiendishly difficult in computational terms and seems unlikely to be capable of sufficient refinement (at least in the short term) to provide the kind of accuracy that body-fixed approaches at present seem to provide.

It is therefore very difficult to see a way forward. All that it seems reasonable to say at the moment is that potential energy surfaces and some sort of molecular structure ideas seem very likely to persist for some time in computational quantum chemistry. As long as these ideas persist, so will the idea that there *ought* to be a quantum mechanical account of molecular structure whether this is, at a deep level, achievable or not.

5 References

1. Ballhausen CJ (1979) J. Chem. Ed. 56: 357
2. Lewis GN (1923) Valence and the structure of atoms and molecules, Chemical Catalog Co., New York
3. Sidgwick NV (1927) The electronic theory of valency, Oxford UP
4. Jammer M (1960) The conceptual development of quantum mechanics, McGraw-Hill, New York
5. Jammer M (1974) The philosophy of quantum mechanics, Wiley, New York
6. Gribbin J (1984) In search of Schrödinger's cat, Bantam
7. Rae A (1986) Quantum physics — Illusion or reality? Cambridge UP
8. de Witt BS (1970) Physics Today 23, 9: 30
9. Mermin ND (1985) Physics Today 38, 4: 38
10. Primas H (1981) Chemistry quantum mechanics and reductionism, Lecture Notes in Chemistry 24, Springer, Berlin Heidelberg New York
11. Sutcliffe BT (1983) J. Mol. Struct. 98: 189
12. Jørgensen CK (1982) Chimia 36: 221
13. von Meyer E (1898) A history of chemistry from earliest times to the present day Tr. (G. McGowan) Macmillan, London.
14. Russell CA (1971) The history of valency, Leicester UP
15. Woolley RG (1976) Adv. Phys. 25: 27
16. Wang SC (1928) Phys. Rev. 31: 579
17. Born M, Oppenheimer JR (1927) Ann. der Physik. 84: 457
18. Simon B (1971) Quantum mechanics for Hamiltonians defined as quadratic forms, Princeton UP
19. Hamermesh M (1962) Group theory, Addison-Wesley, Reading, MA
20. Thomas IL (1969) Phys. Rev. 185: 90; (1970) Phys. Rev. A2: 1200; (1971) Phys. Rev. A3: 565
21. Monkhorst HJ (1987) Phys. Rev. A36: 1544
22. Pettitt BA, Dancura W (1987) J. Phys. B20: 1899
23. Hunter JE, Berry RS (1987) Phys. Rev. A36: 3042
24. Watanabe S, Lin CD (1987) Phys. Rev. A36: 511
25. Sutcliffe BT (1982) In: Carbo R (ed) Current aspects of quantum chemistry, Studies in theoretical chemistry 21, Elsevier, Amsterdam. p 99
26. Handy NC (1987) Mol. Phys. 61: 207
27. Casimir HBG (1931) Koninkl. Ned. Akad. Wetenschap Proc. 34: 844
28. Eckart C (1934) Phys. Rev. 46: 487
29. Hirschfelder JO, Wigner E (1935) Proc. Nat. Acad. Sci. 21: 113
30. Buck B, Biedenharn LC, Cusson RY (1979) Nucl. Phys. A317: 205
31. Eckart C (1935) Phys. Rev. 47: 552
32. Watson JKG (1968) Mol. Phys. 15: 479
33. Wilson EB, Decius JC, Cross PC (1955) Molecular vibrations, McGraw-Hill, New York
34. Biedenharn LC, Louck JD (1981) Angular momentum in quantum physics, Addison-Wesley, Reading, MA
35. Ezra GC (1982) Symmetry properties of molecules, Lecture notes in chemistry 28, Springer, Berlin Heidelberg New York
36. Born M, Huang K (1955) Dynamical theory of crystal lattices App VIII, Oxford UP

37. Slater JC (1927) Proc. Nat. Acad. Sci. 13: 423
38. Czub J, Wolniewicz L (1978) Mol. Phys. 30: 1301
39. Hunter G (1975) Int. J. Quantum. Chem 9: 237
40. Kolos W, Wolniewicz L (1963) Rev. Mod. Phys. 35: 473
41. Schutz B (1980) Geometrical methods of mathematical physics C.U.P., Cambridge
42. Mezey P (1983) I.J.Q.C. Symp. 17: 137
43. Duchiñsky F (1937) Acta Phys. Chem. URSS 7: 551
44. Murrell JN, Carter S, Farantos SC, Huxley P, Varandas AJC (1984) Molecular potential energy functions, Wiley, Chichester
45. Kaplan IG (1975) Symmetry of many electron systems, Academic, London
46. Woolley RG (1982) Structure and Bonding 52: 1
47. Woolley RG (1985) J. Chem. Ed. 62: 1082
48. Woolley RG (1986) Chem. Phys. Letts. 125: 200
49. Woolley RG (1978) J.A.C.S. 100: 1073
50. Claverie P, Diner S (1980) Israel J. Chem. 19: 54
51. Claverie P (1983) In: Maruani J and Serre J (eds) Symmetries and properties of non-rigid molecules, Elsevier, Amsterdam, p 13
52. Sutcliffe BT, Tennyson J (1986) Mol. Phys. 38: 1053
53. Sutcliffe BT, Tennyson J (1987) J. Chem. Soc. Farad. Trans 83: 1663
45. Tennyson J, Sutcliffe BT (1984) Mol. Phys. 51: 887
55. Carney GD, Porter RN (1980) Phys. Rev. Lett. 45: 537
56. Carney GD (1980) Mol. Phys. 39: 923
57. Lathouwers L, van Leuven P (1982) Adv. Chem. Phys. 49: 112

Topology and Properties of Molecules

Oskar E. Polansky†

Max-Planck-Institut für Strahlenchemie, 4330 Mülheim a. d. Ruhr/FRG

The topology of covalent molecules and the metric and topological spaces associated with them are discussed in some detail. It is shown that particular simple connected graphs represent molecular topology. The role which molecular topology plays in the modelling of physical properties is illustrated by the non-empirical rules of topological charge stabilisation (TCS) and topological effects on molecular orbitals (TEMO). The predictions based on these concepts are verified with astonishing fidelity in nature. From this a hierarchic order of molecular topology, matter and space, and energy is postulated and briefly discussed.

1 Introduction

One of the most fascinating problems in chemistry is the relationship between structure and properties of molecules. Although in the last century a large part of chemical research was devoted to that subject, the problem has not lost its fascination and, at present, is still under intensive investigation. About 120 years ago, when the term structure had been introduced into chemistry by Butlerov, its meaning was exhausted by the number and kind of atoms forming a molecule and their mutual connectedness. But subsequently it has come to define more and more and, nowadays, a great deal of detailed information about a particular molecule is subsumed under the term structure. As the connotations of that nomenclature became enriched, so the need for a more basic notion, quite close to the original meaning, has become evident. Such a notion is offered by the ideas of molecular topology.

In the next section this concept is examined and it is shown that the topology of any covalent molecule is represented by the so-called molecular graph which may be uniquely derived from the constitutional formula in question. Section 3 treats the metric and topological spaces associated with covalent molecules and the characteristic and the acyclic polynomials of molecular graphs. In Section 4 the modelling of physico-chemical properties is briefly discussed; with regard to the results presented in the previous sections, three factors, among them a molecular topological factor, are postulated to determine physico-chemical properties of molecules. This scheme is proved by two non-empirical topological rules, namely the rule of topological charge stabilisation (Sect. 5) and the concept of topological effect on molecular orbitals (Sect. 6). Finally, in the last section, the material presented is examined critically and a dominant role of topology in modelling the physico-chemical properties of larger molecules is deduced. As a consequence of that a hierarchic order of molecular topology, matter and space, and energy is postulated.

2 Graphs Representing Molecular Topology

2.1 Complete Molecular Graphs

The chemist's view of molecular topology has been expressed as follows [1]: "In constitutional formulae the atoms are represented by letters and the bonds by lines. They describe the topology of the molecule." Two items of that statement are of particular importance: (i) there is a unique relation between the constitutional formula and the topology of a given molecule and (ii) the constitutional formula is a description but no representation of molecular topology. The essence of the last item becomes evident when the class of covalent diatomic molecules is considered. In total, some thousand diatomic molecules may be constructed and some hundreds of them have been experimentally examined. Among the latter one finds quite stable molecules with single bonds like H_2, HF, etc., or with multiple bonds like N_2, CO, etc., further molecules requiring particular state conditions as NaCl and other alkali halides, and, finally, a large number of short life species like AlS, PN, PbH and many others. Thus, the class of diatomic molecules exhibits an enormous variety in constituent atoms, types of bonds, and properties. However, in spite of all these varieties, all diatomic molecules possess one and the same topology, simply because only a single topology exists for two particles connected by any interaction.

As this example shows, in order to derive the molecular topology from a constitutional formula one has to deport from the nature of atoms and the type of bonds. Depicting each atom of the molecule by a small circle (*vertex*) and connecting two vertices by a line (*edge*) if, and only if, in the constitutional formula a bond is indicated between the corresponding atoms, one arrives at a graph, termed *complete molecular graph*. For example, in the case of isopentane, $(CH_3)_2CHCH_2CH_3$, and trimethylethylene, $(CH_3)_2C{=}CHCH_3$, the graphs $\mathcal{G}_1$ and $\mathcal{G}_2$ are uniquely derived from the respective constitutional formulae; they are shown in Fig. 1. Since in these graphs, neither the nature of atoms nor the types of bonds are specified, they meet the requirements specified above and consequently they represent the respective molecular topology of the molecules considered. This particular result may be generalized as follows: *Molecular graphs depicting constitutional formulae of molecules represent their molecular topology*.

Due to this intimate relationship between molecular topology and molecular graphs, the latter have to be examined in some detail. In doing this, we will apply the terminology used in [2–6], where more details and some proofs also may be found. In what follows molecular graphs will be denoted by $\mathcal{G}$ and symbolically expressed as follows:

$$\mathcal{G} = [\mathcal{V}, \mathcal{E}], \tag{1}$$

where $\mathcal{V}$ and $\mathcal{E}$ stand for the vertex and the edge set, respectively, their cardinalities are denoted by $n = |\mathcal{V}|$ and $m = |\mathcal{E}|$.

Two vertices are called *adjacent* if they are connected by an edge; the *degree* of a vertex expresses the number of its adjacent vertices.

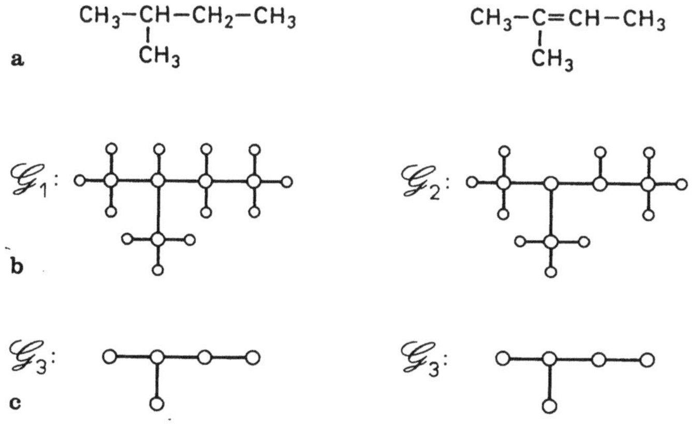

Fig. 1a–c. Molecular graphs ($\mathcal{G}_1$ and $\mathcal{G}_2$) and skeleton graph ($\mathcal{G}_3$) of isopentane and trimethylethylene

Two edges are called *adjacent* if they have an incident vertex in common.

Further, a sequence of pairwise adjacent vertices (without repeating any one) and the edges connecting a pair of adjacent vertices is termed a *path*. The number of edges which belong to a given path is called its *length*.

A graph is said to be *connected* if, for each pair of its vertices, there is at least one such path leading from the one vertex to the other. Since a molecular graph depicts a given molecule [7, 8] in which each atom is bound in some way to the other ones, it must be connected.

Notice, that an edge represents an unordered pair of vertices. Thus, the edge set $\mathscr{E}$ collects all such pairs defined upon the vertex set $\mathscr{V}$. If upon the vertex set of a graph only unordered pairs are defined, then such a graph is called *simple*.

Comparing the molecular graphs of Fig. 1 with the respective constitutional formulae one observes that the coordination number of an atom in the molecule (i.e. the number of directly bonded atoms) is represented in the graphs by the degree of the corresponding vertex. Thus, all hydrogen atoms are depicted by vertices of degree 1 and the carbon atoms by vertices of degree 2, 3, or 4 depending on whether the carbon atoms are involved in triple or double bonds or not.

Taking together all these items, the following list of general properties of a molecular graph $\mathscr{G}$ corresponding to molecule M is obtained:

1. $\mathscr{G}$ is a connected simple graph.
2. The cardinality n of its vertex set equals the number of atoms of M.
3. Since $\mathscr{G}$ is connected, the cardinalities of its vertex and edge sets have to obey the following relation [2–6]:

$$m \geqq n - 1 . \tag{2}$$

4. The degree of a vertex of $\mathscr{G}$ and the coordination number of the corresponding atom in M coincide in their numerical values.

Concerning the topology of a given molecule, the most important information stored in the molecular graph $\mathscr{G}$ is the pattern of its edges, which is very clearly represented by means of the graph's *adjacency matrix* $A(\mathscr{G})$. This is a symmetric square matrix of order n. Its rows and columns correspond to the different vertices of $\mathscr{G}$ and its elements, A_{rs}, are either one or zero depending on whether the vertices r and s are adjacent or not. Thus we have:

$$A(\mathscr{G}) = (A_{rs}) ; \tag{3}$$

$$A_{rs} = \begin{cases} 1 \ldots r \; adj \; s , \\ 0 \ldots \text{otherwise} . \end{cases}$$

For instance, for the molecular graph $\mathscr{G}_4$ corresponding to ethylene one obtains:

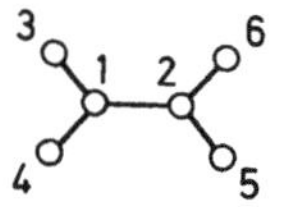

$\mathscr{G}_4$

$$A(\mathscr{G}_4) = \begin{bmatrix} 0 & 1 & 1 & 1 & 0 & 0 \\ 1 & 0 & 0 & 0 & 1 & 1 \\ 1 & 0 & 0 & 0 & 0 & 0 \\ 1 & 0 & 0 & 0 & 0 & 0 \\ 0 & 1 & 0 & 0 & 0 & 0 \\ 0 & 1 & 0 & 0 & 0 & 0 \end{bmatrix} . \tag{4}$$

It is easy to verify that the sum of the entries of one row (column) agrees with the degree of the corresponding vertex. Obviously, the actual form of the adjacency matrix depends on the labeling of the vertices; but whatever the labels, the adjacency matrix of a simple connected graph never takes a block diagonal form with off diagonal zero blocks.

While the relation of a constitutional formula to the corresponding molecular graph is definitely unique, the same is not true for the opposite direction as already seen in the case of diatomic molecules. So, for instance, the molecular graphs derived for dimethylethylsilan or dimethylethylchlorosilan are isomorphic with $\mathcal{G}_1$ and that one derived for trimethylhydrazine is isomorphic with $\mathcal{G}_2$. Two molecules are called *isotopological* if their molecular graphs are isomorphic. Thus, in general, a molecular graph corresponds to a family of isotopological molecules. As it is already illustrated by the examples given above, from one member of such a family another member is obtained when one atom of the molecule is replaced by a different one which has the same valence or coordination number. Because the net electric charge of a molecule has no relevance for its topology, even methane and the ammonium cation are isotopological molecules. Further, from the fact that all stereoisomers of a molecule, if any, agree with a single constitutional formula one may conclude that they form a subfamily of isotopological molecules. As it will turn out later the phenomenon of isotopology gives rise for some unsolved problems in the relation between molecular topology and properties; but in certain cases it could be used as a guide to molecular structure (see for instance Sect. 5).

The phenomena of isotopology and isoelectronicity are not coupled; for instance, trimethylethylene and trimethylhydrazine form a pair of isotopological but not isoelectronic molecules while benzene and borazene are not only isotopological but also isoelectronic.

2.2 Skeleton Graphs

Often in chemical applications of graph theory hydrogen atoms are not depicted. The graphs obtained in such a way are called *hydrogen suppressed graphs* or *skeleton graphs*. Obviously, the skeleton graphs are subgraphs of the complete molecular graphs. Within a defined class of compounds as saturated or fully conjugated hydrocarbons etc. the neglect of hydrogen atoms does not cause any ambiguity; but notice that the skeleton graphs $\mathcal{G}_3$ of saturated iso-pentane and unsaturated trimethylethylene are isomorphic (see Fig. 1). Therefore, in the general treatment of molecular topology, skeleton graphs should be used only with caution.

Of particular interest are the skeleton graphs of fully conjugated systems. They are isomorphic with the *Hückel graphs* of the systems which depict the corresponding π-AO basis sets. They provide the basis for graph theoretical investigations of π-electron systems [9]. Further, they represent the topology of π-electron systems [10].

2.3 Line Graphs

Another type of graph related to the pattern of bonds in molecules is the *line graph of molecular graphs*. They are rarely used in present chemical applications of graph theory. Nevertheless, they are very useful in the discussion of molecular topology and, hence, they will be introduced here.

In order to obtain the line graph, $\mathscr{L}(\mathscr{G})$, of an arbitrary simple graph, $\mathscr{G}$, the edges of $\mathscr{G}$ are depicted as vertices forming the vertex set of $\mathscr{L}(\mathscr{G})$. Two vertices of $\mathscr{L}(\mathscr{G})$ are connected by an edge if and only if the corresponding edges of $\mathscr{G}$ are adjacent, i.e. they have a vertex in common. In such a way the line graphs $\mathscr{G}_5 = \mathscr{L}(\mathscr{G}_1)$ and $\mathscr{G}_6 = \mathscr{L}(\mathscr{G}_2)$ are derived from the molecular graphs $\mathscr{G}_1$ and $\mathscr{G}_2$ shown in Fig. 1; in Fig. 2 they are drawn such that any crossing of edges is omitted. Note that a vertex $v \in \mathscr{G}$ of degree g is incident with g edges which are mutually adjacent. Hence, in $\mathscr{L}(\mathscr{G})$ they are depicted by the complete graph $\mathscr{K}_g$ which is a subgraph of $\mathscr{L}(\mathscr{G})$. Consequently, line graphs may be considered as composed of a series of subgraphs all of which are complete graphs.

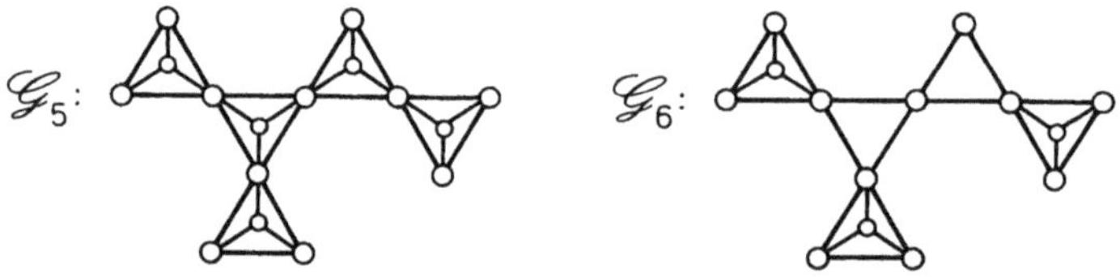

Fig. 2. Line graphs of the molecular graphs of isopentane ($\mathscr{G}_5$) and trimethylethylene ($\mathscr{G}_6$)

The method of construction of line graphs implies that the line graph of a connected graph is connected. Since connectedness is one of the characteristics of molecular graphs, one may conclude: *line graphs of molecular graphs are connected.*

With a single exception it has been shown that the line graphs of non-isomorphic graphs are not isomorphic. The exception from that rule is the complete graph $\mathscr{K}_3$ and the complete bipartite graph $\mathscr{K}_{1,3}$ shown below which have, indeed, iso-

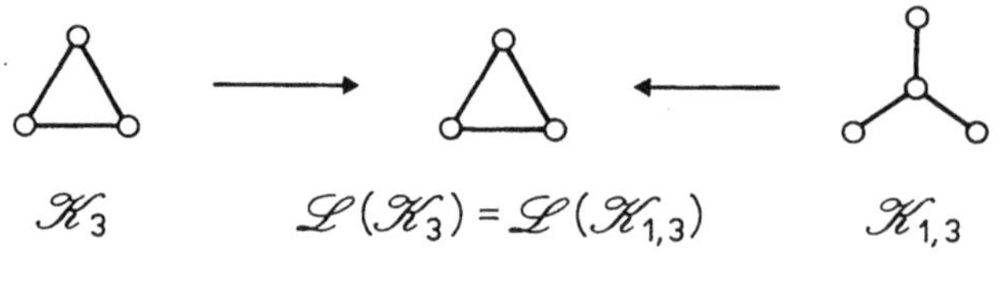

morphic line graphs. Taking this into account, the general rule given above may be denoted as follows:

If $\mathscr{G}' \neq \mathscr{G}''$ and $\{\mathscr{G}', \mathscr{G}''\} \neq \{\mathscr{K}_3, \mathscr{K}_{1,3}\}$ then

$$\mathscr{L}(\mathscr{G}') \neq \mathscr{L}(\mathscr{G}''). \tag{5}$$

The idea of considering the topology of σ-electron systems would normally associate each bond of the molecule with a localized σ- and σ^*-bond orbital (BO). Provided the molecule considered doesn't contain lone pair electrons then the line graph of the molecular graphs depicts both the σ-BO and the σ^*-BO basis sets and consequently it represents their topologies.

In the case of molecules containing lone pairs which are parts of their σ-electron system, the basis graph of σ-BO is obtained by supplementing the line graph with as many vertices as there are lone pairs present together with the appropriate edges; certainly the vertex depicting a lone pair on a given atom is adjacent to those vertices which depict the other bonds in which the atom itself is involved. Details concerning the treatment of the topologies of σ- and σ^*-electron systems may be found elsewhere [5, 11].

2.4 Summary

As has been shown the topology of a molecule is represented by its molecular graph which under certain conditions may be replaced by the corresponding skeleton graph. The skeleton graphs of fully conjugated systems may be considered to be the basis graphs of the π-AO and, hence, they exhibit the topology of the respective π-electron systems. The topology of σ-electron systems is represented either by the line graph of the molecular graph itself or by its appropriate augmented form. In contrast, the topology of a σ^*-electron system is always expressed by the line graph of the molecular graph. As a consequence of the interrelations between these simple connected graphs they offer an alternative description of molecular topologies.

3 Spaces and Polynomials

All simple graphs and, hence, also those discussed in the previous section, are associated with a series of mathematical objects [2–6] such as groups, functions, matrices, numbers, etc. Some of these objects have importance for the discussion of molecular topology and/or the relationship between topological and physicochemical properties of molecules. Hence, a selection of these is introduced in this section, namely, the metric and topological spaces and further, the characteristic and the acyclic polynomials associated with simple connected graphs. Since the following considerations are limited to simple connected graphs only, these graph properties are not always explicitly indicated. Also it should be noted that the two polynomials discussed below are not the only ones which may be associated with graphs; but the polynomials not discussed here are either specific to certain classes of graphs or they have only little relevance in the discussion of molecular topology.

3.1 Metric Spaces

As already explained in the previous section a path consists of a sequence of pairwise adjacent vertices, say $v_0, v_1, \ldots, v_j, \ldots v_l, v_j \, adj \, v_{j-1}$, and those edges which connect adjacent vertices. The number of edges of a path is called its length, l. So, for instance, the path (3, 1, 2, 5) in $\mathcal{G}_4$ has the length 3. Any path leading from vertex r to vertex s or vice versa will be denoted by $\mathcal{P}_{rs}$. All such paths form the set $\{\mathcal{P}_{rs}\}$ which can never be empty in the case of simple connected graphs by definition. The different members of the set $\{\mathcal{P}_{rs}\}$ may have different lengths; those which have shortest length are called geodesic paths and their length is called the distance, $d(r, s)$, between the vertices r and s.

In this way a *distance* is defined for each pair of vertices. As can be easily shown they obey the metric axioms, namely:

$$
\begin{aligned}
&[\mathrm{M1}]: \quad d(r, s) \geqq 0 \quad \text{for all } r, s \in \mathcal{V}(\mathcal{G})\,; \\
&[\mathrm{M2}]: \quad d(r, s) = 0 \quad \text{if and only if } r = s\,; \\
&[\mathrm{M3}]: \quad d(r, s) = d(s, r) \quad \text{for all } r, s \in \mathcal{V}(\mathcal{G})\,; \\
&[\mathrm{M4}]: \quad d(r, s) \leqq d(r, t) + d(s, t) \quad \text{for all } r, s, t \in \mathcal{V}(\mathcal{G})\,.
\end{aligned}
\tag{6}
$$

Hence, the distances defined upon the vertex set $\mathcal{V}(\mathcal{G})$ of a connected simple graph form a *metric*. That metric is discrete because by definition all non-zero distances are expressed by integers. It is said that a set and the metric defined upon it form a metric space; hence, each connected simple graph is associated with such a *metric space*.
Usually the distances between the vertex pairs of $\mathcal{G}$ are described by the *distance matrix* $\boldsymbol{D}(\mathcal{G})$ defined as follows:

$$
\begin{aligned}
\boldsymbol{D}(\mathcal{G}) &= (D_{rs})\,, \\
D_{rs} &= d(r, s)\,.
\end{aligned}
\tag{7}
$$

According to Eq. (7) and [M3] the distance matrix is a symmetric square matrix of order n. For instance, in the case of the molecular graph of ethylene, $\mathcal{G}_4$ (see Eq. (4)), it takes the following form

$$
\boldsymbol{D}(\mathcal{G}_4) = \begin{bmatrix} 0 & 1 & 1 & 1 & 2 & 2 \\ 1 & 0 & 2 & 2 & 1 & 1 \\ 1 & 2 & 0 & 2 & 3 & 3 \\ 1 & 2 & 2 & 0 & 3 & 3 \\ 2 & 1 & 3 & 3 & 0 & 2 \\ 2 & 1 & 3 & 3 & 2 & 0 \end{bmatrix}. \tag{8}
$$

In computer assisted work, the distances may be obtained from the powers of the adjacency matrix since

$$
d(r, s) = \min\{\nu \mid [\boldsymbol{A}^{\nu}]_{rs} \neq 0,\ 0 \leqq \nu \leqq n - 1\}\,. \tag{9}
$$

3.2 Topological Spaces

A topological space is formed by a set and the topological structure defined upon the set [12, 13]. Thus a simple connected graph can be associated with a topological space provided it can be shown that a topological structure is defined upon its vertex set. For the execution of such a proof several equivalent methods are available.

In the first approach [14] to solving that problem an open set formalism was used. Consider a set $\mathscr{R}$ and its subsets $\mathscr{R}_j \subseteq \mathscr{R}, j = 1, 2, 3, \ldots$. Upon the set $\mathscr{R}$ an open set topology is defined if a system of subsets $\boldsymbol{T}_O = \{\mathscr{R}_1, \mathscr{R}_2, \mathscr{R}_3, \ldots\}$ exists which obey the following axioms:

[O1]: each union of members of $\boldsymbol{T}_O$ is a member of $\boldsymbol{T}_O$;
[O2]: each intersection of a finite number of members of $\boldsymbol{T}_O$ is a member of $\boldsymbol{T}_O$;
[O3]: set $\mathscr{R}$ belongs to $\boldsymbol{T}_O$; (10)
[O4]: the empty set $\boldsymbol{\Phi}$ belongs to $\boldsymbol{T}_O$.

It has been shown [14, 15] that the open set formalism may be applied straight forwardly to the vertex sets of bipartite graphs but needs the auxiliary construction of para-spectral duplex [16] in case of non-bipartite graphs.

Another attempt [17] applied the neighbourhood formalism. Consider a set $\mathscr{R}$ upon which a metric is defined and let r, s be arbitrary members of $\mathscr{R}$. For each $r \in \mathscr{R}$ and $\varepsilon > 0$ a ball-neighbourhood of r is defined as the set of all elements of $\mathscr{R}$ whose distance is smaller than ε:

$$\mathscr{U}_\varepsilon(r) = \{s \mid d(r, s) < \varepsilon\}; \tag{11}$$

ε is called the radius of $\mathscr{U}_\varepsilon(r)$. Each ball-neighbourhood contains at least one element, namely r itself. A subset of $\mathscr{R}$ is called a neighbourhood $\mathscr{U}$ of r if for some value $\varepsilon > 0$ it contains a ball-neighbourhood of r, $\mathscr{U} \supseteqq \mathscr{U}_\varepsilon(r)$. Finally, let $\boldsymbol{U}(r)$ be the system of all neighbourhoods of r. Then, upon the set $\mathscr{R}$ a neighbourhood topology or a U-topology (U = Umgebung) $\boldsymbol{T}_U$ is defined if each element r of $\mathscr{R}$ is associated with a system $\boldsymbol{U}(r)$ of neighbourhoods of r obeying the following axioms:

[U1]: $r \in \mathscr{U}$ for $\mathscr{U}$ of $\boldsymbol{U}(r)$;
[U2]: if $\mathscr{U}$ is a member of $\boldsymbol{U}(r)$ and $\mathscr{U}' \supseteqq \mathscr{U}$, then $\mathscr{U}'$ is a member of $\boldsymbol{U}(r)$; $\mathscr{R}$ is a member of $\boldsymbol{U}(r)$;
[U3]: the intersection of two members of $\boldsymbol{U}(r)$ is a member of $\boldsymbol{U}(r)$;
[U4]: for every $\mathscr{U} \in \boldsymbol{U}(r)$ there exists a $\mathscr{U}'' \in \boldsymbol{U}(r)$, such that $\mathscr{U}'' \in \boldsymbol{U}(s)$ for all $s \in \mathscr{U}''$. (12)

In Fig. 3 the application of that formalism to connected graphs is illustrated. The graph considered, $\mathscr{G} = \mathscr{P}_4$, consists of 4 vertices and 3 edges. The set $\mathscr{R}$ used above is identified by the vertex set $\mathscr{V}(\mathscr{G}) = \{1, 2, 3, 4\}$. The neighbourhoods of the vertices are represented by the corresponding subgraphs $\mathscr{G}'_k \subseteq \mathscr{G}$. There are, in total, 10 such subgraphs; together with the empty graph $\mathscr{G}'_0$ they form the system of subgraphs $\boldsymbol{G}$. The neighbourhood systems $\boldsymbol{G}(j)$ of the vertices j, $j = 1, 2, 3, 4$, are subsets of $\boldsymbol{G}$; they are shown in Fig. 3 and the vertex j in question is marked by a full circle. It is easy to prove that the neighbourhoods and the neighbourhood systems of all vertices as exhibited in Fig. 3 obey the axioms [U1]–[U4], hence, they represent

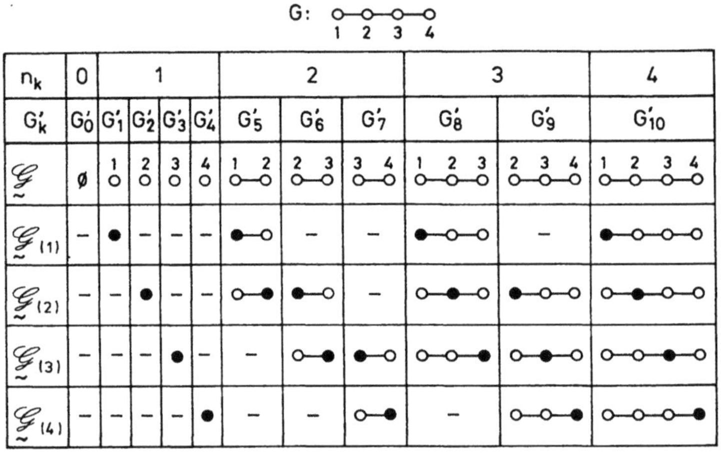

Fig. 3. Neighbourhoods and neighbourhood systems of the vertices of $\mathcal{G} = \mathcal{P}_4$

the T_U-topology of $\mathcal{G}$. Figure 3 further illustrates the formative influence of the edge pattern of the topological space of $\mathcal{G}$.

From these results [14,17] one may conclude that *each connected simple graph $\mathcal{G}$ is associated with a topological space $T(\mathcal{G})$*. Since the molecular graph is uniquely derived from the constitutional formula of the molecule one observes that *each covalent molecule is associated with a topological space via its molecular graph.*

Some properties of these topological spaces may be derived by means of Fig. 3. Suppose we are interested in the topology of another graph $\mathcal{G}'$ consisting of 3 vertices and 2 edges; obviously $\mathcal{G}'$ is isomorphic with subgraph $\mathcal{G}'_8 \subset \mathcal{G}$. The topological structure of $\mathcal{G}' \cong \mathcal{G}'_8$ is obtained from Fig. 3 by striking out the columns referring to subgraphs $\mathcal{G}'_4$, $\mathcal{G}'_7$, $\mathcal{G}'_9$, $\mathcal{G}'_{10}$ and the row representing $G(4)$. This shows that the topological space associated with $\mathcal{G}'$ is a subspace of $T(\mathcal{G})$, $T(\mathcal{G}') \subset T(\mathcal{G})$. In the case that $\mathcal{G}$ represents the molecular graph of a molecule M, then its subgraph $\mathcal{G}'$ corresponds with a distinct moiety of M. Hence, the result obtained may be generalized as follows: *Each moiety of a molecule is associated with a distinct subspace of the topological space associated with the molecule.*

A moiety of special interest is the skeleton of a molecule as represented by skeleton graphs. From the last statement it is immediately apparent that *the topological space associated with a skeleton graph is a subspace of the topological space associated with the molecular graph.* From this the use of skeleton graphs in molecular topological considerations is legitimated in principle; but, as pointed out in the previous section, the use of skeleton graphs requires some caution.

Let us suppose now the molecule M contains two structurally equivalent moieties corresponding to the subgraphs $\mathcal{G}''$ and $\mathcal{G}'''$ of the molecular graph $\mathcal{G}$, respectively; then, obviously, these subgraphs are disjoint and isomorphic, $\mathcal{G}'' \cong \mathcal{G}'''$. In the example of Fig. 3, such a situation is obtained if $\mathcal{G}''$ and $\mathcal{G}'''$ are identified with the subgraphs $\mathcal{G}'_5$ and $\mathcal{G}'_7$, respectively. It is easy to verify that the topological subspaces associated with $\mathcal{G}'_5$ and $\mathcal{G}'_7$, respectively, are isomorphic. Thus, generalizing this result, one may state: *The subspaces associated with structurally equivalent moieties of a molecule are isomorphic.*

It has been shown [12, 13] that the topological structure of a given set does not depend on the method used for its definition. Consequently, the topological spaces derived for a given molecule by means of either the open set [14] or the neighbourhood formalism [17] are equivalent. In general, the neighbourhood formalism is not so readily applicable as the open set formalism because it requires that a metric is defined upon the vertex set. Fortunately this requirement is always met by the vertex sets of connected simple graphs. In the neighbourhood formalism the metric serves as a kind of blueprint showing how the different subsets forming the neighbourhood systems should be constructed. Applying the open set formalism, one also needs such guide; in [14] the transitivity of digraphs corresponding with the simple connected graph is used for that purpose.

In a completely analogous manner the line graphs of molecular graphs are associated with topological spaces too. Provided that the molecular graph is neither $\mathscr{K}_3$ nor $\mathscr{K}_{1,3}$, (see Eq. (5)), the topological space of its line graph, $\boldsymbol{T}(\mathscr{L}(\mathscr{G}))$, also uniquely represents the topology of the molecule considered.

3.3 Characteristic and Acyclic Polynomials

The characteristic polynomial of a graph is defined by means of its adjacency matrix as follows

$$\Phi(\mathscr{G}; x) = \det |\boldsymbol{A}(\mathscr{G}) - x\boldsymbol{I}_n|; \tag{13}$$

therein $\boldsymbol{I}_n$ denotes the unitary matrix of order n and x is the variable of the polynomial. From Eq. (13) one concludes that $\Phi(\mathscr{G}, x)$ is a real polynomial in x of degree n which has n real zeros; they represent the eigenvalues of $\boldsymbol{A}(\mathscr{G})$ and are called the spectrum of graph $\mathscr{G}$.

Consider the expansion theorem for determinants, viz

$$G = |G_{rs}| = \sum_{\{P\}} (-1)^p \mathfrak{P} G_{11} G_{22} \dots G_{jj} \dots G_{nn}, \tag{14}$$

where G and G_{rs} concisely denote the determinant of Eq. (13) and its elements, respectively, $\mathscr{P}$ is a permutation operator acting on the column indices while row indices remain unchanged, p indicates the number of transpositions composing a permutation P, and the summation runs over all $n!$ permutations. By means of this theorem it becomes clear that non-vanishing terms in Eq. (14) are only produced by those transpositions which correspond with edges of the graph [18]. Consequently, each contribution to $\Phi(\mathscr{G}; x)$ may be depicted by a subgraph of $\mathscr{G}$; the complete set of these are called *Sachs graphs* [4–6, 9, 19]. According to the cycle structure of a particular permutation P, the corresponding Sachs graph consists of as much $\mathscr{K}_2$ and $\mathscr{C}_l$, $l = 3, 4, \dots$, as P is composed of (isolated) transpositions and cyclic permutations of length l; $\mathscr{K}_2$ and $\mathscr{C}_l$ denote the complete graph consisting of 2 vertices and 1 edge and the cyclic graph formed by l vertices of degree 2, respectively.

All these details illuminate the intimate relationship between the cofficients of the characteristic polynomial and the edge pattern of a graph which has a formative influence on its topological space as already mentioned above. Nevertheless, there is the phenomenon of cospectral graphs [20] exhibited by particular ensembles of two or more non-isomorphic graphs (all of them have necessarily the same number of vertices) which give rise to identical characteristic polynomials. A detailed discussion of that topic is found elsewhere [19–21]. While Eq. (13) shows that each graph has a unique characteristic polynomial, the phenomenon of cospectrality indicates that this relation is not necessarily unique in the opposite direction.

The characteristic polynomial of graph $\mathcal{G}$ may be expressed in terms of the characteristic polynomials of subgraphs, e.g. as follows:

$$\Phi(\mathcal{G}; x) = \Phi(\mathcal{G} - (rs); x) - \Phi(\mathcal{G} - r - s; x) - 2 \sum_{\{\mathcal{P}_{rs}\}} \Phi(\mathcal{G} - \mathcal{P}_{rs}; x) . \quad (15)$$

Therein $\mathcal{G}$-(rs), $\mathcal{G}$-r-s, and $\mathcal{G}$-$\mathcal{P}_{rs}$ denote the graphs obtained from $\mathcal{G}$ by the deletion of the edge (rs), the vertices r and s (with all incident edges), and the path $\mathcal{P}_{rs}$, respectively. It is easy to verify that the last term in Eq. (15) corresponds with cyclic permutations of length $l > 2$ in Eq. (14).

Another polynomial associated with graph $\mathcal{G}$ is the acyclic (or matching) polynomial defined as follows:

$$\alpha(\mathcal{G}; y) = \sum_{k=0}^{[n/2]} (-1)^k \, m(\mathcal{G}, k) \, y^{n-2k} ; \quad (16)$$

therein $[n/2]$ indicates the largest integer equal or less $n/2$. This polynomial was used for the first time in studying some problems of statistical physics [22]. The coefficients, $m(\mathcal{G}, k)$, of the polynomial are called k-matching; they count the number of different ways k mutually disjoint edges may be selected from graph $\mathcal{G}$. Thus they correspond to those Sachs graphs which do not contain any cycle, a fact to which the name acyclic polynomial refers. If n is an even number then the polynomial possesses an absolute term with the coefficient $m(\mathcal{G}, n/2)$, called perfect matching. In case of Hückel graphs it indicates the number of Kekulé structures. A detailed discussion of acyclic polynomials is found elsewhere [23].

According to the definition of the coefficients $m(\mathcal{G}, k)$ it was assumed, acyclic polynomials have pure combinatorial character. However, it has been shown recently [24] that in principle, an acyclic polynomial $\alpha(\mathcal{G}; y)$ may be considered as the characteristic polynomial of a particular weight matrix, $\boldsymbol{W}(\mathcal{G})$, of the graph, although the explicit notation of their elements (i.e. the edge weights of $\mathcal{G}$) requires in general a more flexible number system than is provided by the set of complex numbers; even the flexibility of quaternions is only sufficient for a limited number of basic polycyclic topologies [24].

In analogy to Eq. (15), acyclic polynomials obey the following relation:

$$\alpha(\mathcal{G}; y) = \alpha(\mathcal{G} - (rs); y) - \alpha(\mathcal{G} - r - s; y) . \quad (17)$$

Notice that Eqs. (15) and (17) formally differ only with respect to cyclic terms not present in case of acyclic polynomials.

Expanding the determinants of acyclic graphs according to Eq. (14), no cyclic terms can be produced because an acyclic graph has no cycles. Hence, the characteristic and the acyclic polynomials of acyclic graphs coincide.

Since the two types of polynomials differ only with respect to the cyclic terms they may be expressed in a unified form, say $\mu(\mathcal{G}; z, t)$, by means of a discriminating parameter, t, which weights the cyclic contributions and takes the value 1 in case of characteristic polynomials but 0 in case of acyclic ones [25]. Thus, these μ-polynomials have to obey the following equalities:

$$\begin{aligned} \mu(\mathcal{G}; z, 1) &= \Phi(\mathcal{G}; z) , \\ \mu(\mathcal{G}; z, 0) &= \alpha(\mathcal{G}; z) . \end{aligned} \tag{18}$$

If t is assumed to be a vector, the elements of which correspond with the different independent cycles of the graph then the topological effects of any cycle or combination of cycles can be elucidated. Applying these ideas to Hückel graphs [25] a series of rules have resulted which complete the Hückel rule for the case of fully conjugated polycyclic systems.

4 Modelling of Physico-Chemical Properties

Molecular and skeleton graphs contain a large amount of detailed information about the topology of the molecule, as for instance, distances of vertex pairs, vertex degrees, the number of paths of a given length, self-returning walks and others. By their means some graph invariants, $\boldsymbol{I}(\mathcal{G})$, may be defined [5] which, to some extent, characterize the global topology of the respective molecules. In chemical applications these numbers are often called topological or graph theoretical indices. They have been frequently used for correlations with a broad variety of physico-chemical data [26–28].

Although the physical and topological background of such correlations and even the physical meaning of the indices used therein are not completely cleared up yet, these correlations indicate the significant role which molecular topology plays in the modelling of physico-chemical properties. But with regard to the phenomenon of isotopology, the kind of atoms used for the realization of a given topology on an actually existing molecule also play a significant role. Finally, the embedding of the constituent atoms in the three-dimensional space must have a contribution too as is indicated by the variation of distinct physico-chemical properties within a family of stereoisomers. As a result of these considerations one has to contemplate at least the following three factors which determine the magnitudes of physico-chemical properties:

[P1] molecular topology;
[P2] nature of constituent atoms and their respective numbers;
[P3] spatial distribution of constituent atoms.

Notice, that the term structure refers to the union of these factors in their entirety. Therefore, [P1]–[P3] can be seen to represent a detailed description of the concept of chemical structure.

From the success achieved in empirical correlation a strong influence of the molecular topological factor [P1] on the modelling of physico-chemical properties may be deduced. This view is supported by two non-empirical rules, the topological charge stabilisation (Sect. 5) and the topological effect on molecular orbitals (Sect. 6).

5 Topological Charge Stabilisation

The concept of topological charge stabilisation (TCS) [29, 30] is based on the observation that within a homonuclear molecule the charge density distribution depends on (i) the topology of the molecule and (ii) the number of electrons present. Suppose, for a given topology and a given number of electrons a charge density distribution results which attributes unequal charge densities to different atoms. In such a homonuclear molecule replacing one atom by another one which has a higher (lower) electronegativity, the position of largest (smallest) charge density is energetically more likely as can be seen from first order perturbation energy

$$E^{(1)} = \langle \psi^\circ | \mathfrak{H}' | \psi^\circ \rangle \,, \qquad (19)$$

$$\mathfrak{H}' = - \sum_{\alpha, j} \Delta Z_\alpha / r_{j,\alpha} \,,$$

wherein ψ° and ΔZ_α denote the unperturbed wave function of the homonuclear molecule and the difference between the nuclear charges of the atoms involved in such a replacement and j labels the electrons. For qualitative considerations it is sufficient to consider valence electrons only, and ΔZ_α may be replaced by the difference of the effective nuclear charges, $\Delta\zeta_\alpha$, which is a rough measure of the relative electronegativities.

The homonuclear molecule and its characteristic charge density distribution is called the *uniform reference frame* [29]. Applying TCS to it, a family of isotopological and isoelectronic molecules is generated; each member represents an energetically favoured hetero-derivative. Thus, in the TCS approach the topological factor [P1] (see Sect. 4) is kept constant. The geometrical factor [P3] will always vary. In the case of cyclic topologies, however, the effect of geometric variations might be negligibly small; in the case of acyclic topologies this effect can be minimized by an appropriate choice of the uniform reference frame (see Sect. 5.3).

In the following TCS is illustrated by means of a few examples taken from [29, 30]; more examples and their detailed discussion is found elsewhere [31–35].

5.1 Planar Conjugated Systems

For pentalene with 8 π-electrons the uniform frame ***1*** is obtained. The structures of the heterocycles **2** and ***3*** [36, 37] agree with the TCS rule applied to ***1***:

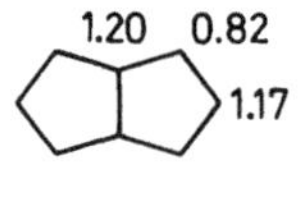

1

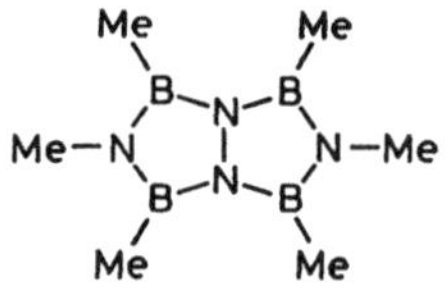

2

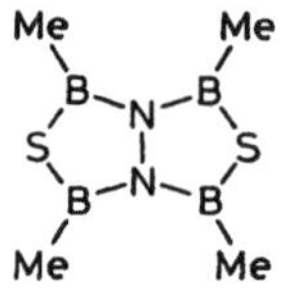

3

In case of 10 π-electrons the uniform frame **4** results. In accordance with TCS the electronegativity pattern of **5** [38] meets the charge density distribution of **4**:

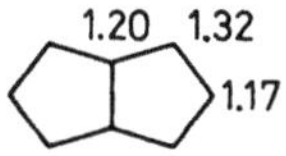

4

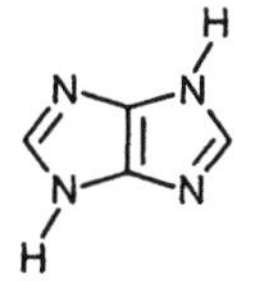

5

In the series of isoelectronic thienothiophenes **6–9** using TCS with reference to **4** one concludes with regard to the relative stabilities that those of **6** and **7** should be

6 7 8 9

comparable, but **8** and **9** should be successively less stable. This prediction agrees with (i) **9** is only known as tetraphenyl-substituted derivative [39]; (ii) selenium analogues of **6–8**, but not of **9** have been obtained [39]; (iii) results of several semiempirical calculations [40–42].

Interesting examples are derived from the uniform reference frames **10** (26 electrons) and **11** (32 electrons) of the porphyrine skeleton [29, 43]:

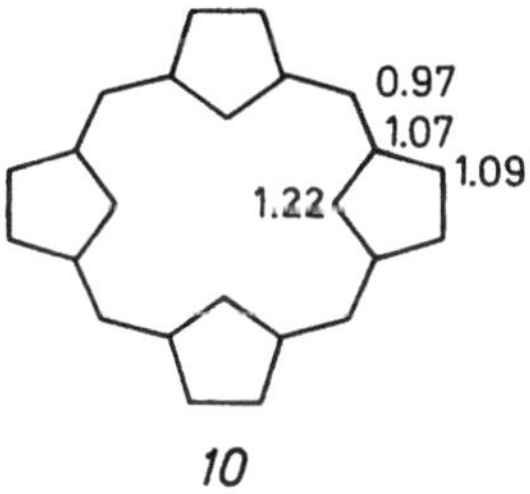

10

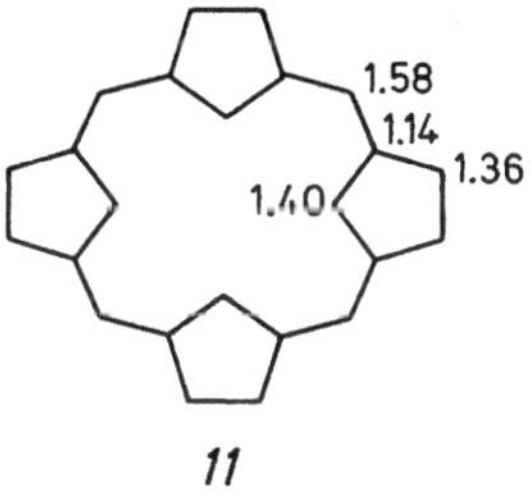

11

In porphyrine anions, isoelectronically with **10**, the nitrogens occupy the positions of largest charge density (1.22). In B_8S_{16}, isoelectronically with **11**, the boron atoms are located at the positions of smallest charge density (1.14).

5.2 Non-Planar Systems

A beautiful example [31] of the operational power of TCS is provided by some molecules which are isotopological and isoelectronic with heptaphosphorous-tri-anion [44], $P_7^{3\ominus}$. The corresponding uniform reference frame **12** has been calculated by means of extended Hückel (EH) method using 3d AOs in the basis set [31]: The largest charge densities are located in the bridging positions, a medium one is attributed to the apex, and smaller ones to the basal sites. The pattern of charges may be normalized by means of

$$q'_r = q_r - Q/N\,, \qquad Q = \sum_r q_r\,, \tag{20}$$

where Q and N denote the total charge and the number of centers of the uniform reference frame, respectively. In such a way from ***12*** the normalized uniform reference frame ***13*** is derived.

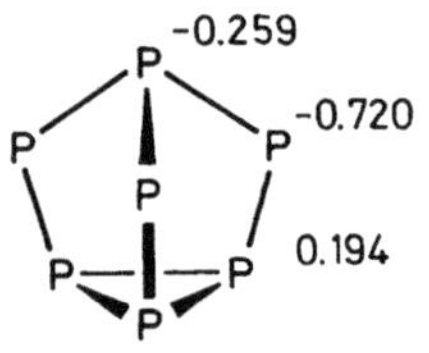

12

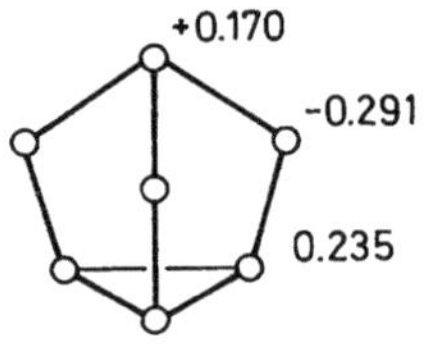

13

The structures of P_4S_3 (***14***) [45], As_4S_3 (***15***) [46], and PS_3As_3 (***16***) [47] agree with those derived from ***12*** by means of TCS.

14

15

16

These compounds are in chemical equilibrium [47], namely

$$P_4S_3 + 3\,As_4S_3 \rightleftarrows 4\,PS_3As_3 \tag{21}$$

which lies at the side of ***15***. This preference of ***15*** has been explained [46] by the greater strength of the PS than the AsS bonds and the increase of their stoichiometric number from 6 to 12; on the other hand, ***15*** is also energetically favoured by TCS, thus, TCS offers an alternative explanation of the driving force of the equilibrium [31].

Numerous applications of TCS to carboranes of the brutto formula $C_2B_{n-2}H_n$, $5 \leqq n \leqq 12$, have been reported: They concern predictions of the relative stabilities among series of isomers [34] and the discussion of the diamond-square-diamond rearrangements [35]. In all these cases the union reference frame is obtained by means of extended Hückel (EH) method (for $C_2B_4H_6$ see also subsection 5.4). The results agreee with experimental observations, if available, and they are supported by ab initio calculations [48].

Very interesting results have been obtained among adamantane-like molecules

[33]. The normalized uniform reference frame *17*, calculated for 56 valence electrons using an appropriate oxygen cluster, exhibits positive charge densities at the bridgehead sites and negative ones at the bridging positions. Thus, according to TCS, in molecules of type A_4B_6, atoms A should be more electronegative than atoms B. This expectation is realized by the structure of P_4O_6, As_4O_6, As_4S_6, Sb_4O_6, $P_4(NMe)_6$, $As_4(NMe)_6$, $(HC)_4S_6$, $(HSi)_4S_6$, and $(HSi)_4Se_6$; however, exceptions

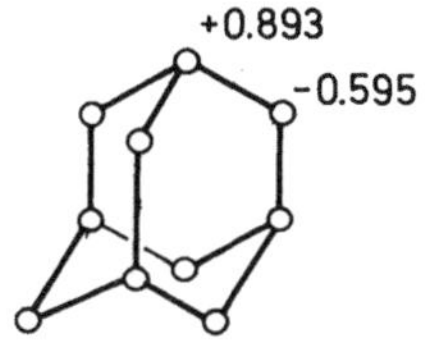

17

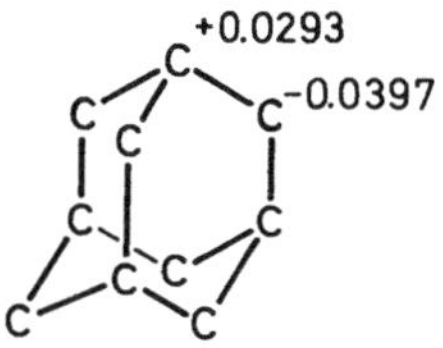

18

are $N_4(CH_2)_6$, $P_4(SiMe_2)_6$, and $P_4(GeMe_2)_6$. The charge density distribution within the skeleton of adamantane itself, as shown by *18*, agrees qualitatively with *17*, as does the normalized uniform reference frame for 44 valence electron systems; the respective normalized charges are —0,077 (bridgeheads) and +0,051 (bridging sites) [33]. In agreement with that pattern are the structures of the molecules $(HC)_4(BR)_6$ with R = Me, Cl, or Br.

For bridgehead substituted adamantanes the normalized uniform reference frame *19* has been obtained from an appropriate oxygen cluster [33]; tetramethyladaman-

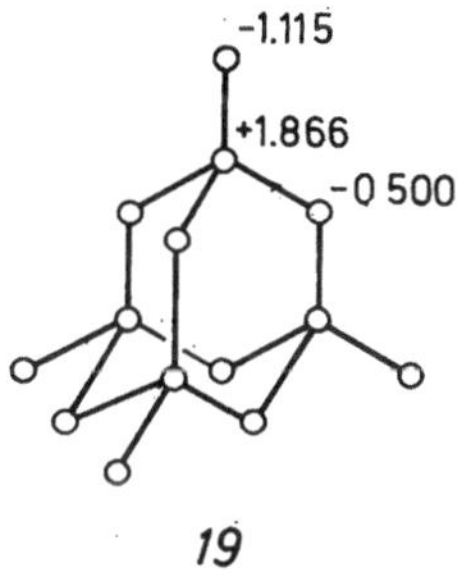

19

tane, $(MeC)_4(CH_2)_6$, represents a homonuclear realisation of that frame. In heteronuclear realisations of *19*, according to the rule of TCS, the electronegativities of the atoms should decrease in the order: exo > bridge > bridgehead. This expectation is met by most bridgehead substituted adamantane analogues of type $A_4B_6C_4$ and $A_{10}C_4$, namely:

(i) type $A_4B_6C_4$:
$P_4S_6O_4$, $P_4O_6S_4$, $Ge_4S_6Br_4$, $Ge_4S_6J_4$, $Ge_4Se_6J_4$, $Si_4(CH_2)_6Cl_4$;

(ii) type $A_{10}C_4$:
P_4O_{10}, P_4S_{10}, P_4Se_{10}, $Si_4S_{10}^{4\ominus}$, $Si_4Te_{10}^{4\ominus}$, $Ge_4S_{10}^{4\ominus}$, $Ge_4Se_{10}^{4\ominus}$, $Sn_4S_{10}^{4\ominus}$, $Sn_4Se_{10}^{4\ominus}$, $B_4S_{10}^{8\ominus}$, $Ga_4S_{10}^{8\ominus}$, $Ga_4Se_{10}^{8\ominus}$, $In_4S_{10}^{8\ominus}$, $In_4Se_{10}^{8\ominus}$

But there are also a few counter-examples, namely [33]:

$$Si_4O_6(t\text{-}Bu)_4,\ Si_4S_6Me_4,\ Ge_4S_6Me_4,\ Sn_4S_6Me_4,\ Sn_4Se_6Me_4;$$

notice, that these structures would agree with TCS predictions derived from reference frame *17* of non-substituted adamantane if the alkyl substituted group IVa atoms are considered as quasi-atoms.

5.3 Linear and Quasi-Linear Systems

As one could observe in the last two subsections, a polycyclic topology very often leads directly to a particular geometry of the uniform reference frame. Then geometrical variations of such a uniform reference frame are very limited. In contrast to that the topology of a chain does not produce such constraints. Because the pattern of the charge density distribution in a chain may vary significantly with its geometry, in applications of TCS to linear systems one must consider a series of geometries of the uniform reference frame. Obviously, this necessity diminishes the pure topological character of TCS in such applications, but the space factor [P3] in the modelling of physico-chemical properties (Sect. 4) can never be eliminated totally and here one has to pay tribute to it.

For symmetric five-atom chains with 24 valence electrons the normalized uniform reference frames **20** and **21** are obtained by means of extended Hückel (EH) method [30]. In agreement with these patterns OCCCO is almost linear, but OBOBO and

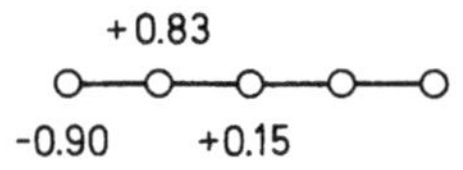

20

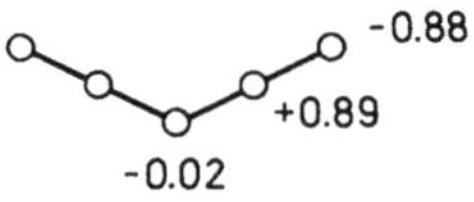

21

NCSCN are V-shaped [30]. Further reference frames are **22** (28 valence electrons), **23** (30 valence electrons), **24** (34 valence electrons), and **25** (36 valence electrons).

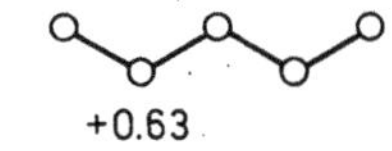

22

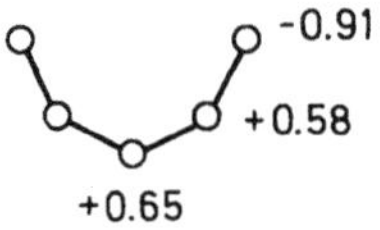

23

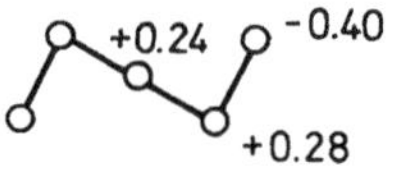

24

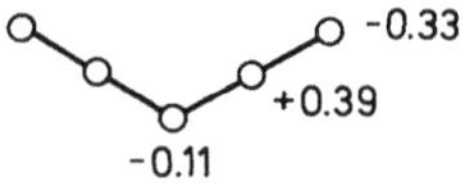

25

The planar W-shaped conformation of N_2O_3 agrees with **22**, the cations $J_5^{\oplus}$, $J_3Cl_2^{\oplus}$, $J_3Br_2^{\oplus}$ and the anions $J_5^{\ominus}$, $J_2Cl_3^{\ominus}$, $J_2Br_3^{\ominus}$ are in accord with **24** and **25**, respectively, and $Xe_2F_3^{\oplus}$ also agrees with **25**. The planar U-shaped anion $ClSNSCl^{\ominus}$, however, does not agree with **23**; at present it represents the only significant deviation from the rule of TCS in this series [30].

5.4 Uniform Reference Frames with Even Charge Density Distribution

In a few cases it may happen that all centers of the uniform reference frame are equivalent. Then for any number of electrons corresponding with a closed shell system, all centers have the same charge density and, hence, the rule of TCS cannot directly be applied. This difficulty is easily overcome by introducing a heteroatom of larger (smaller) electronegativity into one of the positions.

So, for instance, in the case of benzene as a uniform reference frame, the introduction of a nitrogen atom gives a pyridinium cation which has negative charges at the β-positions. The structures ***26–28*** are in agreement with that pattern of charge distribution [30]:

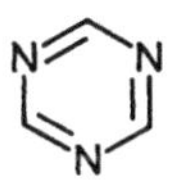
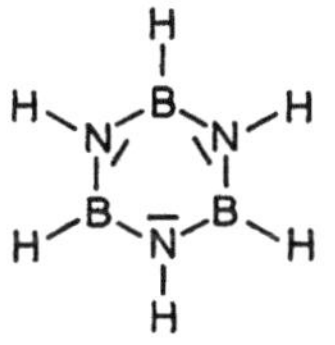
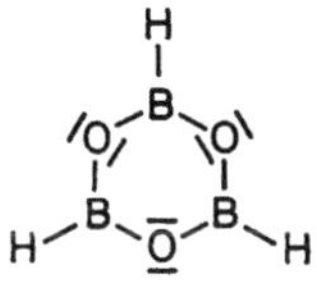

26 *27* *28*

A similar situation arises in the case of the $C_6^{2\ominus}$ octahedral reference frame. ***29*** shows the charge density distribution for $C_5N^{\ominus}$, normalized by means of Eq. (20), from which one concludes that the *trans*-isomer, 1,6-$C_2B_2H_6$, should be more stable than the *cis*-isomer, 1.2-$C_2B_4H_6$.

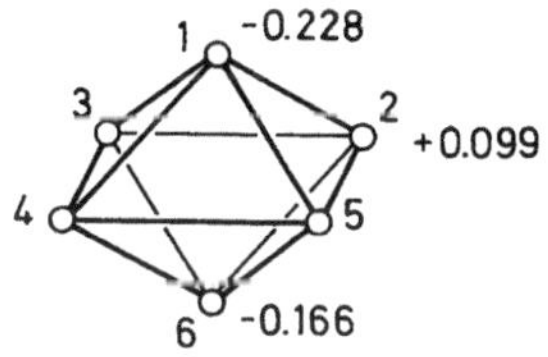

29

6 Topological Effect on Molecular Orbitals

6.1 Topomers

A completely different approach for the utilization of molecular topology is represented by the topological effect on molecular orbitals (TEMO) [49–52]. As pointed out in Sect. 4, there are three factors determining physico-chemical properties of molecules; for convenience they are listed here again:

[P1] molecular topology;
[P2] nature of constituent atoms and their respective numbers;
[P3] spatial distribution of constituent atoms.

A possibility of proving the relevance of the molecular topological factor [P1] is provided by comparing the properties of two topologically different molecules, say S and T, which are chosen such that the influence of the other two factors, [P2] and [P3], is the same or nearly the same for both molecules. This is easily achieved in case of [P2] if S and T are isomers, The spatial factor [P3], however, will always differ for unequal molecules but the difference of this factor for S and T will be small if the discrepancy between their topologies is not too large. Such a situation may be achieved if the isomers S and T are formed from only two or three fragments (moieties) which are connected differently in S and in T. Then the topological spaces

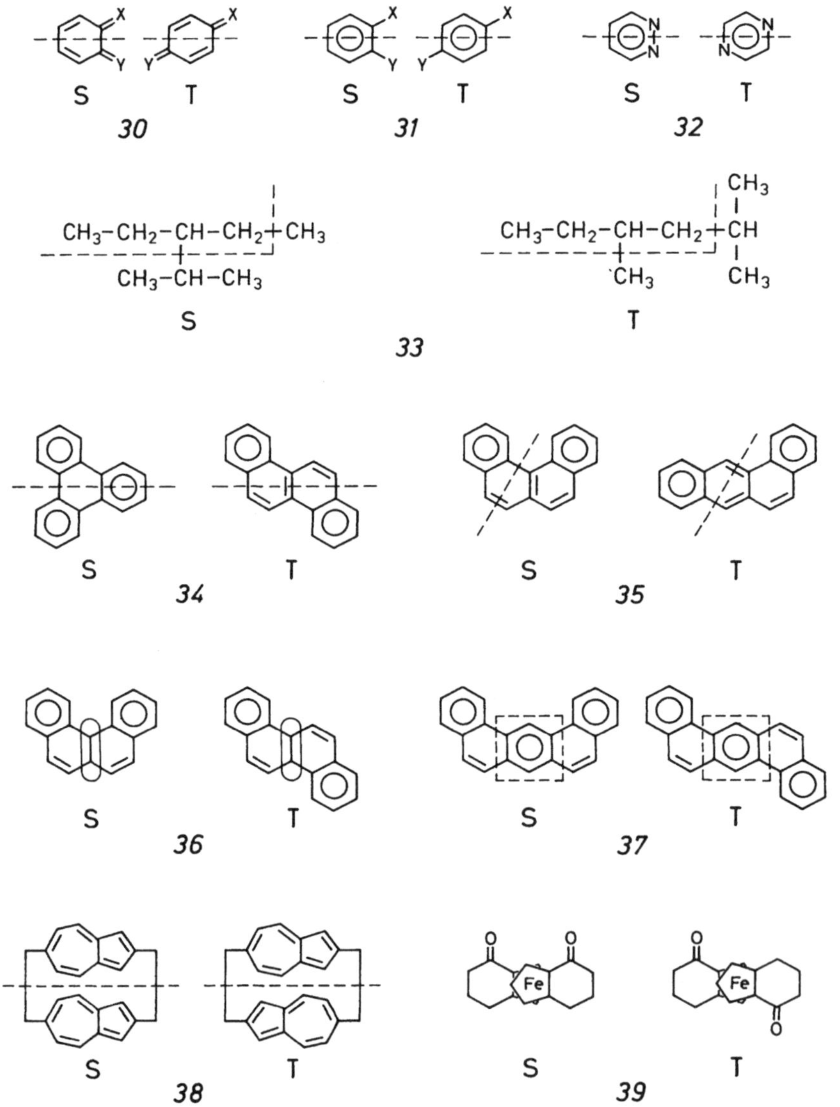

Fig. 4. Some examples for pairs of topomers ($X = Y$ or $X \neq Y$)

$T(\mathcal{S})$ and $T(\mathcal{T})$, associated with these molecules, consist of pairwise isomorphic subspaces. Isomers which meet this condition have been called *topologically related isomers*, or, more concisely, *topomers*. In Fig. 4 some examples for such pairs are given.

In order to obtain different topologies for S and T, the fragments from which they are formed, must contain at least two non-equivalent sites for their mutual connection; thus, the valency of the used fragments, l, must satisfy the condition $l \geqq 2$. In all examples shown in Fig. 4 except the pair **32** the connection sites are topologically non-equivalent. In the case of **32**, pyridazine and pyrazine, however, these centers are topologically equivalent but, in the molecules, they correspond to different elements, carbon and nitrogen, respectively. As shown in Fig. 4 the topomers may be constructed from two or three (**36**, **37**) moieties. The fragments may be saturated (**33**) or unsaturated, homonuclear (**33–38**) or heteronuclear; both fragments may be equal (**32**, **34**, **38**, **39**) or different; one of the fragments may even be disconnected (**33**).

The basis of this great variety in the construction of topomers is furnished by three models shown in Fig. 5. In these the fragments used for the formation of S and T are depicted by small boxes designated A, B, and C. The sites used for the mutual connection of the fragments are denoted by $k, l, \ldots \in$ A, $p, q, \ldots, \in$ B, $a, b, \ldots, e, f, \ldots, \in$ C. For the sake of simplicity in Fig. 5 only topological models for the case of $l = 2$ are depicted; analogous forms for $l > 2$ are easy to draw.

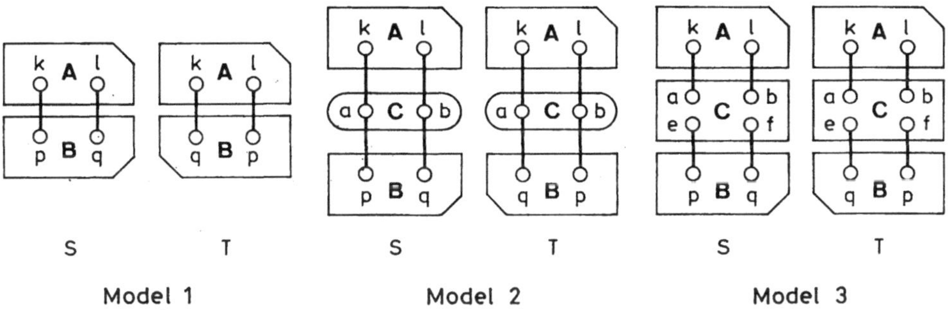

Fig. 5. Some models for the construction of pairs of topomers (for $l = 2$)

As an illustration the application of model 1 (Fig. 5) to the pair **35** (Fig. 4) is shown graphically below.

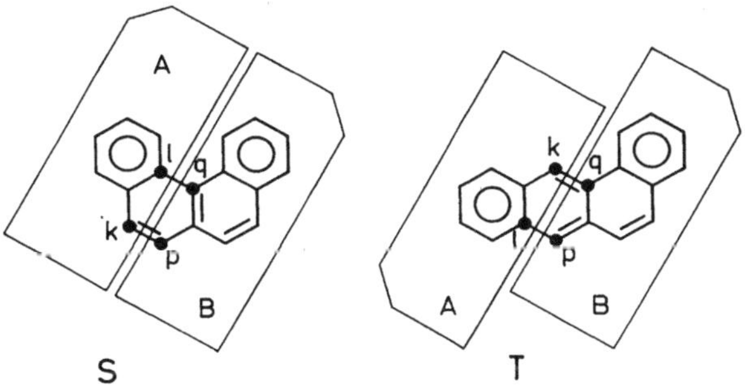

35

As already stated above the vertices k and l and also p and q are topologically non-equivalent.

As perusal of Fig. 5 shows, model 1 uses two, model 2 and 3 use three moieties for the formation of the topomers; the sites of the central moiety C are bivalent in model 2, but univalent in model 3.

It should be noticed that besides these three models a series of others has been studied [52–58]; some of them have been constructed for particular purposes [57, 58].

For the sake of brevity in that which follows only model 1, $l = 2$, is treated in some detail. Significant results will be supplemented by the corresponding ones obtained for other models.

6.2 Interlacing Theorem

In Sect. 3 it was shown that the topological structure of the vertex set of a connected simple graph $\mathcal{G}$ is represented by the pattern formed by the edges of the graph and, hence, determines the topological space $\boldsymbol{T}(\mathcal{G})$ associated with that graph. As exemplified by Eq. (4) this pattern is stored in the adjacency matrix $\boldsymbol{A}(\mathcal{G})$, by means of which the characteristic polynomial $\Phi(\mathcal{G})$ is defined, Eq. (13). From this relationship one may conclude that the comparison of the topological spaces of topomers, $\boldsymbol{T}(\mathcal{S})$ and $\boldsymbol{T}(\mathcal{T})$, may be performed by a comparison of the respective characteristic polynomials, $\Phi(\mathcal{S})$ and $\Phi(\mathcal{T})$.

According to Fig. 5, the characteristic polynomials of the molecular graphs $\mathcal{S}$ and $\mathcal{T}$ can be expanded in terms of the characteristic polynomials of the fragment graphs $\mathcal{A}$ and $\mathcal{B}$ and some of their subgraphs [59]. In case of model 1, $l = 2$, one obtains

$$\begin{aligned}\Phi(\mathcal{S}) = {} & \Phi(\mathcal{A})\,\Phi(\mathcal{B}) - \Phi(\mathcal{A}-k)\,\Phi(\mathcal{B}-p) - \\ & - \Phi(\mathcal{A}-l)\,\Phi(\mathcal{B}-q) + \Phi(\mathcal{A}-k-l)\,\Phi(\mathcal{B}-p-q) - \\ & - 2\sum_{\{\mathcal{P}_{kl}\}}\sum_{\{\mathcal{P}_{pq}\}} \Phi(\mathcal{A}-\mathcal{P}_{kl})\,\Phi(\mathcal{B}-\mathcal{P}_{pq}), \\ \Phi(\mathcal{T}) = {} & \Phi(\mathcal{A})\,\Phi(\mathcal{B}) - \Phi(\mathcal{A}-k)\,\Phi(\mathcal{B}-q) - \\ & - \Phi(\mathcal{A}-l)\,\Phi(\mathcal{B}-p) + \Phi(\mathcal{A}-k-l)\,\Phi(\mathcal{B}-p-q) - \\ & - 2\sum_{\{\mathcal{P}_{kl}\}}\sum_{\{\mathcal{P}_{pq}\}} \Phi(\mathcal{A}-\mathcal{P}_{kl})\,\Phi(\mathcal{B}-\mathcal{P}_{pq}),\end{aligned} \tag{22}$$

where $\mathcal{A} - k$ and $\mathcal{A} - \mathcal{P}_{kl}$ denote those subgraphs of $\mathcal{A}$ which are obtained by removing the vertex k (with all its incident edges) and the path $\mathcal{P}_{kl}$, respectively; all other notations are analogous.

Three of the terms of $\Phi(\mathcal{S})$ and $\Phi(\mathcal{T})$ are pairwise identical; hence, they cancel each other out in the difference polynomial, $\Delta(x)$, defined as follows:

$$\Delta(x) = \Phi(\mathcal{T}) - \Phi(\mathcal{S})\,. \tag{23}$$

After some simple algebra one arrives at

$$\Delta(x) = [\Phi(\mathcal{A}-k) - \Phi(\mathcal{A}-l)]\,[\Phi(\mathcal{B}-p) - \Phi(\mathcal{B}-q)]\,. \tag{24}$$

Let n_A and n_B denote the respective number of vertices of $\mathscr{A}$ and $\mathscr{B}$, then the degree of $\Delta(x)$ is equal or less than $n_A + n_B - 6$. While all characteristic polynomials must have real zeros, this is not necessarily true for $\Delta(x)$ because its factors are differences of two such characteristic polynomials.

No conclusion can be drawn from Eq. (24) regarding the sign of $\Delta(x)$. However, this situation changes when the moieties A and B are equal, i.e. the fragment graphs $\mathscr{A}$ and $\mathscr{B}$ are isomorphic, $\mathscr{A} \cong \mathscr{B}$. Then the two factors of $\Delta(x)$ are equal and one obtains

$$\Delta(x) = [\Phi(\mathscr{A} - k) - \Phi(\mathscr{A} - l)]^2 \geqq 0 \,. \tag{25}$$

Substituting in Eq. (23) implies that, in this very case, the characteristic polynomials of $\mathscr{S}$ and $\mathscr{T}$ are related as follows:

$$\Phi(\mathscr{T}) \geqq \Phi(\mathscr{S})\,, \qquad x \in (-\infty, +\infty)\,; \qquad \mathscr{A} \cong \mathscr{B} \tag{26}$$

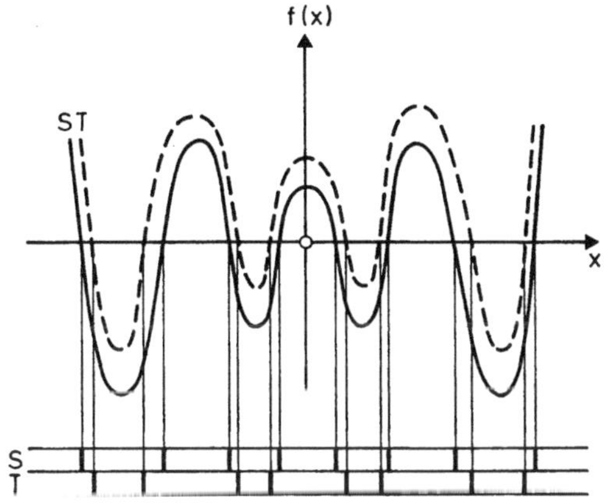

Fig. 6. Schematic illustration of the interlacing theorem

The situation characterized by Eq. (26) is schematically depicted in Fig. 6: In the complete range of the variable x the characteristic polynomial $\Phi(\mathscr{T})$ lies above $\Phi(\mathscr{S})$. As a consequence of that the zeros of these two polynomials, i.e. the eigenvalue of the topomeric graphs $\mathscr{S}$ and $\mathscr{T}$, interlace in a characteristic manner (see Fig. 6):

(i) the largest and most negative eigenvalue belongs to the spectrum of $\mathscr{S}$;

(ii) in the ranges determined by two subsequent eigenvalues of $\mathscr{S}$ there are alternantly two or no eigenvalues of $\mathscr{T}$.

This interlacing may be formally expressed as follows:

$$x_1^S \leqq x_1^T \leqq x_2^T \leqq x_2^S \leqq \dots \leqq x_{2j-1}^S \leqq x_{2j-1}^T \leqq x_{2j}^T \leqq x_{2j}^S \leqq \dots\,, \tag{27}$$

where the eigenvalues of $\mathscr{S}$ and $\mathscr{T}$ are labeled in non-decreasing order.

This result, as expressed by Eq. (27), has been called the *interlacing theorem*. To the best of our knowledge it was not obvious in the literature. Therefore the theorem was proved rigorously for $\mathscr{S}$ and $\mathscr{T}$ according to model 1, $l = 2$, $\mathscr{A} \cong \mathscr{B}$ (see

Fig. 5). One proof [60] applies Cauchy inequalities to $\Phi(\mathscr{S})$, $\Phi(\mathscr{T})$ and an appropriate set of further topological functions closely related to $\Phi(\mathscr{S})$ and $\Phi(\mathscr{T})$. In the other proof [61] the properties of two polynomials obeying the interlacing theorem are examined at first and then it is shown that $\Phi(\mathscr{S})$ and $\Phi(\mathscr{T})$ form a pair belonging to this class of polynomials. As a result of these proofs another sequence of eigenvalues which might agree to Eq. (26) but differs from Eq. (27) is excluded.

What kind of eigenvalue spectra should one expect in the case of non-isomorphic fragment graphs $\mathscr{A} \neq \mathscr{B}$, Eq. (24)? This situation is schematically depicted in Fig. 7. Let us denote the topomers S and T so that their polynomials obey the relation

$$\lim_{x \to \infty} [\Phi(\mathscr{T}) - \Phi(\mathscr{S})] > 0 . \tag{28}$$

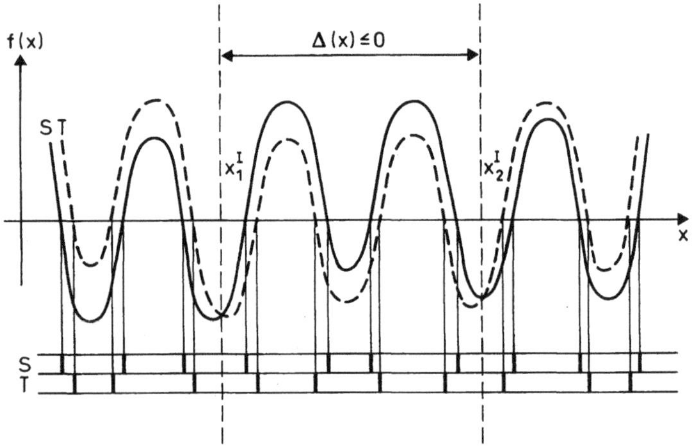

Fig. 7. Inversion of the interlacing in case of $\Delta(x) \leqq 0$

Then in those intervals of the variable x where $\Delta(x) \geqq 0$, the polynomial $\Phi(\mathscr{T})$ lies above $\Phi(\mathscr{S})$ and the sequence of eigenvalues accords with Eq. (27). But in those intervals of x, in which $\Delta(x) < 0$, the polynomials $\Phi(\mathscr{S})$ and $\Phi(\mathscr{T})$ interchange their mutual locations and within such an interval interlacing occurs in an inverted order. Only in the neighbourhood of the real zeros of $\Delta(x)$ is the sequence of Eq. (27) perturbed.

The real zeros of $\Delta(x)$ are called *inversion points*. Let x_k^I denote such an inversion point and supposing it lies between x_{2j-1}^{T} and x_{2j}^{T}, then the sequence of eigenvalues is perturbed as follows:

$$\ldots \leqq x_{2j-2}^{\mathrm{T}} \leqq x_{2j-2}^{\mathrm{S}} \leqq x_{2j-1}^{\mathrm{S}} \leqq x_{2j-1}^{\mathrm{T}} \leqq x_k^I \leqq x_{2j}^{\mathrm{S}} \leqq x_{2j}^{\mathrm{T}} \leqq \ldots . \tag{29}$$

Because no conclusions can be drawn concerning the real eigenvalues of $\Delta(x)$, as already mentioned above, the non-isomorphism $\mathscr{A} \neq \mathscr{B}$ is a necessary but not sufficient condition for the appearance of inversion points. Further, it may also happen that $\Phi(\mathcal{S})$ and $\Phi(\mathcal{T})$ have no eigenvalues within an interval of $\Delta(x) < 0$, i.e. the inversions are not traced out in the spectra of $\mathcal{S}$ and $\mathcal{T}$.

In the case of model 2, $l = 2$, $\mathcal{A} \cong \mathcal{B}$, the difference polynomial takes the following form

$$\Delta(x) = [\Phi(\mathcal{A} - k) - \Phi(\mathcal{A} - l)]^2 \, \Phi(\mathcal{C} - a - b) , \tag{30}$$

provided, $\Phi(\mathcal{C} - a) = \Phi(\mathcal{C} - b)$. Thus the sign of $\Delta(x)$ agrees with that of $\Phi(\mathcal{C} - a - b)$. If $\mathcal{C}$ consists of two vertices only, then $\Phi(\mathcal{C} - a - b) = 1$ and $\Delta(x) \geqq 0$ for all values of x.

For model 3, $l = 2$, $\mathcal{A} \cong \mathcal{B}$, one obtains

$$\Delta(x) = [\Phi(\mathcal{A} - k) - \Phi(\mathcal{A} - l)]^2 \, [\Phi(\mathcal{C} - a - f) - \Phi(\mathcal{C} - a - e)] , \tag{31}$$

provided $\Phi(\mathcal{C} - a) = \Phi(\mathcal{C} - b) = \Phi(\mathcal{C} - e) = \Phi(\mathcal{C} - f)$, etc. [52]. Once again the sign $\Delta(x)$ depends on that of its linear factor. For the C-moiety of *37*, Fig. 4, that factor equals 1, thus, the eigenvalues of the topomers of pair *37* interlace according to Eq. (27). A blue print for the construction of C-moieties such that $\Delta(x)$, Eq. (31), is non-negative may be found elsewhere [56].

If $l \geqq 3$, for all models $\Delta(x)$ consists of more than one term with partly opposite sign [5, 52]; thus, in all these cases the appearance of inversions cannot be excluded.

6.3 Physical Interpretation of the Interlacing Theorem: TEMO

The interlacing theorem represents a pure mathematical formalism relating the spectra of the graphs $\mathcal{S}$ and $\mathcal{T}$ to each other. Although these molecular graphs possess chemical meaning, the physical relevance of the interlacing theorem is not necessarily self-evident and, hence, must be proved. The basis for such proofs is provided by a physical interpretation of the eigenvalues of the topomers $\mathcal{S}$ and $\mathcal{T}$. It seems quite reasonable to identify them with the molecular orbitals of either the π-electron or the σ-electron system. As shown in Sect. 2 the basis graphs of these electron systems correspond with the skeleton and the line graph of the molecular graph, respectively, hence, their topologies are uniquely determined by that of the molecular graph. In what follows the eigenvalues of $\mathcal{S}$ and $\mathcal{T}$ are considered as π-MO's, as they may result from HMO calculations; a comment on the topological behaviour of σ-electron systems is given at the end of this subsection.

In Fig. 8 the eigenvalue pattern of an S/T-pair according to model 1, $l = 2$, $\mathcal{A} \cong \mathcal{B}$, is schematically depicted. Because $\mathcal{A} \cong \mathcal{B}$ it follows $n_A = n_B$ and, hence, S and T have an even number, $n = 2n_A$, of π-MO's. In the case of $n = 4\nu + 2$ ($n = 4\nu$) an odd (even) number of π-MO's, $m = 2\nu + 1$ ($m = 2\nu$), are doubly occupied in S and T. This is indicated by an horizontal bar in Fig. 8. From the eigenvalue pattern provided by the interlacing theorem and the interpretation of the eigenvalues as π-MO's, conclusions can be drawn concerning the relative energies of HOMO, ε_m, and LUMO, ε_{m+1}, of the topomers S and T. Interpreting these MO's as usual [62], some predictions immediately follow concerning the relative magnitude of first ionization potentials IP_1, electron affinities, EA, and HOMO-LUMO transition energies, $\Delta\varepsilon$, of the topomers S and T. All these relations are listed in Table 1 and they are augmented by a relation for the total π-electron energies, E_π, recently [53] derived.

It should be noticed that all these predictions are expressed by inequalities reflecting the qualitative character of the interlacing theorem. Thus, this theorem

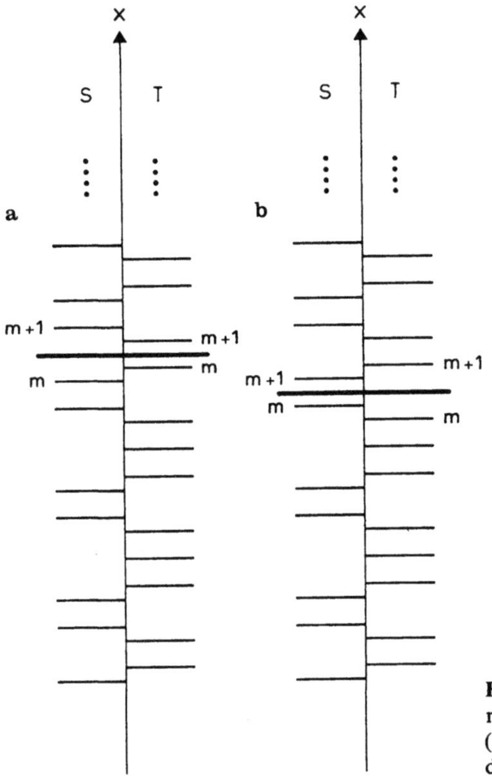

Fig. 8. MO pattern of an S/T-pair according to model 1, $l = 2$, $\mathscr{A} \cong \mathscr{B}$, with (a) $n = 4\nu + 2$ (b) $n = 4\nu$ π-electrons, respectively, and $m = n/2$ doubly occupied MO's

Table 1. Consequences of the interlacing theorem applied for π-electrons

	$n = 4\nu + 2$	$n = 4\nu$
HOMO	$\varepsilon_m^S \leqq \varepsilon_m^T$	$\varepsilon_m^S \geqq \varepsilon_m^T$
	$IP_1^S \geqq IP_1^T$	$IP_1^S \leqq IP_1^T$
LUMO	$\varepsilon_{m+1}^S \geqq \varepsilon_{m+1}^T$	$\varepsilon_{m+1}^S \leqq \varepsilon_{m+1}^T$
	$EA^S \leqq EA^T$	$EA^S \geqq EA^T$
HOMO-LUMO	$\Delta\varepsilon^S \geqq \Delta\varepsilon^T$	$\Delta\varepsilon^S \leqq \Delta\varepsilon^T$
E_π	$E_\pi^S \geqq E_\pi^T$	$E_\pi^S \geqq E_\pi^T$

does not replace the determination of an appropriate molecular constant by experiment and/or theoretical calculation but it elucidates the dependence of its magnitude on the topological features of the molecules under consideration and illuminates the interplay between the physico-chemical properties and the topology of molecules in a general manner.

All the physical consequences of the interlacing theorem as listed in Table 1 are included in the term *topological effect on molecular orbitals (TEMO)*.

The relationships collected together in Table 1 provide a clue for the proof of the physical relevance of the interlacing theorem by means of experimental data. These verifications and some support by quantum chemical data at the SCF-HF-level are the subjects of the next subsection.

The pattern of interlacing MO's without inversions (as depicted in Fig. 8 with reference to model 1, $l = 2$, $\mathscr{A} \cong \mathscr{B}$) is also obtained for model 2 and 3, $l = 2$, $\mathscr{A} \cong \mathscr{B}$, provided the linear factor appearing in Eqs. (30) and (31), respectively, is non-negative. Certainly in those cases the number of occupied π-MO's is given by $m = n_A + n_C/2$. As a consequence of this the predictions listed in Table 1, except the relation of total π-electron energies, E_π^S and E_π^T, are also valid in these cases. For model 2, $l = 2$, $\mathscr{A} \cong \mathscr{B}$, $\Phi(\mathscr{C} - a - b) \geqq 0$, one has obtained [53] that $E_\pi^S \leqq E_\pi^T$, but no such relation has been derived for model 3.

A simple method useful in the study of the topological behaviour of σ- and σ^*-electron systems has been outlined recently [5, 52]. The basis for this approach is provided by localized σ- and σ^*-bond orbitals (BO) as they are depicted by the line graph of the molecular graph. Since this graph is also simple and connected (see Sect. 2.3), the line graph is also associated with a topological space closely related to $\boldsymbol{T}(\mathscr{G})$ as pointed out in Sect. 3.2. It has been shown [52] that the characteristic polynomials of the σ-electron systems of S and T behave similarly to those for their π-electron systems. This means that in the case of topomers constructed by use of model 1, 2, or 3, $l = 2$, $\mathscr{A} \cong \mathscr{B}$, the topological σ-MO's of S and T interlace in the same manner as is schematically depicted in Fig. 8 for π-MO's. However, in contrast to the π-electron system, all σ-MO's are doubly occupied. In the case of model 1, $l = 2$, $\mathscr{A} \cong \mathscr{B}$, the number of σ-MO's is even, hence, the highest occupied σ-MO (lowest unoccupied σ^*-MO) of S lies above (below) that of T. From that, similar conclusions are drawn as listed in Table 1 for π-electrons. For topomers formed by means of model 2 or 3, $l = 2$, $\mathscr{A} \cong \mathscr{B}$, the parity of the number of their σ-MO's is determined by the number of σ-MO's of their central moiety C; therefore in these cases no general conclusions other than the interlacing of the σ-MO's can be drawn.

If either $\mathscr{A} \neq \mathscr{B}$ or $l \geqq 3$, in the pattern of the σ-MO's of S and T inversions may appear.

6.4 Verification of TEMO by Experimental and Quantum Chemical Data

In this subsection the TEMO predictions listed in Table 1 as well as the interlacing of MO's itself are examined by means of various experimental and quantum chemical data. For the sake of brevity, in that what follows, only a small part of the material available [52] will be presented. No attempt is made to verify TEMO by means of HMO calculations because as a consequence of the topological character of the Hückel (HMO) method these results must always agree with TEMO [49].

6.4.1 Energy Content

In zeroth order approximation the energy content of a molecule can be considered as the sum of the contributions of the σ-electrons and the π-electrons, E_σ and E_π, respectively, thus for one topomer, X = S, T, one may write

$$E^X = E_\sigma^X + E_\pi^X\,, \qquad X = S, T\,. \tag{32}$$

In the case of topomers constructed by means of any model described in Sect. 6.1, the contribution of σ-electrons to the energy content of S and T will be nearly equal, $E_\sigma^S \approx E_\sigma^T$, thus the difference of the energy contents of topomers will be mainly determined by the contributions of the π-electrons as follows

$$E^S - E^T \approx E_\pi^S - E_\pi^T\,. \tag{33}$$

For topomers according to model 1, $l = 2$, $\mathscr{A} \cong \mathscr{B}$ it has been deduced [53], that $E_\pi^S \geqq E_\pi^T$, thus, one may further conclude that the S-topomer of such a pair is never less stable than the T-topomer.

This conclusions may be proved by looking at the heat of combustion. The data obtained for phenanthrene (*40*S) and anthracene (*40*T) are collected in Table 2 [63, 64]. As they show, in accordance with the TEMO prediction [53], phenanthrene (*40*S) is energetically stabilized by an amount of 56.5 kJ mol^{-1} in comparison to anthracene (*40*T). This particular intriguing example has given rise to a series of speculations about its origin; as can be easily recognized, it is a simple consequence of molecular topology.

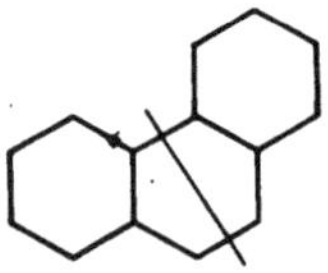
S

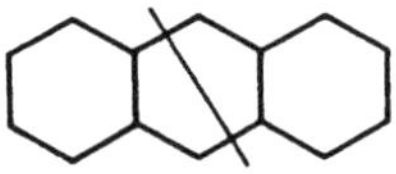
T

40

Table 2. Heat of combustion and mesomeric energies of phenanthrene (*40*S) and anthracene (*40*T) in kJ mol^{-1}

	Heat of combustion		Mesomeric energy
	Exp.	Calc.	
Phenanthrene (*40*S)	7100,1	7515,7	415,6
Anthracene (*40*T)	7153,8	7512,9	359,1

Another example of that kind is provided by the heat of formation of the topomers dibenzotetraazapentalenes ***41***, S and T, which are formed according to model 2, $l = 2$, $\mathscr{A} \cong \mathscr{B}$. In contrast to model 1, for model 2 $E_\pi^S \leqq E_\pi^T$ has been derived [53] and the heats of formation [65], $\Delta H_f^S = 552.7 \pm 6.3$ kJ mol^{-1} and $\Delta H_f^T = 597.5 \pm 5.4$ kJ mol^{-1}, are in agreement with that prediction.

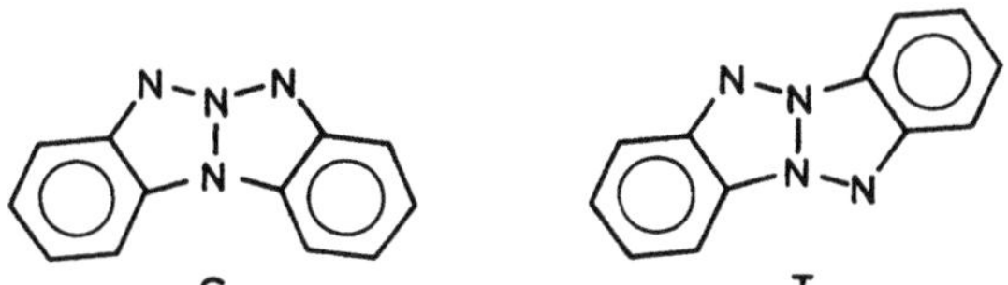

S T

41

6.4.2 HOMO-LUMO-Transition Energies

Some UV-absorption bands may be related to HOMO-LUMO-transitions as, for example, the para-bands in the absorption spectra of polycyclic aromatic hydrocarbons (PAH). In Tables 3 and 4 the energies of that transition in the spectra of some pairs of topomeric PAH's with $4\nu + 2$ and 4ν π-electrons are given.

As indicated in Table 1, in the case of the examples collected in Table 3, one concludes from TEMO that $\Delta\varepsilon^S \geqq \Delta\varepsilon^T$. This expectation is verified by all pairs of Table 3 except the pair in line 8 where the S-topomer is sterically overcrowded. For the 4ν π-electron systems collected in Table 4 $\Delta\varepsilon^S \leqq \Delta\varepsilon^T$ is verified by all the pairs.

Another example of absorption bands related to HOMO-LUMO transitions is the longest wavelength band in the absorption spectra of intramolecular coupled polymethines. For that class of compounds the following TEMO rule has been derived [58]:

Within a series of isomeric intramolecular coupled polymethines the symmetric chromophor absorbs at the shortest (longest) wavelength if it contains $4\nu + 2$ (4ν) *π-electrons.*

A verification of this rule is given by the following 10 π-electron example:

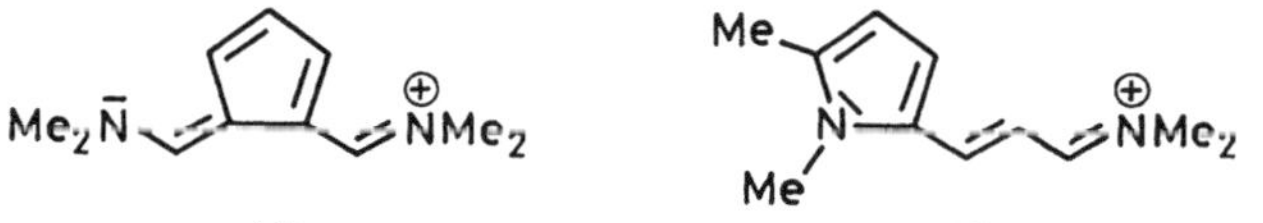

42 *43*

$\lambda_{max} = 409$ nm [70] $\lambda_{max} = 553$ nm [71]

Table 3. Energies of p-bands, $\Delta\varepsilon$ in [eV], of topomeric PAHs with $4\nu + 2$ π-electrons constructed by means of model 1, $l = 2$, $\mathscr{A} \cong \mathscr{B}$

No.	*S*	*T*	$\Delta\varepsilon^S$	$\Delta\varepsilon^T$	Ref.
1	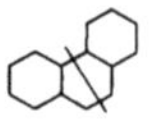	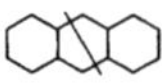	4.32 	 3.39	[66] [66]
2	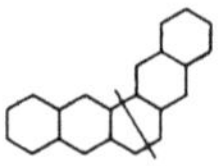	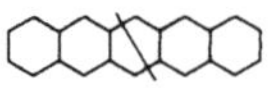	3.59 	 2.30	[66] [66]
3	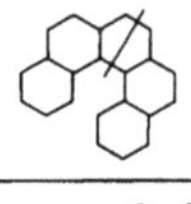	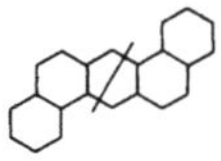	3.88 3.71 	 3.65	[67] [66] [66]
4	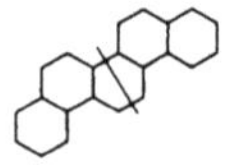	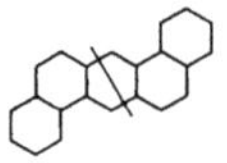	3.92 	 3.65	[66] [66]
5	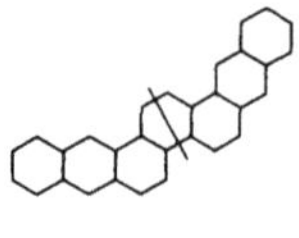	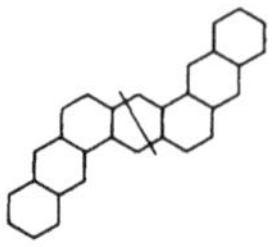	3.53 3.39 	 3.35	[67] [66] [66]
6	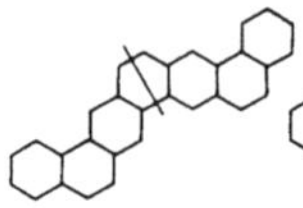	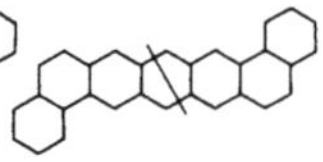	3.52 	 2.53	[66] [66]
7		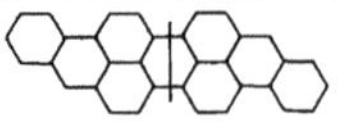	2.71 2.67 	 2.56	[68] [67] [68]
8		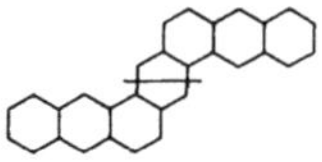	 2.74	3.35 	[66] [67]
9	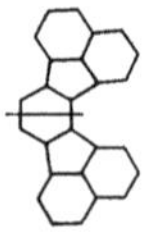	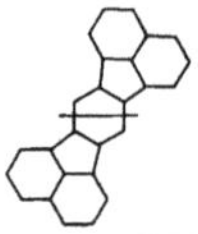	3.24 	 3.12	[67] [67]

6.4.3 Electron Affinities

The energy change due to the gas phase reaction

$$M + e^{\ominus} \rightarrow M^{\ominus} \tag{34}$$

is called electron affinity (*EA*) of molecule M and accordingly defined by

$$EA = E(M) - E(M^{\ominus}) \tag{35}$$

whereby the kinetic energy of the electron is assumed to be zero. Negative (positive) signs of *EA* indicate that the total energy of the molecule is lower (higher) than that

Table 4. Energies of p-bands, $\Delta\varepsilon$ in [eV], of topomeric PAHs with 4ν π-electrons constructed by means of model 1, $l = 2$, $\mathscr{A} \cong \mathscr{B}$

No.	S	T	$\Delta\varepsilon^S$	$\Delta\varepsilon^T$	Ref.
1	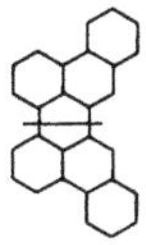	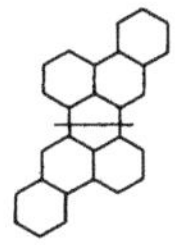	2.97	3.01	[69] [69]
2		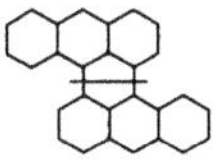	2.37	2.38	[69] [69]
3	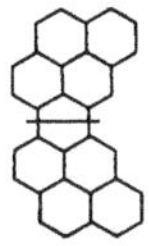		2.53	2.68	[67] [67]

Table 5. Electron affinities, EA in [eV], of some pairs of topomers with $n = 4\nu + 2\pi$-electrons

No.	S_1	T	S_2	EA^{S_1}	EA^T	EA^{S_2}	Ref.
1	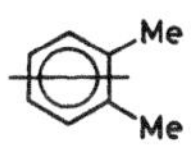	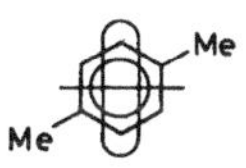	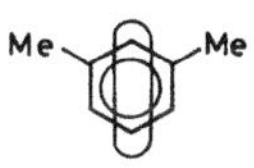	-1.12	-1.07	-1.06	[75] [75] [75]
2		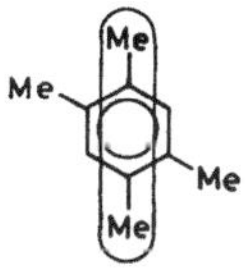	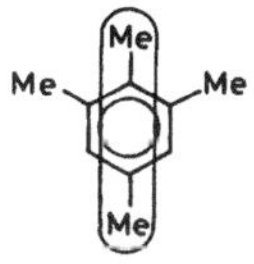		0.067	0.108	[76] [76]
3		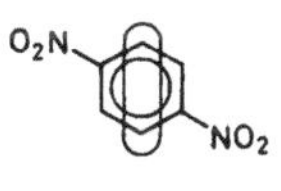	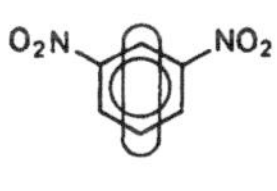		1.89	1.57	[77] [77]
4	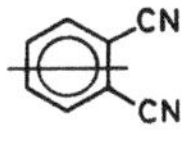	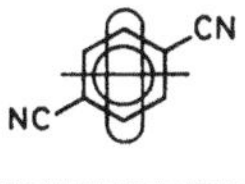	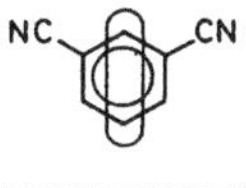	0.95	1.10	0.90	[78] [78] [78]
5	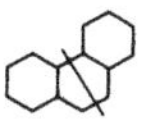	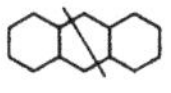		0.305	0.520		[76] [76]
6	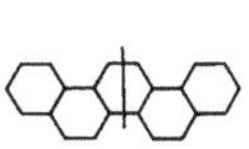	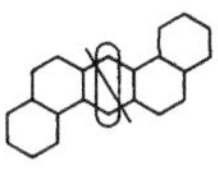	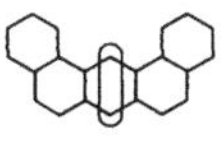	0.542	0.595	0.591	[73] [73] [73]

of the molecule anion. Provided Koopmans theorem [72] is valid, the electron affinity may be expressed by the energy of LUMO [62, 73, 74] as follows

$$EA = -\varepsilon_{m+1} . \tag{36}$$

Deviations from Eq. (36) are interpreted as being the result of a reorganisation of the charge distribution in the anion and an effect of electron correlation which may be approximately described by additive constants [74].

From Table 1 and Eq. (36) one would expect that for pairs of topomers formed according to either model 1, $l = 2$, $\mathscr{A} \cong \mathscr{B}$, or model 2, $l = 2$, $\mathscr{A} \cong \mathscr{B}$, $\Phi(\mathscr{C} - a - b) \geqq 0$,

$$\begin{aligned} EA^{S} &\leqq EA^{T} \quad \text{if} \quad n = 4\nu + 2 , \\ EA^{S} &\geqq EA^{T} \quad \text{if} \quad n = 4\nu . \end{aligned} \tag{37}$$

Table 6. Electron affinities, EA in [eV], of some pairs of topomers with $n = 4\nu$ π-electrons

No.	S	T	EA^S	EA^T	Ref.
1	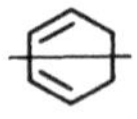	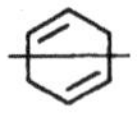	-0.8	-1.75	[79] [79]
2	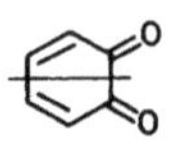	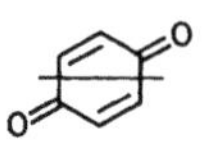	-0.31	-0.51	[80] [80]
3	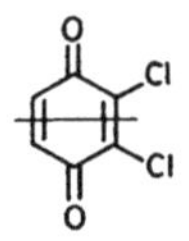	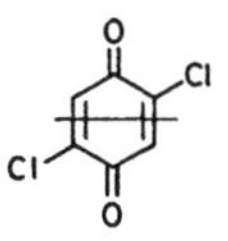	2.19	2.30	[78] [78]
4	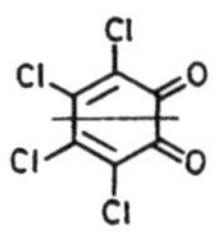	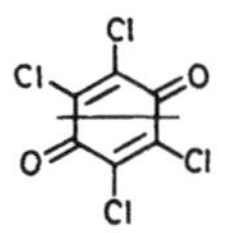	2.57	2.48	[78] [78]
5	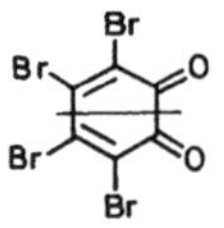	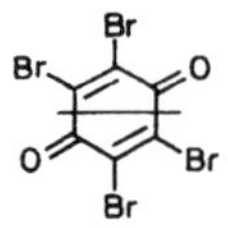	2.62	2.48	[78] [78]
6	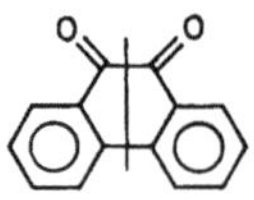	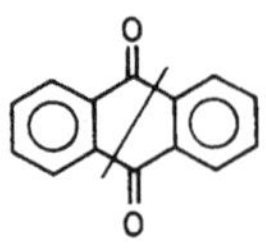	1.77	1.57	[78] [78]

These predictions are verified, in general, by the examples collected in Tables 5 and 6. In Table 5 the pairs S_1/T and S_2/T are formed by model 1 and 2, respectively, and have the T-topomer in common. There are two violations of the TEMO rule: The first one (Table 5, line 1, EA^{S_2}) ranges within the experimental error; the other one (Table 6, line 3) may be caused by the interaction of the two vicinal chlorine substituents in the S-topomer.

6.4.4 Ionization Potential — PE-Spectra

Provided, Koopmans theorem may be applied, the PE spectra exhibit the energies of the highest occupied molecular orbitals. Thus by means of PE spectroscopy not only the TEMO relation for the first ionization potentials (see Table 1) but also the interlacing of the MO's itself may be proved.

Since the topological spaces associated with the σ- and the π-electron system are different, the interlacing rule as expressed by Eq. (27) must be separately applied to these two electron systems (whereby lone pair electrons are accounted for by σ-electrons). For this reason, in the following tables, σ-MO's, if any, are listed separately from π-MO's.

In Figure 9, the PE-spectra of some pairs of topomeric PAHs are depicted [66] exhibiting the interlacing of the 5 to 10 highest occupied π-MO's which are labelled

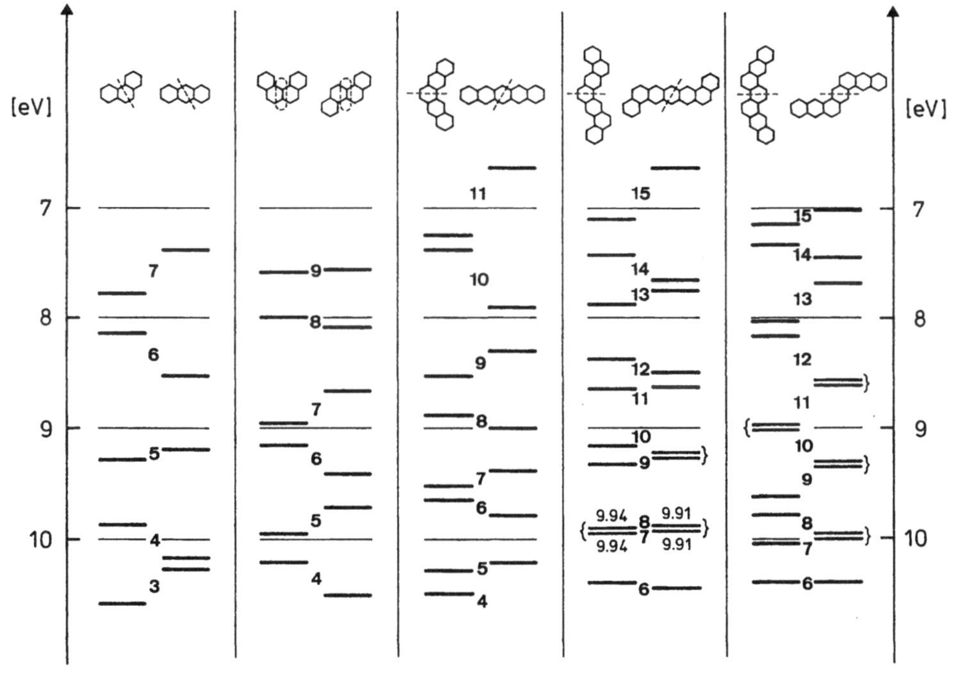

Fig. 9. PE-spectra of some pairs of topomeric PAHs [66]. The π-MO's are labeled by j in increasing order

in energetically increasing order by $j = 1, 2, 3, \dots$. As an inspection of Fig. 9 shows the TEMO rule is perfectly obeyed, except for the energies of the doubly degenerated levels at 9.94 (S) and 9.91 (T) eV of the pair where dibenzopentacene is the T-topomer. According to Eq. (27) their levels should have the same energy in S and T. Certainly the discrepancy of 0.03 eV ranges within the experimental error, but even a more pronounced difference could be attributed to the spatial factor [P3], see Sect. 4, which cannot be neglected completely.

The TEMO predictions concerning energy content, HOMO-LUMO transition energies, and electron affinities can be proved only by means of topomers for which the appearance of inversions is exluded because otherwise the statements could be reversed to their very opposites. The use of PE-spectra in order to prove the interlacing of MO's does not succumb to such limitations. In Table 7, the PE-spectra of some further pairs of topomeric PAHs are listed; since in their case the building moieties are different, A ≠ B, inversions may occur. As shown by Table 7, the PE-spectra of two topomeric pairs only exhibit one inversion each. The assignment of the topomers as S and T, respectively, is in agreement with Eq. (28). Notice, that tetraphene is part of each pair presented in Table 7.

Table 7. PE-spectra [66] of topomeric PAHs with unequal fragments, A ≠ B. All energies in [eV]

j	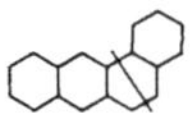 S	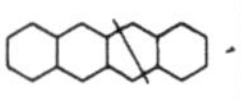 T	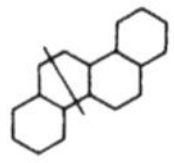 S	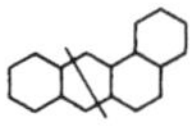 T	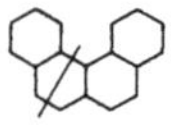 S	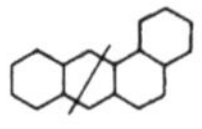 T
9		6.97		7.41		7.41
	7.41		7.59		7.60	
8	8.04			8.04	8.02	
		8.41	8.10			8.04
7		8.41	8.68			8.86
	8.86			8.86	8.98	
6	9.38			9.38	9.18	
		9.56	9.43			9.38
5		9.70	9.72			9.91
	9.91			9.91	9.96	
4		10.25		10.36	10.22	
	10.36		10.52			10.36

In Fig. 10 the PE-spectra [66] of those topomeric PAHs are depicted which are constructed by means of model 1 or 2, $l = 2$, $\mathscr{A} \cong \mathscr{B}$, and have dibenzanthracene (a, e) as a T-topomer in common; they consist of the π-energy levels ε_j^X, $j = 4, 5, \dots, 11$, $X = S, T$. According to Eq. (27) the energy levels of the S-topomers, ε_9^S and ε_8^S should range within the interval $[\varepsilon_9^T, \varepsilon_8^T]$; ε_9^S of dibenzphenanthrene and picene violate this expection seriously, the deviation of ε_8^S of picene is within the experimental error. Thus from about 30 energy levels only two ones show significant deviations.

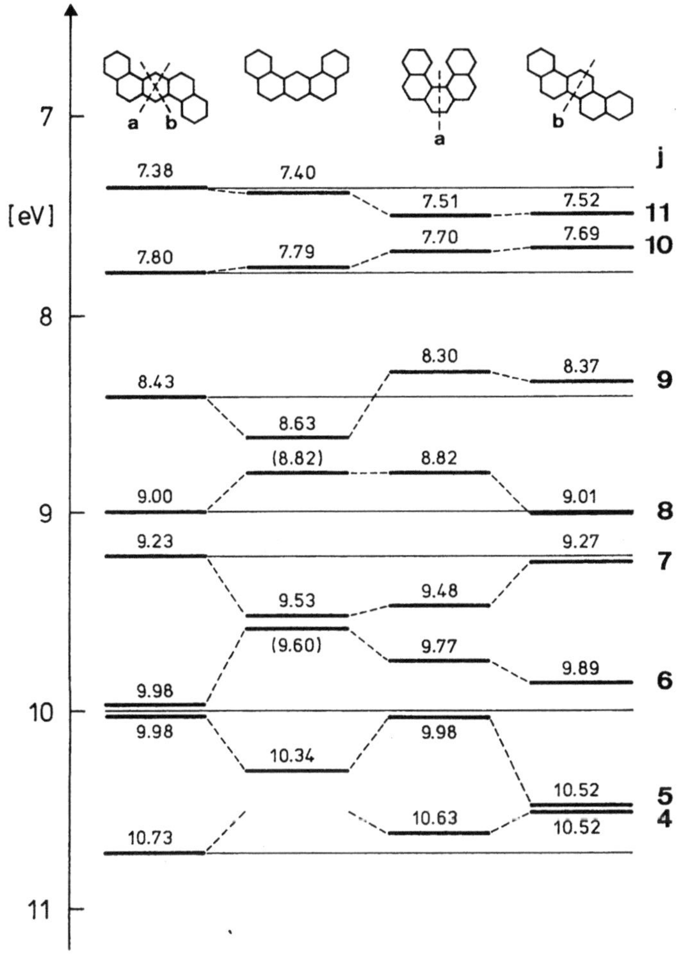

Fig. 10. PE-spectra [66] of some pairs of topomeric PAHs having the T-topomer in common. The π-MO's are labeled by j in increasing order

The PE-spectra of the methyl derivatives of the topomeric triazine pair ***44*** are collected in Table 8. Because of non-isomorphic moieties, A $\neq$ B, inversions cannot be excluded, but the spectra exhibit perfect interlacing of the π- and σ-MO's.

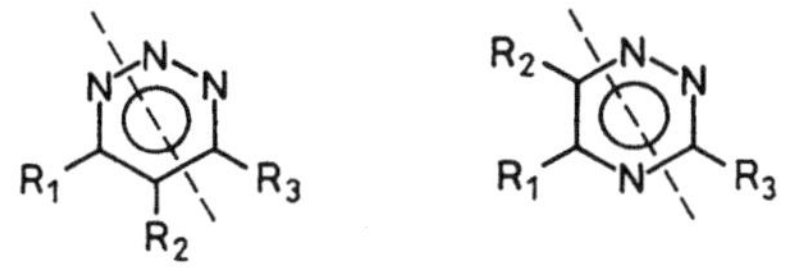

S T

44

	44a	***44b***	***44c***	***44d***	***44e***
R_1;	H	H	Me	Me	Me
R_2:	H	Me	Me	H	Me
R_3:	H	H	H	Me	Me

All the σ-MO's shown in Table 8 are lone pairs. The PE-spectra of cyclohexadiene-(1.3) and cyclohexadiene-(1.4), listed in Table 9, illustrate the perfect interlacing of ordinary σ-MO's.

Table 8. PE-spectra (in [eV]) of topomeric triazine derivatives ***44a–e***

	44a		***44b***		***44c***		***44d***		***44e***	
	S [81]	T [82]	S [81]	T [82]	S [81]	T [82]	S [81]	T [82]	S [81]	T [82]
π		11,30		10,65		10,33		10,27		9,86
	11,6		11,0		10,6		10,8		10,3	
	12,0		11,8		11,3		11,0		10,9	
		12,43		12,17		11,7		11,7		11,4
σ		9,61		9,35		9,15		9,02		8,84
	10,0		9,8		9,5		9,5		9,4	
	10,4		10,3		9,9		9,9		9,7	
		11,82		11,53		11,27		11,2		11,0
		12,42		11,82		11,7		11,7		11,4
	13,1		12,8		12,6		12,5		12,4	

Table 9. PE-spectra of cyclohexadiene-(1,3) and cyclohexadiene-(1,4) [83], in [eV]

	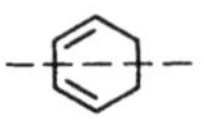 S	 T
π	8.25	
		8.82
		9.88
	10.7	
σ		11.0
	11.3	
	11.8	
		12.0
		12.0
	12.7	

6.4.5 Quantum Chemical Data

Besides experimental data, the results of ab initio SCF calculations may also be used in order to prove the physical relevance of TEMO. Certainly such calculations must be kept at the SCF level for fear of damaging the MO picture. As an example of such results in Table 10 the π- and σ-MO energies of pyridazine (**32**S) and pyrazine (**32**T), see Fig. 5, are given [84, 85]. The MO pattern of π-electrons is without inversions. Both highest σ-MO's have lone pair character (*l*). Within the

Table 10. π- and σ-MO energies of pyridazine (*32*S) and pyrazine (*32*T) in a. u. [84, 85]; STO3G basis sets, standard geometries

		S	T	
π			−0.317710	3
	3	−0.324026		
	2	−0.328821		
			−0.329795	2
			−0.495559	1
	1	−0.498333		
σ			−0.320050	(*l*) 12
	12 (*l*)	−0.323930		
		---i---		
			−0.393363	(*l*) 11
	11 (*l*)	−0.396884		
		---i---		
			−0.489412	10
	10	−0.515740		
	9	−0.548695		
			−0.573421	9
			−0.604389	8
	8	−0.605193		
	7	−0.630982		
			−0.639649	7
			−0.658568	6
	6	−0.674291		
		---i---		
			−0.804231	5
	5	−0.814511		
	4	−0.839386		
			−0.844117	4
			−0.996887	3
	3	−1.039821		
	2	−1.044507		
			−1.114881	2
			−1.212398	1
	1	−1.247275		

pattern of σ-MO's there are two inversion intervals to which the σ-MO's 6–10 and 12, respectively, belong; the inversion points are indicated by a broken line (---i---).

Because σ-MO's are not so well exposed by PE-spectra then π-MO's, the results of SCF [49, 52, 84–88] and CNDO/2 calculations [89] are of particular interest. However, as it has been observed in the case of *o-/p*-benzoquinone (S/T), the appearance of inversions within the calculated TEMO pattern may be caused by some shortness of the basis sets used [86, 88].

By means of an especially adapted version [85, 86, 88] of variational perturbation theory, PV-RR [90, 91], the physical reasons for the appearance of inversions can be traced out, namely perturbations of the topologically induced MO pattern by (i) the variation of the potential within the molecule according to its constituent atoms and its geometry and (ii) the interaction between non-adjacent centres. In case of *o-/p*-benzoquinodimethane (S/T) the inversion of the lowest π-MO's is found to be caused by interactions between non-adjacent centres [88]. It is interesting to notice that the PE-spectra of these compounds [92, 93] also exhibit an inversion of π-MO's, however, here all π-MO's are inverted except the highest ones. All the above mentioned data are collected in Table 11.

Table 11. Calculated π-MO energies in a. u. (4-31G basis set, optimized geometry [88]) and PE-spectra [92 (S), 93 (T)] in eV of *o-/p*-benzoquinodimethane (S/T)

j	Calc. [a. u.]: S	T	Exp. [eV]: S[92]	T[93]
	−0.268155		7,70	
4				
		−0.274594		7,87
			------------i	------------
		−0.372965	9,6	
3				
	−0.388363			9,7
	−0.392576			9,8
2		−0.409358	10,05	
	------------i	------------		
	−0.508766		10,49	
1				
		−0.510804		— ? —

6.4.6 Zero Field Splitting Parameters

Even in the absence of an external magnetic field the lowest triplet state T_1 is split into three zero field levels (ZFL), t_x, t_y, and t_z, by magnetic dipole-dipole interaction of the unpaired electrons. Empirically the zero field splitting is described by the molecular parameters D and E as illustrated by Fig. 11.

This corresponds with a spin-Hamiltonian, $\mathfrak{H}_S$, as follows [94]:

$$\mathfrak{H}_S = D\left(\mathfrak{S}_z^2 - \frac{1}{3}\mathfrak{S}^2\right) + E(\mathfrak{S}_x^2 + \mathfrak{S}_y^2)\,, \tag{38}$$

wherein D and E have the dimension of energy and $\mathfrak{S}_x$, $\mathfrak{S}_y$, $\mathfrak{S}_z$ are the components of the dimensionless spin operator $\mathfrak{S}$. Provided the zero field splitting is solely

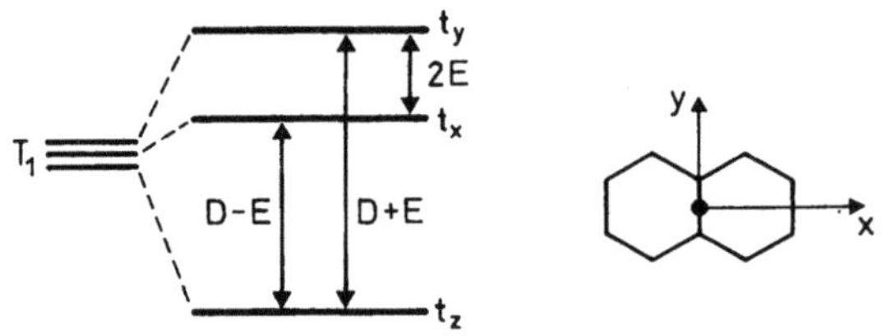

Fig. 11. Zero field splitting of the lowest triplet state into the zero field levels t_x, t_y, and t_z

generated by magnetic dipole-dipole-interactions of both the unpaired electrons, the parameters D and E are expressed as follows:

$$D = \frac{3g^2\beta^2(r_{12}^2 - 3z_{12}^2)}{4r_{12}^5}, \tag{39}$$

$$E = \frac{3g^2\beta^2(y_{12}^2 - x_{12}^2)}{4r_{12}^5}, \tag{40}$$

where g and β denote the gyromagnetic constant and Bohr's magneton of the electron, respectively, and $\vec{r}_{12} = (x_{12}y_{12}z_{12})^T$ stands for the distance vector of the unpaired electrons e_1 and e_2. The absolute magnitude of the parameters D and E is very small; D ranges in the order of 10^{-1} [cm^{-1}]. Nevertheless they may be very precisely determined by double resonance methods [95, 96].

Table 12. Experimentally determined zero field splitting parameters D [cm^{-1}] for pairs of topomers (model 1, $l \geqq 2$)

No.	S	T	D [cm^{-1}] S	D [cm^{-1}] T	Ref.
1	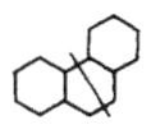	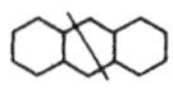	0.100	0.072	[97] [97]
2	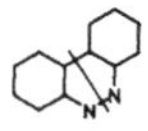	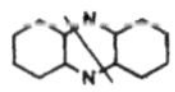	0.129	0.0751	[98] [99]
3	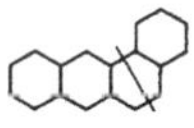	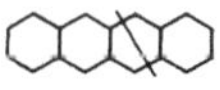	0.079	0.057	[97] [100]
4	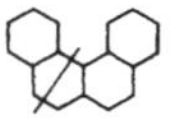	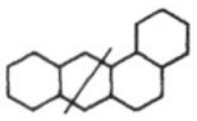	0.093	0.079	[101] [97]
5	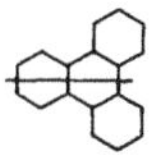	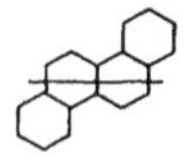	0.134	0.095	[102] [97]
6	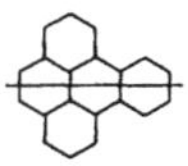	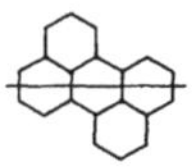	0.097	0.076	[97] [97]
7	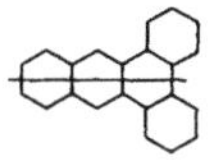	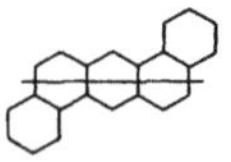	0.098	0.090	[100] [97]

By inspection of the available data extracted from literature, one finds as a kind of empirical rule that

$$D^{S} \geqq D^{T} \tag{41}$$

for pairs of topomers. As an illustration, in Table 12 and 13 the D parameters for some pairs of topomers constructed by means of model 1 or 2, respectively, $l = 2$, $\mathscr{A} \cong \mathscr{B}$, are listed. There is only a single violation of this rule (see line 4 in Table 13). Further examples are found elsewhere [52].

Table 13. Experimentally determined zero field splitting parameters D [cm^{-1}] for pairs of topomers (model 2, $l = 2$)

No.	S	T	D [cm^{-1}] S	D [cm^{-1}] T	Ref.
1			0.173	0.162	[98] [98]
2	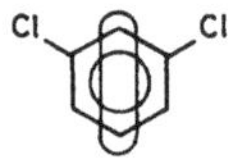	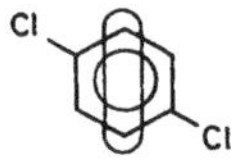	0.154	0.144	[99] [99]
3	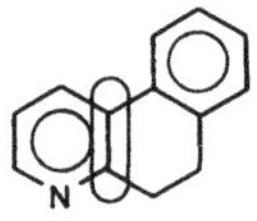	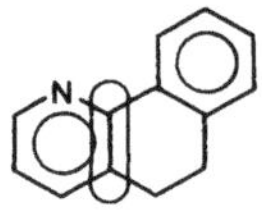	0.103	0.102	[103] [103]
4	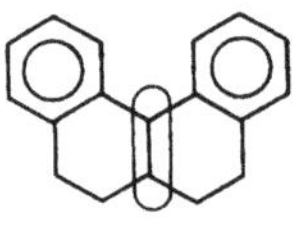	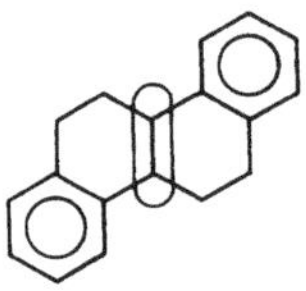	0.093	0.095	[101] [101]

At present there is no theoretical explanation for the topology-dependence of the D parameter. However, a relation between the structure and the magnitude of the D parameter of polycyclic aromatic hydrocarbons (PAH) has already been observed earlier [104] resulting in an empirical method [101] for estimating the numerical value of D parameter as follows:

$$D = \Sigma\, D_{L} K_{L} / \Sigma\, K_{L}\,, \tag{42}$$

where D_{L} and K_{L} denote the D parameter and the number of Kekulé structures of fragment L, respectively; the choice of fragments L is constrained by some particular rules [102].

7 The Role of Molecular Topology

The relevance of molecular topology is already verified by empirical correlations [26–28] and by the rule of topological charge stabilisation (TCS). But, as the topological effect on molecular orbitals (TEMO) indicates, molecular topology plays a very *dominant* role in the modelling of physico-chemical properties of molecules. This suggests a *hierarchic order* of matter, space, molecular topology, and energy as follows

(1) molecular topology,
(2) matter and space,
(3) energy.

At the first glance, one might be bothered by the fact that a formal (i.e. a non-physical) concept, such as topology, occupies the top-level within that hierarchy. But, indeed, this is none other than a straightforward consequence of the mutually related magnitudes of certain properties as, for instance, expressed by the inequalities (27), (37), (41), and those listed in Table 1, which all have purely topological origin and are numerously verified by suitable data. These inequalities and their verification show that the actual values of such properties are obviously not allowed to increase to arbitrary amounts, on the contrary, they seem indeed to be bound by some limits resulting solely from molecular topology. Although to the best of our knowledge such a hierarchic order has not been postulated elsewhere with such clarity, it agrees in principle with some previous quantum chemical conclusions [105].

Certainly, a marked influence of topology may be anticipated only for molecules above a critical size. Obviously, in the case of diatomic molecules, all having the same topology, the significance of molecular topology cannot be manifested. The minimum number of atoms of which a molecule should at least consist in order to exhibit topological effects is an open question at present. However, TEMO is plainly recognizable in the case of topomeric triazines ***44a*** (see Table 8) formed from only 9 atoms (in [105] a minimum of 6 atoms is suggested). Thus the hierarchic order governs the greater number of known molecules.

Disregarding small molecules and speaking casually, the molecular topology seems to provide some kind of framework within and only within which physical properties of molecules are realized.

Its prominent position within the hierarchic order may be a result of the rigorous abstractions necessarily performed in order to obtain the topology of a molecule. Consequently, molecular topology itself is no object of direct experience but it might be better considered as the general architecture of a certain type of molecules.

However, even if the architecture is known, the molecule itself does not start to exist until the given design is realized by the proper kind and number of atoms. But obviously at the same time as the molecule is materially realized it also fills distinct parts of space. Hence, matter and space seem to be non-separable with regard to the hierarchic order and, consequently, they should occupy one and the same level as postulated above.

Within the hierarchic order, the energy plays a selective role in that the geometries corresponding to relative energy minima are determined, and further, the various possible positional isomers and/or steric configurations, if any, may be ordered into a sequence of species with increasing energy content.

All these considerations show that in spite of its unfamilarity the above postulated hierarchic order is in accordance with the established picture of nature we have. Being aware of the significance of topology one possesses a novel point of view from which deeper insights into nature might be gained [51].

Acknowledgement: Fruitful discussions with Dr. G. Mark and Mrs. E. Currell and the technical assistance of Mrs. I. Heuer, Mrs. I. Schneider, and Mrs. R. Speckbruck is appreciated.

8 List of Symbols

A	TEMO subunit
A_{rs}	element of $\boldsymbol{A}(\mathscr{G})$
$\mathscr{A}$	graph of TEMO subunit A
$\boldsymbol{A}(\mathscr{G})$	adjacency matrix of graph $\mathscr{G}$
adj	adjacent
a	vertex
B	TEMO subunit
$\mathscr{B}$	graph of TEMO subunit B
b	vertex
$\mathscr{C}_l$	cycle of length l
D, D^X	zero field splitting parameter (of molecule X = S, T)
D_{rs}	element of $\boldsymbol{D}(\mathscr{G})$
$\boldsymbol{D}(\mathscr{G})$	distance matrix of graph $\mathscr{G}$
$d(r, s)$	distance between the vertices r and s
$d(r\vert\mathscr{G})$	sum of all distances to vertex r in graph $\mathscr{G}$
E	zero field splitting parameter
E, $E(\mathrm{M})$	total energy (of molecule M)
$E^{(1)}$	first order perturbation energy
E_π^X, E_σ^X	total energy of π- (σ-) electrons of molecule X = S, T
EA^X	electron affinity of molecule X = S, T
$\mathscr{E}$, $\mathscr{E}(\mathscr{G})$	edge set of graph $\mathscr{G}$
e	vertex
f	vertex
G	secular determinant of graph $\mathscr{G}$
G_{rs}	element of determinant G
$\mathscr{G}$	graph (also molecular or skeleton graph)
g	gyromagnetic constant of electron
$g(r)$	degree of vertex r
$\mathfrak{H}_S$	spin Hamiltonian
$\mathfrak{H}'$	perturbation operator
$I(\mathscr{G})$	invariant of graph $\mathscr{G}$
IP_j^X	j-th ionization potential of molecule X = S, T

$\boldsymbol{I}_n$	unitary matrix of order n
K	number of Kekulé structures
$\mathscr{K}_n$	complete graph on n vertices
$\mathscr{K}_{n,n'}$	complete bipartite graph on n and n' differently coloured vertices
k	vertex
$\mathscr{L}(\mathscr{G})$	line graph of graph $\mathscr{G}$
l	vertex
l	valency of a TEMO subunit
M	covalent molecule
m	cardinality of the edge set $\mathscr{E}(\mathscr{G})$
m	number of occupied MO's
$m(\mathscr{G}, k)$	number of k-matchings of the graph $\mathscr{G}$
N	number of centers of the uniform reference frame
n	cardinality of the vertex set $\mathscr{V}(\mathscr{G})$
n	number of MO's
P	permutation
$\mathfrak{P}$	permutation operator
$\mathscr{P}_{rs}$	path connecting vertices r and s
p	vertex
p	number of transpositions composing a permutation P
Q	total charge of a union reference frame
q	vertex
q_r	charge density on center r
q'_r	charge density on center r within a normalized union reference frame
$\mathscr{R}$	set (of points)
r	vertex
$r_{jj'}$	distance vector of electrons j and j'
$r_{j,\alpha}$	distance vector of electron j from nucleus α
S, S_j	member of a pair (an ensemble) of topomers
$\mathfrak{S}$	spin operator
$\mathfrak{S}_x$, $\mathfrak{S}_y$, $\mathfrak{S}_z$	spin component operators
$\mathscr{S}$, $\mathscr{S}_j$	graphs of S (S_j)
s	vertex
TCS	topological charge stabilisation
TEMO	topological effect on molecular orbitals
T	member of a pair (an ensemble) of topomers
$\boldsymbol{T}_O$	open set topology
$\boldsymbol{T}_U$	neighbourhood topology
$\boldsymbol{T}(\mathscr{G})$	topological space associated with graph $\mathscr{G}$
$\mathscr{T}$	graph of T
t	vertex
t	discriminator in μ-polynomials
t_x, t_y, t_z	zero field levels
$\boldsymbol{U}(r)$	system of neighbourhoods $\mathscr{U}(r)$
$\mathscr{U}(r)$	neighbourhood of vertex r
$\mathscr{U}_\varepsilon(r)$	ball neighbourhood of vertex r
$\mathscr{V}$, $\mathscr{V}(\mathscr{G})$	vertex set of graph $\mathscr{G}$

v, v_j	vertex
x_j^S, x_j^T	j-th eigenvalue in the spectrum of $\mathscr{S}$ and $\mathscr{T}$, respectively
x_k^I	k-th inversion point
Z_α	nuclear charge of atom α
$\alpha(\mathscr{G}; y)$	acyclic (matching) polynomial of graph $\mathscr{G}$
β	Bohr's magneton
$\Delta(x)$	difference polynomial
$\Delta\varepsilon^X$	HOMO-LUMO separation for topomeric molecules X = S, T
ε_m^X	HOMO of molecule X = S, T
ε_{m+1}^X	LUMO of molecule X = S, T
ζ_α	effective nuclear charge of atom α
μ	cyclomatic number of graph $\mathscr{G}$
$\mu(\mathscr{G}; z, t)$	μ-polynomial of graph $\mathscr{G}$
$\Phi(\mathscr{G}; x)$	characteristic polynomial of graph $\mathscr{G}$

9 References

1. Prelog V (1975/76) J. Mol. Catalysis 1: 163
2. Harary F (1969) Graph theory, Addison-Wesley, Reading, MA
3. Balaban AT (1976) Chemical applications of graph theory, Academic, London
4. Trinajstic N (1983) Chemical graph theory (2 Vols.), CRC Press, Boca Raton
5. Gutman I, Polansky OE (1986) Mathematical concepts in organic chemistry, Springer, Berlin Heidelberg New York
6. Polansky OE (1989) in: Bonchev D, Rouvray D (eds) Graph theory and its application to chemistry, Gordon and Breach, London, vol 1, in press
7. Catenanes and rotaxanes [8] are considered as aggregates composed of several topologically disjoint moieties; the aggregate has in total three translational degrees of freedom.
8. Schill G (1971) Catenanes, rotaxanes, and knots, Academic, New York
9. Graovac A, Gutman I, Trinajstic N (1977) Topological approach to the chemistry of conjugated molecules Springer, Berlin Heidelberg New York (Lecture Notes in Chemistry, vol 4)
10. Ruedenberg K (1954) J. Chem. Phys. 22: 1878; Ruedenberg K (1958) J. Chem. Phys. 29: 1232; Ruedenberg K (1961) J. Chem. Phys. 34: 1884
11. See Sect. 13.4 in [5]
12. Dugundji J (1966) Topology, Allyn and Bacon, Boston
13. Rinow W (1975) Lehrbuch der Topologie, Deutscher Verlag der Wissenschaften, Berlin
14. Merrifield RE, Simmons HE (1980) Theor. Chim. Acta 55: 55
15. Merrifield RE, Simmons HE (1983) in: King RB (ed) Chemical applications of topology in graph theory, Elsevier, Amsterdam, p 1
16. Merrifield RE, Simmons HE (1979) Chem. Phys. Lett. 62: 235
17. Polansky OE (1986) Z. Naturforsch. 41 a: 560
18. Sachs H (1963) Publ. Math. (Debrecen) 11: 119
19. Cvetkovic DM, Doob M, Sachs H (1980) Spectra of graphs — theory and applications, Academic New York
20. Collatz L, Sinogowitz U (1957) Abh. Math. Sem. Univ. Hamburg 21: 63
21. Herndon WC, Ellzey ML, jr. (1986) Match (Math. Chem.) 20: 53
22. Heilmann OJ, Lieb EH (1972) Commun. Math. Phys. 25: 190
23. Godsil CD, Gutman I (1981) J. Graph Theory 5: 137
24. Polansky OE, Graovac A (1986) Match (Math. Chem.) 21: 93
25. Gutman I, Polansky OE (1981) Theor. Chim. Acta 60: 203

26. Kier LB, Hall LH (1976) Molecular connectivity in chemistry and drug research, Academic, New York
27. Charton M, Motoc I (1983) Steric effects in drug design, Springer, Berlin Heidelberg New York (Topics in Current Chemistry vol 114)
28. Bonchev D (1983) Information theoretic indices for characterisation of chemical structures, Research Studies Press, Chichester
29. Gimarc BM (1983) J. Amer. Chem. Soc. 105: 1979
30. Gimarc BM, Ott JJ (1986) in: Trinajstic N (ed) Mathematics and computational concepts in chemistry, Ellis Horwood, Chichester, p 74
31. Gimarc BM, Joseph PJ (1984) Angew. Chem. 96: 518; (1984) Angew. Chem. Int. Ed. Engl. 23: 506
32. Gimarc BM, Juric A, Trinajstic N (1985) Inorg. Chim. Acta 102: 105
33. Gimarc BM, Ott JJ (1986) J. Amer. Chem. Soc. 108: 4298
34. Ott JJ, Gimarc BM (1986) J. Amer. Chem. Soc. 108: 4303
35. Gimarc BM, Ott JJ (1986) Inorg. Chem. 25: 83; Gimarc BM, Ott JJ (1986) Inorg. Chem. 25: 2708; Gimarc BM, Ott JJ (1987) J. Amer. Chem. Soc. 109: 1388
36. Nölle D, Nöth H (1972) Z. Naturforsch. 27b: 1425
37. Nöth H, Ullmann R (1975) Chem. Ber. 108: 3125
38. Ferris JP, Antonucci FR (1972) J. Chem. Soc., Chem. Comm. 126; Ferris JP, Antonucci FR (1974) J. Amer. Chem. Soc. 96: 2010
39. Litinov VP, Gol'dfark YL (1976) Adv. Heterocycl. Chem. 19: 123
40. Dewar MJS, Trinajstic N (1970) J. Amer. Chem. Soc. 92: 1453
41. Hess BA Jr, Schaad LJ, Holyoke CW Jr (1972) Tetrahedron 28: 3657; Hess BA Jr, Schaad LJ, Holyoke CW Jr (1975) Tetrahedron 31: 295
42. Gutman I, Milun M, Trinajstic N (1977) J. Amer. Chem. Soc. 99: 1692
43. Gimarc BJ, Zhu Ji-Kang (1983) Inorg. Chem. 22: 479
44. Schmering von HG, Menge G (1981) Z. Anorg. Allg. Chem. 481: 33
45. Askin PA, Rambidi NG, Ezov SY (1960) Z. Neorg. Khim. 5: 747
46. Whitfield J (1970) J. Chem. Soc. A: 1800
47. Blachnik R, Wickel U (1983) Angew. Chem. 95: 313; (1983) Angew. Chem. Int. Ed. Engl. 22: 317
48. Ott JJ, Gimarc BJ (1986) J. Comput. Chem. 7: 673
49. Polansky OE, Zander M (1982) J. Mol. Struct. 84: 361
50. Polansky OE (1984) J. Mol. Struct. 113: 281; (1986) in: Trinajstic N (ed) Mathematics and computational concepts in chemistry, Ellis Horwood, Chichester, p 262
51. Zander M, Polansky OE (1984) Naturwiss. 71: 623
52. Polansky OE, Mark G, Zander M (1987) Der topologische Effekt an Molekülorbitalen (TEMO) — Grundlagen und Nachweis Schriftenr. des Max-Planck-Instituts für Strahlenchemie, Mülheim a. d. Ruhr Nr. 31
53. Graovac A, Gutman I, Polansky OE (1984) Mh. Chemie 115: 1
54. Graovac A, Polansky OE (1984) Croat. Chem. Acta 57: 1595
55. Hoxha J, Graovac A, Polansky OE (1986) Croat. Chem. Acta 59: 591
56. Kruszewski J, Polansky OE (1986) Match (Math. Chem.) 19: 243, 267
57. Polansky OE, Mark G (1985) Match (Math. Chem.) 18: 249
58. Dähne S, Graovac A, Polansky OE (1987) J. Molec. Struct. 151: 61
59. Heilbronner E (1953) Helv. Chim. Acta 36: 170
60. Graovac A, Gutman I, Polansky OE (1985) J. Chem. Soc., Faraday Trans. II 81: 1543
61. Gutman I, Graovac A, Polansky OE (1985) Chem. Phys. Lett. 116: 206; (1988) Discret Appl. Math. 19: 195
62. Streitwieser A Jr (1961) Molecular orbital theory for organic chemists, Wiley, New York
63. Klages F (1950) Chem. Ber. 82: 4278
64. Franklin JL (1950) J. Amer. Chem. Soc. 72: 4278
65. Chia JT, Simmons HE (1967) J. Amer. Chem. Soc. 89: 2638
66. Schmidt W (1977) J. Chem. Phys. 66: 828
67. Clar E (1964) Polycyclic hydrocarbons, Academic, London and Springer, Berlin Göttingen Heidelberg
68. Clar E, Schmidt W (1978) Tetrahedron 34: 3219

69. Clar E, Schmidt W (1977) Tetrahedron 33: 2093
70. Hafner E, Völpel KH, Ploss G, König C (1963) Liebigs Ann. Chem. 661: 52
71. Strell M, Kreis F (1954) Chem. Ber. 87: 1011
72. Koopmans T (1934) Physica 1: 104
73. Becker RS, Chen E (1966) J. Chem. Phys. 45: 2403
74. Jounkin JM, Smith LJ, Compton RN (1976) Theor. Chim. Acta 41: 157
75. Jordan KD, Michejda JA, Burrow PD (1976) J. Amer. Chem. Soc. 98: 1295
76. Wojnarovits L, Földiak G (1981) J. Chromatogr. 206: 511
77. Grimsrud EP, Caldwell G, Chowdhury S, Kebarle F (1985) J. Amer. Chem. Soc. 107: 4627
78. Chen, ECM, Wentworth WE (1975) J. Chem. Phys. 63: 3183
79. Jordan KD, Michejda JA, Burrow PD (1976) Chem. Phys. Lett. 42: 227
80. Peover ME (1962) J. Chem. Soc. 4540
81. Gleiter R, Spanget-Larsen J, Bartetzko R, Neunhoeffer H, Clausen M (1983) Chem. Phys. Lett. 97: 94
82. Gleiter R, Kobayashi M, Neunhoeffer H, Spanget-Larsen J (1977) Chem. Phys. Lett. 46: 231
83. Bieri G, Burger F, Heilbronner E, Maier JP (1977) Helv. Chim. Acta 60: 2213
84. Polansky OE, Zander M, Motoc I (1983) Z. Naturforsch. 38a: 196
85. Motoc I, Silverman JN, Polansky OE (1983) Phys. Rev. A 28: 3673
86. Motoc I, Silverman JN, Polansky OE (1984) Chem. Phys. Lett. 103: 285
87. Motoc I, Polansky OE (1984) Z. Naturforsch 39b: 1053
88. Motoc I, Silverman JN, Polansky OE, Olbrich G (1985) Theor. Chim. Acta 67: 63
89. Fabian W, Motoc I, Polansky OE (1983) Z. Naturforsch. 38a: 916
90. Silverman JN, Sobouti Y (1978) Astron. Astrophys. 62: 355
91. Silverman JN (1983) J. Phys. A 16: 3417
92. Kreile J, Münzel N, Schulz R, Schweig A (1984) Chem. Phys. Lett. 108: 609
93. Koenig T, Southworth S (1977) J. Amer. Chem. Soc. 99: 2807
94. McGlynn SP, Azumi T, Kinoshita M (eds) (1969) Molecular spectroscopy of the triplet state, Prentice Hall, Englewood Cliffs, p 330
95. El-Sayed MA (1975) Ann. Rev. Phys. 26: 235
96. Clarke RH (1982) Triplet state ODMR spectroscopy, Wiley, New York
97. Brinen JJ, Orloff MK (1966) J. Chem. Phys. 45: 4747
98. Burland DM, Schmidt J (1971) Mol. Phys. 22: 19
99. Kothandaraman G, Tinti DS (1975) Chem. Phys. Lett. 19: 225
100. Clarke RH, Frank HA (1976) J. Chem. Phys. 65: 39
101. Bräuchle C, Voitländer J (1982) Tetrahedron 38: 279
102. van der Waals JH, Malen G (1964) Mol. Phys. 8: 301
103. Chodkowskaja A, Grabowski ZR (1974) Chem. Phys. Lett. 24: 11
104. Bräuchle C, Kabzda H, Voitländer J (1980) Chem. Phys. 48: 369
105. Woolley RG (1976) Adv. Phys. 25: 27

Symmetry in Molecules

Jens Peder Dahl

Chemical Physics, Chemistry Department B, The Technical University of Denmark, DTH 301, DK-2800 Lyngby, Denmark

This article reviews and illustrates our understanding of the interplay between group theory and the theory of atomic and molecular electronic structure. It begins by demonstrating how the quest for group theory already emerged from the classical chemistry and crystallography of the nineteenth century, and goes on to describe the geometrical and dynamical symmetries met in the quantum mechanical one-electron problem. Next, detailed attention is paid to atoms and molecules with two or more electrons and hence to the permutation symmetry induced by the indistinguishability of electrons. Many-electron wave functions must span the antisymmetric representation of the permutation group, and the problem of constructing such functions is considered from various angles. Thus, contact is made to modern methods of determining many-electron wave functions by large scale computations. The article addresses itself to the student of chemistry who has but little knowledge of group theory, and hence it also includes a brief discussion of the theory of group representations.

---

* Further address: Department of Chemistry, The University of Chicago, 5735 S. Ellis Avenue, Chicago, Illinois 60637, U.S.A.

1 Introduction

The present paper deals with symmetry in molecules, a subject about which an immense number of articles and a great many books have been written. Accordingly, this is not a comprehensive review of the subject. It is rather like an essay in which we try to illustrate the fundamental role which symmetry plays in our understanding of the molecular world at various levels of description.

To us as scientists symmetry is a mathematical idea which finds its exact expression through the theory of groups. This implies that whenever an object possesses a certain symmetry then group theory enables us to make important assertions about some of the object's properties. In simple cases, such assertions may often be made without the formal methods of group theory. We then say that a symmetry argument has been applied, but it is understood that such an argument may always be rephrased in group theoretical terms.

Since symmetry finds its exact expression through the theory of groups our contribution might as well be entitled Group Theory and Molecules. But we prefer to retain the word Symmetry because of the associations this word gives occasion for. Symmetry is of general occurrence in the natural world and in the creations of man, and it has always been a highly valued concept. This is beautifully described by Hermann Weyl [1] in his little book Symmetry, which is based on the notion symmetry = harmony of proportions. And in a very recent book [2], which also includes a contribution by the editor of the present volume, symmetry is described as a concept unifying human understanding. Thus, symmetry is present everywhere and we tend to connect it with order, beauty, and understanding. So also in our description of the molecular world.

The present paper addresses itself to the student of chemistry who has but little knowledge of group theory. It begins with a discussion of the three-dimensional molecule (Section 2) and the geometric symmetry of molecules and crystals (Section 3). This discussion is essentially historic and aims at demonstrating how the quest for group theory naturally emerges from the classical chemistry of the nineteenth century. Sections 4 and 5 serve as an introduction to descriptive group theory and group representations, thus providing the foundation for the discussion in the remaining sections, which are devoted to the electronic structure of atoms and molecules as described by quantum mechanics.

Sections 6–10 are devoted to the one-electron atom and the one-electron molecule. They describe the geometrical and dynamical symmetries of orbitals and the function spaces which orbitals define. Detailed attention is also paid to the description of the electron's spin. Thus, these sections allow the student to appreciate the interplay between group theory and molecular orbital theory which plays such an important role in almost all branches of modern chemistry.

In Section 11 we turn to atoms and molecules with two or more electrons. We discuss the drastic implication of the fact that electrons are indistinguishable, and introduce the permutation group to describe these implications. The permissible N-electron wave functions must span the antisymmetric representation of the permutation group, and in Section 12 we discuss how such functions may be constructed by means of Slater determinants. In Section 13 we describe an alternative approach which draws more heavily on the representation theory of the permutation group. This approach is, in particular, of great value in connection with the large scale computations of today's quantum chemistry.

Section 14 is a brief discussion of the so-called unitary group approach to the quantum mechanical N-electron problem. The role played by group theory is here somewhat different from that of the previous sections. For the unitary group does not directly refer to the symmetry of a Hamiltonian. It refers to the geometry of whatever one-electron space one uses in the construction of many-electron wave functions.

Section 15 contains our concluding remarks and points to other topics that with equal right might have been included in a paper on symmetry in molecules.

2 The Three-Dimensional Molecule

As a generally accepted concept, the picture of a molecule as a three-dimensional arrangement of atoms is a fairly young one, not much more than a hundred years old. It played no role in the long and monumental development of chemistry which led to the final acceptance of Avogadro's hypothesis and the idea of valence in the late 1850s and to the establishment of the periodic law laid down in Mendeleev's periodic table of 1869 [3]. It was not until 1874 that the structural formulas of chemistry were extended into space. This was the year when van't Hoff and le Bel independently documented that the valencies of the carbon atom are directed toward the corners of a tetrahedron of which the carbon itself occupies the center.

The original papers by van't Hoff and le Bel are reprinted as English translations in the book by Benfey [4]. Both authors based their arguments on the existence of a correlation between a spatial arrangement of atoms and optical activity. As discussed by Benfey, the necessity of such a correlation had been pointed out earlier by Pasteur (1860) and Wislicenus (1869), and a tetrahedral carbon model had also been considered as a possibility by Wollaston (1808), Butlerov (1862), and Kekulé (1867).

Nevertheless, it was the systematic analysis by van't Hoff and le Bel that made the acceptance of a three-dimensional arrangement of atoms an unavoidable one.

Van't Hoff and le Bel's work is the foundation of the stereochemistry of carbon compounds. It is the directions of the four valencies of the carbon atom that determine the positions of the atoms bound to it. The maximum number of atoms bound to a carbon atom is accordingly four.

When we go beyond the carbon compounds we encounter great difficulties if we try to tie stereochemistry to valencies alone. These difficulties were overcome by Werner (1893) through the introduction of a new concept, the coordination number. Werner's most significant papers have been translated into English by Kauffman [5], and we quote from the first of these papers:

"The valence number indicates the maximum number of monovalent atoms which can be bound directly to the atom in question without the participation of other elementary atoms.

The coordination number indicates the maximum number of atoms and groups which can be bound directly to the atom in question.

For carbon, the numerical values for these two number concepts are the same; on the other hand, for almost all other elements, as far as can be judged today, they are different from each other."

By accepting the coordination number concept we obtain a general rationale for describing the possible arrangements of atoms in space, i.e. for describing the atomic positions in molecules, complexes, and crystals.

The correlation between the spatial arrangement of atoms and optical activity played a decisive role in van't Hoff and le Bel's arguments. It was understood that optical enantiomers are the mirror images of each other and hence, that a molecule which is superimposable on its mirror image cannot rotate the plane of polarized light.

Thus, the development of stereochemistry was greatly influenced by symmetry considerations.

The validity of the symmetry considerations hinges, however, on an important assumption not mentioned by van't Hoff, but spelled out by le Bel:

"In the reasoning which follows, we shall ignore the asymmetries which might arise from the arrangement in space possessed by the atoms and univalent radicals; but shall consider them as spheres or material points, which will be equal if the atoms or radicals are equal, and different if they are different."

This passage reflects the fact that the scientists of the nineteenth century were unable to explain what an atom really was and how it should be visualized. By the end of the century not only stereochemistry, but also spectroscopy and the kinetic theory of gases were well developed subjects, and no one had been able to combine the chemical properties of the elements, the characteristic spectral lines of the elements, and the kinetic properties of atoms and molecules into a coherent picture of the atom. As is well known, there were even heavy debates about the very reality of atoms.

It was this lack of knowledge about the nature of the atom that le Bel made allowance for in his exposition.

3 The Geometric Symmetry of Molecules and Crystals

At the time stereochemistry had been well developed, by the end of the last century, another important branch of science had reached a point of culmination, namely crystallography. Crystallography is the science of crystals, and its historical development is well described by Phillips [6]. As an exact science it began when Nicolaus Steno (Niels Stensen) published his measurements of the angles between corresponding faces on crystals of quartz (1669). He found these angles to be constant, and thereby opened the road to the Law of Constancy of Angles which was finally confirmed as a general law by Romé de l'Isle a hundred years later (1772–83).

The external regularities of crystals which this law describes were by many considered to reflect an internal regularity, and in 1784 Abbe Haüy suggested that continued cleavage of a crystal would ultimately lead to a smallest possible unit, a molécule intégrante, by a repetition of which the whole crystal is built up.

Haüy also formulated the important Law of Rational Indices concerning the relative slopes of crystal surfaces. On the basis of this law he made important deductions about the internal symmetries of crystals. The study of internal symmetries continued during the following hundred years and reached a complete clarification by 1890 with the enumeration of the 230 space groups, by Fedorov, Schoenflies, and Barlow. The internal structure of a crystal, as described by its unit cell and the arrangement of structural units within the unit cell, must necessarily conform with one of the 230 space groups. A particular space group specifies the combination of translations, rotations, reflections, and inversion which leave the corresponding internal structure invariant.

Group theory arose in the first half of the ninetenth century, and it was in the study of crystals that it found its first physical application. With the arrival of the picture of the three-dimensional molecule one could identify the structural units within the unit cell with atoms and/or molecules, and thus the basis was laid for one of the most important uses of groups in chemistry whatsoever, namely, in X-ray crystallography, a science of the twentieth century. The determination of a crystal's space group is the first and very essential step in a modern crystal structure analysis by X-ray diffraction. The second step is the accurate determination of the atomic positions within the unit cell. With the atoms belonging to a molecule it leads to a precise determination of the three-dimensional structure of the molecule.

The geometric symmetry of a crystal is given by its space group. The geometric symmetry of a molecule is, in turn, given by its point group. The point group specifies the combinations of rotations, reflections, and inversion which leave the molecular geometry invariant.

Just knowing to which point group a given molecule "belongs" is of considerable importance in many respects. One of the simplest applications concerns the optical activity of the molecule. A necessary condition for a molecule to be optically active is that its point group contains no reflections. Such a molecule is called chiral and may still have a considerable amount of symmetry.

The 230 space groups are listed and described in the International Tables for Crystallography [7]. Each space group has one of a set of 32 point groups associated with it. These so-called crystallographic point groups are also described in the tables. The number of possible point groups is, however, unlimited. An excellent description of the classification of the general point groups is given by Landau and Lifshitz [8].

Point groups are named by symbols describing the symmetry elements of the group, i.e. the set of axes, planes, and inversion center with respect to which the symmetry operations are carried out. Crystallographers prefer the so-called Hermann-Mauguin notation, whereas the older Schoenflies notation is the one most commonly used in molecular work. Accordingly, we shall adhere to the Schoenflies notation in what follows.

A useful little handbook of point group properties is the book by Atkins et al. [9]. It gives, for example, the Schoenflies symbols for all point groups and, in addition, the Hermann-Mauguin symbols for the 32 crysrallographic point groups. Let us take a look at one of the smaller groups.

4 Descriptive Group Theory

The structure of a finite group is given by its multiplication table, and two groups whose elements satisfy the same multiplication table are said to be isomorphic. Any group is accordingly isomorphic to an abstract group, i.e. a group to whose elements we attach no other properties than those that follow from the multiplication table.

There is one or more abstract groups of any order, g, the order being defined as the number of elements of a group. (Lomont [10] lists the number of abstract groups for $g \leqq 215$.) The smallest non-commutative group is of order 6, and this group is accordingly a favoured text-book example. Its multiplication table is reproduced in Table 1. The point groups C_{3v} and D_3 are isomorphic with this group and so is S_3, the group of permutations of three objects.

Table 1. The multiplication table for the smallest non-commutative group

	a_1	a_2	a_3	a_4	a_5	a_6
a_1	a_1	a_2	a_3	a_4	a_5	a_6
a_2	a_2	a_3	a_1	a_6	a_4	a_5
a_3	a_3	a_1	a_2	a_5	a_6	a_4
a_4	a_4	a_5	a_6	a_1	a_2	a_3
a_5	a_5	a_6	a_4	a_3	a_1	a_2
a_6	a_6	a_4	a_5	a_2	a_3	a_1

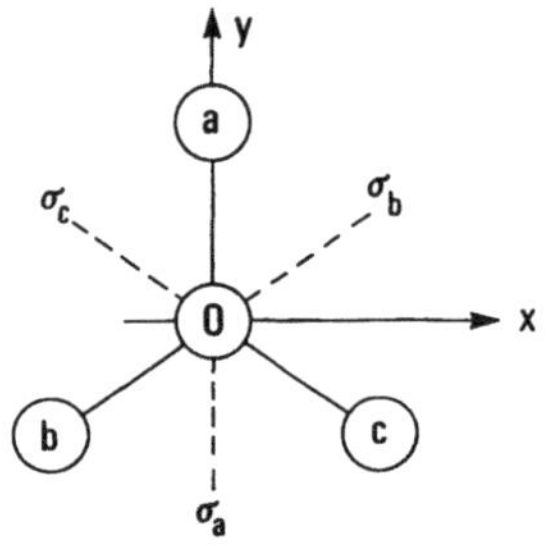

Fig. 1. C_{3v} molecule. a, b, and c are equivalent atoms (e.g. hydrogen) in the xy-plane. The central atom 0 (e.g. nitrogen) lies above the plane, along the positive z axis. σ_a, σ_b, and σ_c are vertical reflection planes. The z axis is a threefold rotation axis

Figure 1 shows the symmetry elements of a C_{3v} molecule. The symmetry operations are:

R_1 = identity, E.
R_2 = rotation through $2\pi/3$, C_3.
R_3 = rotation through $4\pi/3$, C_3^2.
R_4 = reflection in the σ_a plane, σ_a.
R_5 = reflection in the σ_b plane, σ_b.
R_6 = reflection in the σ_c plane, σ_c.

The rotations are understood to be positive (counterclockwise) rotations about the z axis. The numbering of the symmetry operations corresponds to the numbering in Table 1. We have also added Schoenflies-like symbols (E, C_3, C_3^2, σ_a, σ_b, σ_c) for the symmetry operations.

Stereochemistry and the great importance of X-ray crystallography have caused the kind of descriptive group theory which we have just given for C_{3v} to be part of the general knowledge of most chemists. A great many chemists have, however, had to go one step further to make themselves acquainted with the concept of group representations. There are essentially two practical reasons for this, apart from the obvious theoretical ones.

The first of these reasons is that many forms of spectroscopy have been indispensable tools of analytic chemistry for many years; infrared and Raman spectroscopy, visible and UV spectroscopy, ESR and NMR spectroscopy, to mention the most common ones. Spectral lines and spectral bands represent quantum transitions, and the selection rules that govern such transitions depend on the symmetries of the quantum states involved. This dependence can only be understood through representation theory.

The second reason is that the availability of path ways for chemical reactions in many cases is governed by the symmetries of the highest occupied molecular orbitals (HOMOs) and the lowest unoccupied molecular orbitals (LUMOs). The role played by the symmetries of these orbitals is sufficiently simple and general that it can be used in the synthetic chemist's arguments at the same level as all his other rules. The very classification of orbital symmetries is, however, a problem of representation theory.

5 Group Representations

A (matrix) representation of a group $\{a_1, a_2, \ldots\}$ is a set of square nonsingular matrices $\{D(a_1), D(a_2) \ldots\}$ such that

$$D(a_i)\,D(a_j) = D(a_i a_j) \tag{1}$$

for any pair of group elements. For a finite group this implies that the matrices of a representation satisfy the group-multiplication table. $D(a_i)$ is said to represent the group element a_i. The identity element is, in particular, represented by the unit matrix. The number of rows and columns of the matrices is called the dimensionality of the representation.

In connection with the study of molecules we may in general assume that a matrix representation is "carried" by a function space, by which we mean the following. With n denoting the dimensionality of the representation there exists a set of n linearly independent functions $\{\psi_1, \psi_2, \ldots, \psi_n\}$ such that

$$R_i\psi_r = \sum_{s=1}^{n} \psi_s D_{sr}(R_i), \qquad \text{for all } i\,. \tag{2}$$

The operators $\{R_1, R_2, ...\}$ are supposed to form a group isomorphic with the group $\{a_1, a_2, ...\}$, and we have accordingly written $D(R_i)$ instead of $D(a_i)$ in Eq. (2). The set of all functions of the type

$$\psi = \sum_{r=1}^{n} c_r \psi_r , \tag{3}$$

with the coefficients c_r being arbitrary complex numbers, are said to form a linear function space, or vector space, Ω_n.

Let S be a nonsingular $n \times n$ matrix. The function

$$\psi'_r = \sum_{s=1}^{n} \psi_s S_{sr} , \qquad r = 1, 2, \ldots, n \tag{4}$$

will then also be linearly independent, and any function of the type (3) may equally well be written as

$$\psi = \sum_{r=1}^{n} c'_r \psi'_r . \tag{5}$$

Thus, the new functions define an alternative basis in Ω_n.

Elementary manipulations show that

$$R_i \psi'_r = \sum_{s=1}^{n} \psi'_s D'_{sr}(R_i) , \qquad \text{for all i} . \tag{6}$$

The relations (2) and (6) are similar. We say that the matrices $D(R_i)$ and $D'(R_i)$ are connected by a similarity transformation and find that

$$D'(R_i) = S^{-1} D(R_i) S , \qquad \text{for all i} . \tag{7}$$

The primed matrices also form an n-dimensional representation of our group. The two representations are called equivalent.

The relations (2) and (6) may be said to imply that Ω_n is invariant under the group $\{R_1, R_2, ...\}$. By a suitable choice of basis, i.e. by a suitable choice of the matrix S it may happen that we can split Ω_n into smaller subspaces, each of which is again invariant under our group. We shall then have relations similar to (2) satisfied for each of the subspaces, and thus each subspace defines its own representation of the group. If Ω_n has been split into subspaces none of which can be split further, then we call the representations carried by the subspaces irreducible. If the dimension of a subspace is larger than 1 we still can vary the basis in the subspace. Hence, an irreducible representation is only defined up to a similarity transformation.

A group can in a unique way be divided into (equivalence) classes, with no two classes having any elements in common. The elements in the class of a_i are determined as $a_k^{-1} a_i a_k$ when a_k is allowed to scan the group. We find, for instance, that the group C_{3v} contains the classes E, $2\,C_3 = \{R_2, R_3\}$, and $3\sigma_v = \{R_4, R_5, R_6\}$.

Representation theory tells us that the number of (inequivalent) irreducible representations of a finite group equals the number of classes. It also tells us that if we square the dimension of each irreducible representation and add the resulting numbers, then the sum equals the order of the group. Applying these rules to C_{3v} shows us that this group has three irreducible representations, with the dimensions 1, 1, and 2, respectively. These are of course also the irreducible representations of any group isomorphic with C_{3v}.

Given an irreducible representation of a group we may take the trace of each matrix of the representation. The resulting set of g numbers is called the character of the representation. We shall also refer to the trace of a matrix as the character of the corresponding group element. It is now easy to show that all the elements of a given class have the same character, and hence all information about the characters of the irreducible representations can be collected in an array with q rows and columns, where q is the number of classes of the group. Such an array is called a character table.

Table 2. Character table for the group C_{3v}

C_{3v}	E	$2\,C_3$	$3\sigma_v$
A_1	1	1	1
A_2	1	1	−1
E	2	−1	0

The character table for the group C_{3v} is given in Table 2. The two one-dimensional representations are denoted A_1 and A_2, the two-dimensional representation is called E. The E matrices are only defined to within a similarity transformation; the trace of a matrix is, however, invariant under such a transformation. The E character is thus uniquely defined.

Table 3. Standard matrices for the E representation of C_{3v}

R_1	R_2	R_3	R_4	R_5	R_6
$\begin{bmatrix} 1 & 0 \\ 0 & 1 \end{bmatrix}$	$\begin{bmatrix} -\frac{1}{2} & -\frac{\sqrt{3}}{2} \\ \frac{\sqrt{3}}{2} & -\frac{1}{2} \end{bmatrix}$	$\begin{bmatrix} -\frac{1}{2} & \frac{\sqrt{3}}{2} \\ -\frac{\sqrt{3}}{2} & -\frac{1}{2} \end{bmatrix}$	$\begin{bmatrix} -1 & 0 \\ 0 & 1 \end{bmatrix}$	$\begin{bmatrix} \frac{1}{2} & \frac{\sqrt{3}}{2} \\ \frac{\sqrt{3}}{2} & -\frac{1}{2} \end{bmatrix}$	$\begin{bmatrix} \frac{1}{2} & -\frac{\sqrt{3}}{2} \\ -\frac{\sqrt{3}}{2} & -\frac{1}{2} \end{bmatrix}$

In Table 3 we give a set of matrices for the E representation of C_{3v}. We shall refer to these as the standard matrices of the representation. They have been chosen to be unitary, in accordance with the lemma that any irreducible representation of a finite group is equivalent to a representation by unitary matrices.

With reference to Fig. 1 we may easily specify a function space which carries the E representation, with a basis corresponding to the matrices of Table 3. To this end we consider a general function f(x, y, z) defined with respect to the coordinate system of Fig. 1 and operate on this function with a symmetry operator R_i. This results in a new function which we denote $R_i f(x, y, z)$. Per definition its contour surfaces shall be obtained by subjecting the contour surfaces of f(x, y, z) to the operation R_i. If r denotes the distance from the point (x, y, z) to the origin we find in particular that

$$R_i f(r) = 1 \times f(r)\,, \qquad \text{for all } i\,. \tag{8}$$

This implies that f(r) "spans" the A_1 representation. Now let us introduce

$$p_x = f(r)\,x\,, \qquad p_y = f(r)\,y\,. \tag{9}$$

It is then readily seen that

$$R_2[p_x, p_y] = [p_x, p_y]\begin{bmatrix} -\frac{1}{2} & -\frac{\sqrt{3}}{2} \\ \frac{\sqrt{3}}{2} & -\frac{1}{2} \end{bmatrix}, \tag{10}$$

and we recognize the matrix representing R_2 in Table 3. Subjecting p_x and p_y to the other symmetry operations of C_{3v} are found to generate the remaining matrices of Table 3. Hence p_x and p_y define a two-dimensional function space carrying the E representation, and they also define the standard basis in this space.

We have now introduced the general language of representation theory, with the C_{3v} group as a simple example. We shall need some further elements of the theory, but these will be introduced as we go along by referring to already existing and comprehensive presentations by, e.g. Wigner [11], Tinkham [12], and Hamermesh [13]. We are, however, sufficiently prepared for a discussion of the role played by symmetry in the description of the interior world of atoms and molecules. This role is probably best understood by moving from smaller to larger systems, and this will accordingly be our line of approach.

6 The Three-Dimensional Atom

As described in Section 2, stereochemistry was as highly developed science by the end of the nineteenth century. It was based on the concept of atoms, yet nobody understood what an atom really was. It took thirty years to reach such an understanding, from the discovery of the electron in 1897 by Thomson, over Rutherford's model of the planetary atom of 1911, to the discovery of the Schrödinger equation in 1926. Those years saw the rise of quantum mechanics and the theory of relativity. The rise of these sciences was much tied to the search for an understanding of the nature of atoms and radiation. It was the era of the Bohr-Sommerfeld-Pauli theory of the atom, and the era of the valence theories by Kossel, Langmuir, and Lewis. It was cemented that an atom with atomic number Z consists of a heavy, compact nucleus carrying the electric charge Ze, and of Z electrons each carrying the charge —e. It was also understood that the states of atoms and ions were quantized, and that the absorption and emission of electromagnetic radiation (photons) resulted in transitions between the possible quantum states. The lower states of an atom or ion could be accounted for in an empirical way by the *aufbau* principle, based upon a set of one-electron quantum numbers and Pauli's exclusion principle.

Symmetry also had a part to play in the atomic models of the period, as I have described elsewhere [14]. However, we shall not dwell on the models of early quantum mechanics, but concentrate instead on the development initiated by Schrödinger's discovery.

What Schrödinger showed was that the problem of quantization can be formulated as an eigenvalue problem [15]. The stationary states of a system are determined by an equation of the form

$$H\Psi = E\Psi , \tag{11}$$

where H is the Hamiltonian, Ψ the wave function, and E the energy. For bound states it must be possible to normalize Ψ such that

$$\langle\Psi|\Psi\rangle = 1 , \tag{12}$$

and this condition limits the possible values of E to a discrete spectrum. There is also a continuous eigenvalue spectrum corresponding to scattering states, but we shall not consider such states in the present paper.

The wave function of a given atomic state contains all information about the state. It leads to a three-dimensional picture of an atom as a well defined distribution of electronic charge, but indeed it gives us a much richer picture than that. Many aspects of this richness are, however, incomprehensible in terms of classical concepts. This is due to the fact that a wave function primarily is an entity in an abstract function space (a Hilbert space), and that the laws that determine its form and possible change of form are laws relating to that space.

Subjecting an atom or any other quantum system to a symmetry operation will induce a particular transformation on the system's Hilbert space. Thus, Hilbert space becomes the carrier of representations of the symmetry group involved.

Let us illustrate this by considering the symmetry properties of the simplest atom in some detail.

7 The Hydrogen-Like Atom

By a hydrogen-like atom we understand a one-electron atom with nuclear charge Ze. The quantum description of this system is of course discussed in many text books (see, e.g. Refs. 16–19) and needs no elaboration here. For simplicity, we neglect the center of mass motion and assume that the nucleus is kept fixed at the origin of a rectangular coordinate system. A point $\vec{r} = (x, y, z)$ in this system will then define a possible position of the electron. Alternatively, we may specify the electron's position by the spherical polar coordinates (r, θ, φ), defined by the well-known relations

$$\begin{cases} x = r \sin\theta \cos\varphi\,, & 0 \leqq r < \infty \\ y = r \sin\theta \sin\varphi\,, & 0 \leqq \theta \leqq \pi \\ z = r \cos\theta\,, & 0 \leqq \varphi < 2 \end{cases} \tag{13}$$

The Hamiltonian has the form

$$H = -\frac{\hbar^2}{2\mu}\nabla^2 - \frac{Ze^2}{4\pi\varepsilon_0}\frac{1}{r} \tag{14}$$

where the first term represents the kinetic energy of the electron and the second its potential energy in the field of the nucleus. μ is the mass of the electron. Solving the Schrödinger equation (11) leads to the Balmer expression for the allowed energy levels, i.e.

$$E_n = -\left(\frac{Ze^2}{4\pi\varepsilon_0}\right)^2 \frac{\mu}{2\hbar^2 n^2}\,, \qquad n = 1, 2, \ldots \tag{15}$$

There are n^2 linearly independent wave functions corresponding to the energy E_n. Any linear combination of these is also a solution to the Schrödinger equation. In other words, the wave functions corresponding to the n'th energy level define an n^2-dimensional function space $\Omega(n)$. The degeneracy of the level is said to be n^2.

A standard basis in $\Omega(n)$ is given by the functions

$$\varphi_{nlm_l}(r, \theta, \varphi) = R_{nl}(r)\, Y_{lm_l}(\theta, \varphi) \tag{16}$$

with $l = 0, 1, \ldots, n-1$ and $m_l = -l, -l+l, \ldots, l$. $R_{nl}(r)$ is a radial function and $Y_{lm_l}(\theta, \varphi)$ is a spherical harmonic. We assume the form of these functions to be well known. The $2l + 1$ functions corresponding to a fixed value of l define a function space $\Omega(n, l)$. The space $\Omega(n)$ is thus the direct sum of n subspaces, i.e.

$$\Omega(n) = \sum_{l=0}^{n-1} \Omega(n, l)\,. \tag{17}$$

We shall now tie the nature of these spaces to the symmetry of the problem.

With this end in view, let A be some operator that commutes with the Hamiltonian, i.e.

$$[A, H] \equiv AH - HA = 0\,. \tag{18}$$

Application of A to both sides of Eq. (11) gives

$$AH\Psi = EA\Psi\,. \tag{19}$$

E is a real number, and we assume that

$$AE\Psi = EA\Psi\,, \tag{20}$$

a relation that will be satisfied whenever A is either linear or antilinear [11]. Applying Eqs. (18) and (20) to Eq. (19) shows that

$$HA\Psi = EA\Psi\,, \tag{21}$$

i.e. the function $A\Psi$ is an eigenfunction of H with the same E value as Ψ.

This familiar result implies that if the elements of some group all commute with the Hamiltonian (14), then each of the function spaces $\Omega(n)$ must be invariant under the group, i.e. each $\Omega(n)$ carries a representation of the group. The representation carried by an $\Omega(n)$ may either be reducible or irreducible. "Understanding" the degeneracies of our problem from the point of view of symmetry is tantamount to finding a group such that each $\Omega(n)$ carries an irreducible representation of that group.

It is easy to find a whole class of groups whose elements commute with the Hamiltonian (14). Any point group that leaves the origin (the nucleus) invariant is in fact a candidate. An important example is the group $C_i = \{E, I\}$, where E is the identity and I the inversion in the origin. Its irreducible representations are given in Table 4.

Table 4. Character table for the group C_i

C_i	E	I
A_g	1	1
A_u	1	−1

As is well known, functions spanning the A_g representation are called even (*gerade*) and are said to have the parity 1, functions spanning the A_u representation are called odd (*ungerade*) and are said to have the parity −1.

Since the group C_i only has one-dimensional irreducible representations it can in no way account for the degeneracies of the hydrogenic energy levels. Its presence does, however, guarantee us that the wave function for the non-degenerate ground state has a definite parity (it is even). It also allows us to choose the wave functions for other states such that each has a definite parity. The functions (16) have in fact been so chosen, the parity being $(-1)^l$. Actually, the parity is the same for all functions belonging to a chosen $\Omega(n, l)$ subspace, but it alternates in the direct sum (17) that leads to $\Omega(n)$.

The group of all rotations about any axis through the origin is the three-dimensional rotation group O(3). It has an infinite number of elements. The elements may, however, be labelled by three continuous parameters such as the three components of the vector $\theta\hat{n}$, where $\hat{n}$ is a unit vector specifying the axis of rotation, and θ is the actual angle of rotation. The form of the rotation operators is [11–13]:

$$R(\theta\hat{n}) = \exp\left(-i\theta\hat{n}\cdot\vec{l}/\hbar\right)\,, \tag{22}$$

where

$$\vec{l} = (l_x, l_y, l_z) \tag{23}$$

is the angular momentum vector operator. Its components satisfy the familiar commutation relations

$$[l_x, l_y] = i\hbar l_z\,, \qquad \text{cyclic} \tag{24}$$

as well as the Jacobi identity

$$[u, [v, w]] + [v, [w, u]] + [w, [u, v]] = 0\,, \tag{25}$$

where u, v, w are any of the three components of $\vec{l}$.

All operators of the form

$$A = \alpha_1 l_x + \alpha_2 l_y + \alpha_3 l_z\,, \tag{26}$$

with α_1, α_2, α_3 being arbitrary real numbers, define a linear space. This space is closed under an antisymmetric binary operation (the commutator) which satisfies Jacobi's identity. These properties of the space characterises it as a real Lie algebra [13, 20]. It is denoted o(3). The group O(3) is a Lie group, and because its parameters (the three components of $\theta\vec{n}$) are bound to vary over a finite domain, it is called compact.

A linear function space which is invariant under the operators (26) defines a matrix representation of the algebra o(3). Reducibility and irreducibility are defined in a similar way as for groups, and it is obvious, from a comparison of the operators (22) and (26), that a function space which carries an irreducible representation of O(3) also carries an irreducible representation of o(3).

Standard angular momentum theory [16–18] addresses itself to the problem of determining the irreducible representations of o(3). In so doing one considers o(3) as an abstract algebra with "generators" j_x, j_y, j_z satisfying commutation relations similar to those of Eq. (24). It is then found that the representations of this abstract algebra may be labelled by the eigenvalues of the bilinear operator

$$\vec{j}^2 = j_x^2 + j_y^2 + j_z^2 \tag{27}$$

which commutes with all elements of the algebra. The possible eigenvalues of $\vec{j}^2$ are $j(j + 1)\hbar^2$, where $j = 0, {}^1/_2, 1, {}^3/_2, \ldots$ The abstract vector space which carries the irreducible representation corresponding to a fixed value of j has the dimension $2j + 1$.

By returning to the l_x, l_y, l_z operators in order to represent the carrier space vectors as functions of r, θ, φ one finds that such functions only exist for j (which we now call l) being an integer. When l is an integer the $2l + 1$ functions in question have the form of an arbitrary radial function times the spherical harmonics $Y_{lm_l}(\theta, \varphi)$. This set of functions span an irreducible representation of the group O(3) as well as of the algebra o(3).

Thus, we have arrived at the conclusion that each of the function spaces $\Omega(n, l)$, Eq. (17), carries an irreducible representation of O(3). The degeneracy with respect to the quantum number m_l Eq. (16), is accordingly a consequence of the fact that the elements of O(3) commute with the Hamiltonian (14). Actually, the elements of O(3) commute with the kinetic and potential energy operators separately, and the same is true for the inversion operator. This spherical symmetry of the Hamiltonian is naturally called a geometrical symmetry.

The spherical symmetry of the Hamiltonian (14) accounts for the degeneracy with respect to m_l for a given value of l. But it does not account for the degeneracy with respect to l for a given value of n. This remaining denegeracy, which only occurs when the potential energy is of the Coulomb type, has accordingly been called accidental.

It is not just in quantum mechanics that the Coulomb potential has an exceptional position. In classical mechanics it leads to the closed elliptic orbits of the Kepler problem, and hence to the periodic motion of the earth that causes each year to have the same number of days. It turns out that this situation can be accounted for by noting that the motion in a Coulomb potential not only conserves the angular momentum vector, but also a vector perpendicular to the angular momentum. This is the so-called Runge-Lenz vector. Its presence allows the classical equations of motion to be solved in an extremely elegant way [21, 22].

Pauli [23] demonstrated that the corresponding quantum mechanical vector operator, likewise called the Runge-Lenz vector, commutes with the Hamiltonian (14) and he utilized this in an elegant derivation of the energy level spectrum (15). The form of the Runge-Lenz vector is

$$\vec{M} = \frac{1}{2\mu}(\vec{p} \times \vec{l} - \vec{l} \times \vec{p}) - \frac{Ze^2}{4\pi\varepsilon_0}\frac{\vec{r}}{r}. \tag{28}$$

Subsequently, Fock [24] studied the Schrödinger equation in momentum space and showed that the Coulomb problem is intrinsically four-dimensional. Finally Bargmann [25] showed that the three components of the vector

$$\vec{M}' = \left(-\frac{\mu}{2E}\right)^{1/2} \vec{M}, \tag{29}$$

together with the three components of $\vec{l}$, define a Lie algebra isomorphic with o(4), i.e., the Lie algebra of the four-dimensional rotation group O(4).

The function spaces $\Omega(n)$, Eq. (17), carry irreducible representations of O(4), and thus the degeneracy of the levels (15) has been completely accounted for by symmetry. The components of $\vec{M}$ do not commute with the kinetic and potential energy parts of (14) separately, and hence the O(4) symmetry is referred to as a dynamical symmetry.

For a detailed discussion of the O(4) symmetry we refer to Schiff [16] and Bander and Itzykson [26]. The latter authors also show that it is possible to introduce a larger group, O(1, 4), in such a way that the whole Hilbert space of bound states carries a single irreducible unitary representation of this group. One of the O(1, 4) group parameters cannot be restricted to a finite range, and hence O(1, 4) is called a non-compact group. It can be shown [20] that any unitary representation of a non-compact group is of infinite dimension.

By allowing the coefficients α_1, α_2, α_3 in Eq. (26) to be complex we obtain the so-called complexification of o(3). This complex Lie algebra contains the shift operators $l_x \pm il_y$ which transform the various basis functions of $\Omega(n, l)$ into each other. Similarly, shift operators can be defined from o(4). The complexification of the Lie algebra o(1, 4) contains shift operators that transform all bound state wave functions into each other. For this reason, we may refer to o(1, 4) as a spectrum generating algebra.

The discussion of the present section serves several purposes. Firstly, it introduces a set of notions for continuous groups and their algebras. Secondly, it illustrates that symmetry may have dynamical origins beyond the more obvious geometrical ones. Thirdly, it introduces the concept of a spectrum generating algebra. And last, but not least, it gives an overview of the bound states of a hydrogen-like atom.

In view of the fact that the Coulomb potential is the potential that nature actually prefers, it is tempting to ask if the O(4) symmetry of the Coulomb problem can be tied to other of nature's symmetries in a natural manner. The only attempt to answer this question seems to be a classical mechanical investigation by the author [27] in which it is shown that the Runge-Lenz vector occurs as a part of the generator of a Lorentz transformation, in the proper two-body problem. An extension of this approach to the quantum mechanical domain must also include the spins of the particles, so although such an extension would be interesting, it would also be exceedingly difficult.

8 The Electron Spin

The necessity of assuming that the electron has an intrinsic angular momentum, the so-called spin, was established by Uhlenbeck and Goudsmit in 1925 [28]. The component of the spin in any chosen direction is restricted to two values, $+\hbar/2$ and $-\hbar/2$, and this implies according to the standard theory of angular momentum, that the j quantum number (now called s) must be 1/2. It also implies that the wave function (the state vector) associated with the spin must belong to a two-dimensional vector space, $\Omega_{1/2}$, which carried an irreducible representation of the o(3) algebra.

A priori, it might be natural to assume that the spin reflects the presence of an internal axis whose direction may be described by spherical polar angles θ', φ'. A spin wave function would then be a function of these angles. But in accordance with the previous section it is impossible to construct functions of spherical polar angles corresponding to half-integral values of j. Hence the picture of an electron as a (point) particle that specifies a direction must be abandoned.

Following Pauli [27] one therefore introduces a purely formal spin variable σ restricted to just two values, say $+1$ and -1. A normalized spin function, $\eta(\sigma)$, is then subject to the interpretation that $|\eta(1)|^2$ is the probability of measuring a spin projection $\hbar/2$ along the z-axis. Similarly, $|\eta(-1)|^2$ gives the probability of measuring the projection $-\hbar/2$. The eigenfunctions of s_z corresponding to the eigenvalues $+\hbar/2$ and $-\hbar/2$ are denoted $\alpha(\sigma)$ and $\beta(\sigma)$, respectively.

Working on a general spin function, $\eta(\sigma)$, with the rotation operator (22), but with $\vec{l}$ replaced by $\vec{s}$, leads to a change of sign when θ is put equal to 2π. The $\Omega_{1/2}$ function space does not accordingly carry a proper representation of the group O(3). The representation is a double-valued one.

To understand this double-valuedness one notes that o(3) is isomorphic with su(2) which is the Lie algebra corresponding to the group SU(2) [11–13]. SU(2) is the group of 2 by 2 unitary matrices with determinant 1, and it is the so-called covering group of O(3). There is a 2 to 1 relation between the two groups, such that one element of O(3) corresponds to two distinct elements of SU(2). The irreducible representations of SU(2) are all proper single-valued representations [11–13]. The representations corresponding to integral values of j are also single-valued representations of O(3), but those corresponding to half-integral values of j are double-valued representations of O(3).

Double-valued representations are acceptable in quantum mechanics because all physical predictions involve the formation of bilinear combinations of wave functions, and such combinations are evidently left unchanged by a rotation through 2π. The sign change of a spin 1/2 wave function may, however, give rise to interference effects in suitably conducted experiments. Such experiments have now been carried out [30, 31], and the sign change has been amply confirmed.

It is interesting to note that although it is common practice to work with the formal spin variable σ, it is in fact possible to construct an electron model in which σ is replaced by angle variables [32]. Such a model is obtained by realizing that although it is impossible to consider an electron as a point particle that specifies a direction, it is possible to consider it as a point rotor, i.e. as a particle that specifies an orientation. The orientation is that of a set of coordinate axes attached to the electron and is given by three Euler angles α, β, γ. Considering a spin function as a function of α, β, γ leads to the following expressions for the spin operators s_x, s_y, s_z:

$$\begin{cases} s_x = i\hbar\left(\sin\alpha\,\dfrac{\partial}{\partial\beta} + \cot\beta\cos\alpha\,\dfrac{\partial}{\partial\alpha} - \dfrac{\cos\alpha}{\sin\beta}\,\dfrac{\partial}{\partial\gamma}\right) \\ s_y = i\hbar\left(-\cos\alpha\,\dfrac{\partial}{\partial\beta} + \cot\beta\sin\alpha\,\dfrac{\partial}{\partial\alpha} - \dfrac{\sin\alpha}{\sin\beta}\,\dfrac{\partial}{\partial\gamma}\right) \\ s_z = -i\hbar\,\dfrac{\partial}{\partial\alpha}\,. \end{cases} \tag{30}$$

In order not to interfere with the designation for the Euler angles we call the α and β spin functions χ_1 and χ_2 instead. They are

$$\begin{cases} \chi_1 = (8\pi^2)^{-1/2}\cos\dfrac{\beta}{2}\,e^{i\alpha/2}\,e^{i\gamma/2} \\ \chi_2 = (8\pi^2)^{-1/2}\sin\dfrac{\beta}{2}\,e^{-i\alpha/2}\,e^{i\gamma/2}\,. \end{cases} \tag{31}$$

A similar pair of functions are

$$\begin{cases} \chi_3 = -(8\pi^2)^{-1/2} \sin \frac{\beta}{2} e^{i\alpha/2} e^{-i\gamma/2} \\ \chi_4 = (8\pi^2)^{-1/2} \cos \frac{\beta}{2} e^{-i\alpha/2} e^{-i\gamma/2} . \end{cases} \tag{32}$$

The half angles in these expressions make it immediately clear that a spin function is multiplied by -1 when it is rotated through the angle 2π.

An advantage of the point rotor model is that it is directly transferable to the relativistic domain. Dirac's relativistic equation [33] makes an electron wave function a so-called four-component spinor. The rotor equivalent is a wave function of the form

$$\psi(x, y, z; \alpha, \beta, \gamma) = \sum_{i=1}^{4} \varphi_i(x, y, z)\, \chi_i(\alpha, \beta, \gamma) , \tag{33}$$

where $\varphi_1, \varphi_2, \varphi_3, \varphi_4$ are the components of the Dirac spinor, and $\chi_1, \chi_2, \chi_3, \chi_4$ are the functions specified above. For further details we refer to Ref. 32.

9 Atomic Spin-Orbitals

We must now synthesize the contents of the two preceding sections. In so doing we shall generalize the discussion by allowing the Coulomb field of the nucleus to be replaced by a "screened" Coulomb field. Such a field occurs, for instance, in the SCF (self-consistent field) description of an N-electron atom or ion [14, 34], in which each electron is assumed to move in a central field which is the sum of the Coulomb field from the nucleus and a spherically averaged field from $N - 1$ electrons.

Thus we replace the Hamiltonian (14) with the similar Hamiltonian

$$H = -\frac{\hbar^2}{2\mu} \nabla^2 + V(r) , \tag{34}$$

where V(r) describes the generalized central field. The bound state eigenfunctions of H are atomic orbitals. They can still be chosen in the form (16), with $R_{nl}(r)$ varying with V(r), but the energy is no longer a function of n alone (except in the Coulomb case, whose special features we now ignore). Instead, the energy is a function of both n and l, and each $\Omega(n, l)$ function space has its own energy, $E(n, l)$. The degeneracy of an energy level is $2l + 1$. It is completely explained by the spherical symmetry of the problem, i.e. by the presence of the C_i and O(3) symmetry groups.

Even though an atomic orbital is the solution of a one-electron Schrödinger equation, it is not a genuine electron wave function. The integrity of the electron requires that we combine it with a spin function $\eta(\sigma)$. By taking η to be either $\eta_{1/2} = \alpha$, or $\eta_{-1/2} = \beta$, we get instead of Eq. (16):

$$\psi_{nlm_lm_s}(x) = \varphi_{nlm_l}(\vec{r})\,\eta_{m_s}(\sigma)\,, \tag{35}$$

with $m_s = \pm\frac{1}{2}$. We have collected the space coordinates in the position vector $\vec{r}$ and denoted the combined space and spin coordinates by x.

For fixed values of n and l, Eq. (35) specifies $2(2l + 1)$ functions and hence a $2(2l + 1)$ dimensional linear function space. Since it is the direct product of two vector spaces, i.e. a tensor space, we denote it by $\Omega(n, l) \otimes \Omega_{1/2}$. The normalized functions of this space, in particular the basis functions (35), are called spin-orbitals. They are all eigenfunctions of the Hamiltonian (34) with the energy E(n, l). The inclusion of the spin has doubled the degeneracy of the energy level.

The Hamiltonian (34) commutes separately with the operators that rotate the spatial orbitals (atomic orbitals) and the operators that rotate the spin functions. This is, of course, the reason for the doubled degeneracy of an E(n, l) level. The Hamiltonian (34) is, however, only approximately correct. The most important term that must be added to improve the description of an atomic electron is the spin-orbit interaction

$$H' = \xi(r)\,\vec{l}\cdot\vec{s}\,. \tag{36}$$

The function $\xi(r)$, which is proportional to $\partial V/\partial r$, is, for instance, specified in Ref. 12.

The operator (36) does not commute with the components of $\vec{l}$ and $\vec{s}$ separately, but only with the compbnents of $\vec{j} = \vec{l} + \vec{s}$. Thus, we must write $\vec{j}$ instead of $\vec{l}$ in Eq. (22) to obtain the rotation operators that commute with H'. The relevant rotation operators are accordingly those that honour the integrity of the electron by simultaneously rotating the spatial part and the spin part of a wave function. These operators form the group SU(2). For $l \neq 0$ the space $\Omega(n, l) \otimes \Omega_{1/2}$ carries a reducible representation of this group. By a change of basis we may split the space into two linear spaces, carrying irreducible representation corresponding to $j = l + \frac{1}{2}$ and $j = l - \frac{1}{2}$, respectively. These two spaces will correspond to different values of the spin-orbit energy. The product space is irreducible for $l = 0$, and the spin-orbit energy is zero for this case.

The reduction of the representation for $l \neq 0$ is most easily performed by using angular momentum theory, i.e. by working with the complexification of the su(2) algebra, and we refer to standard text books, for instance Ref. 18, for a detailed treatment of this problem.

We have now completed our discussion of the symmetries associated with the central field problem. In the following section we shall consider the description of an electron moving in a molecular field.

10 Molecular Orbitals

In Sect. 2 we discussed the chemist's picture of a molecule as a three-dimensional arrangement of atoms. The quantum mechanical basis for this picture was first discussed by Born and Oppenheimer [35]. They introduced a molecule-fixed coordinate system in which the nuclei were supposed to perform small oscillations about their "equilibrium" positions, in a potential determined by the quantum state of the electrons. Rotation and translation of the molecule as a whole means rotation and translation of the molecule-fixed coordinate system. These motions, as well as the nuclear vibrations, are of course supposed to be quantized.

This is not the place for a detailed discussion of the Born-Oppenheimer approximation, or of its extension, the adiabatic approximation [35–38]. It will suffice to recall that the electronic states of a molecule are obtained by solving the Schrödinger equation for the motion of the electrons in the electrostatic field created by the nuclei. An electronic wave function depends parametrically on the positions of the nuclei, and its energy, considered as a function of the nuclear geometry, defines, in turn, the potential energy surface for the actual motion of the nuclei. The original Born-Oppenheimer approximation assumes that the potential energy surface has a well defined minimum, and it is this minimum that defines the above mentioned "equilibrium" positions of the nuclei. Present studies of molecular quantum states include, however, quite general potential energy surfaces, and this allows one to study non-rigid, or floppy, molecules as well as dissociating and reacting molecules ([39–42] and references therein).

In any case, the solution of the electronic Schrödinger equation for a fixed nuclear geometry is of paramount importance, and it is the symmetry associated with this problem that we shall discuss in the following.

We begin by considering the one-electron problem in which a single electron moves in the electrostatic field of a set of nuclei, and we generalize immediately to the SCF situation where the field from the nuclei has been augmented by an averaged field from $N - 1$ electrons. We call the resulting field $V(\vec{r})$ and assume that it has been averaged so as to have the same geometrical symmetry as the nuclear geometry. The resulting one-electron Hamiltonian is

$$H = -\frac{\hbar^2}{2\mu}\nabla^2 + V(\vec{r}) . \tag{37}$$

The bound state eigenfunctions of this Hamiltonian are molecular orbitals, $\varphi_i(\vec{r})$.

The molecular orbitals will, according to the general discussion following Eq. (18), be symmetry orbitals, i.e. the orbitals corresponding to the same energy will span a representation of the group whose elements commute with the Hamiltonian (37). This group will in almost all cases coincide with the geometrical symmetry group defined by the nuclear geometry. Dynamical .symmetry is rare in molecular problems. However, it does occur in the diatomic one-electron molecule, as a left-over from the dynamical symmetry in the hydrogen-like atom. We shall ignore this situation in the present discussion, and refer the reader to [43] and references therein. Accordingly, we now assume that the representation spanned by the molecular orbitals of a given energy is an irreducible representation of the geometrical symmetry group.

The exact eigenfunctions of the Hamiltonian (37) are in general out of reach. Hence, we are obliged to look for approximate molecular orbitals. The most common description represents such approximate orbitals as LCAOs (linear combinations of atomic orbitals), i.e.

$$\varphi_i(\vec{r}) = \sum_{s=1}^{M} \chi_s(\vec{r})\, c_{si}\,, \tag{38}$$

with the atomic orbitals $\chi_s(\vec{r})$ appropriately chosen. They may be solutions to a one-electron atomic problem, but need not be so. The coefficients c_{si} are determined by the variational method, as amply described in standard text books [17, 18, 44].

To ensure that the approximate molecular orbitals be of the same symmetry as the exact ones, the set of atomic orbitals must be chosen such, that if $\chi_s(\vec{r})$ is a member of the set, so is any function that can be generated from it by a symmetry transformation. With this condition fulfilled, the set is called symmetry restricted.
SCF set of molecular orbitals for ammonia, NH_3. With reference to Fig. 1 and its
As a simple, and often used example, let us discuss the form of an approximate legend, we choose the "minimal basis set" consisting of the following orbitals:

χ_{1s}, χ_{2s}:	the 1s and 2s atomic orbitals of nitrogen,
$\chi_{2x}, \chi_{2y}, \chi_{2z}$:	the three 2p orbitals of nitrogen,
χ_a, χ_b, χ_c:	hydrogen 1s orbitals centered at a, b, c.

Obviously, this is a symmetry restricted set.

The symmetry group of NH_3 is C_{3v} whose irreducible representations we described in Sect. 5 and listed in Tables 2 and 3. From the discussion of Sect. 5 we know that χ_{1s}, χ_{2s}, and χ_{2z} each span the A_1 representation, and that χ_{2x} and χ_{2y} together span the E representation. The three hydrogen orbitals span a reducible representation, and hence it is advantageous to replace them by three "symmetry-adapted" orbitals which span irreducible representations of C_{3v}. The way to do this is well described elsewhere [9, 45], and the new orbitals are easily found to be:

$$\begin{cases} \chi_{h1} = \sqrt{\dfrac{1}{3}}\,(\chi_a + \chi_b + \chi_c) \\[2ex] \chi_{hx} = \sqrt{\dfrac{1}{2}}\,(\chi_c - \chi_b)\,, \\[2ex] \chi_{hy} = \sqrt{\dfrac{1}{6}}\,(2\chi_a - \chi_b - \chi_c)\,. \end{cases} \tag{39}$$

The χ_{h1} orbital spans the A_1 representation. The orbitals χ_{hx} and χ_{hy} span the E representation, with the standard matrices of Table 3. The square-root factors in front of the orbitals have been chosen such that the orbitals are normalized for infinite separation of the hydrogen atoms. Any other choice would be equally valid.

We can now conclude that the molecular orbitals of NH_3 must have the form:

$$\begin{cases} \varphi(ka_1) = c_{k1}\chi_{1s} + c_{k2}\chi_{2s} + c_{k3}\chi_{2z} + c_{k4}\chi_{h1}\,, & k = 1, 2, 3, 4\,, \\ \varphi(me_x) = d_{m1}\chi_{2x} + d_{m2}\chi_{hx}\,, & \\ \varphi(me_y) = d_{m1}\chi_{2y} + d_{m2}\chi_{hy}\,, \quad m = 1, 2\,, & \end{cases} \tag{40}$$

where the coefficients must be variationally determined. The quantum numbers k and m are supposed to be so chosen that they increase with the energy of the molecular orbitals. We expect that $\varphi(1a_1)$ is almost pure χ_{1s}, and that $\varphi(2a_1)$, $\varphi(3a_1)$, and $\varphi(4a_1)$ contain very little of χ_{1s}. The variational method ensures that the molecular orbitals be mutually orthogonal, and the coefficients d_{11} and d_{12} of the bonding $\varphi(1e)$ orbitals will in accordance with this be of the same sign, the coefficients d_{21} and d_{22} of the antibonding $\varphi(2e)$ orbitals of opposite sign.

It is important to note that LCAO molecular orbitals will be symmetry orbitals whether or not we symmetrize the atomic orbital basis set before the variational method is applied, provided the basis set is symmetry restricted. However, we must solve an 8×8 secular problem for NH_3 if we do not symmetrize. With symmetrization we only need to solve a 4×4 and a 2×2 secular problem. The final result is, of course, independent of the procedure chosen.

In concluding the NH_3 example, we note that Pauli's exclusion principle assigns the electronic configuration $(1a_1)^2\,(2a_1)^2\,(3a_1)^2\,(1e)^4$ to the NH_3 molecule in its ground state. The double occupancy of the molecular orbitals reflects the fact that each molecular orbital may be multiplied by an α or a β spin function to give a molecular spin-orbital.

Just as for atoms it may be necessary to add the spin-orbit interaction to the Hamiltonian (37). This does not change the symmetry of H, but its eigenfunctions will be spin-orbitals of the general form $\varphi_1(\vec{r})\,\alpha(\sigma) + \varphi_2(\vec{r})\,\beta(\sigma)$, and will only be symmetry orbitals under the simultaneous transformation of the spatial and the spin parts. Due to the change of sign experienced by a spin function under a rotation through 2π, it now becomes necessary to double the number of elements in the original symmetry group. Thus we obtain a so-called double group.

Double groups were first introduced by Bethe [46], and their representations are well known. We must, however, refrain from entering the fascinating world of these groups in the present paper. The reader is referred to the occasionally colourful literature on the subject [12, 13, 47, 48].

The interplay between group theory and molecular orbital theory plays a large and important role in almost all branches of modern chemistry. This is reflected in a large number of monographs and review articles, far too numerous to be listed here. As to the general acceptance of this role I would, however, like to emphasize the special influence which the broad chemical community assigns to the pedagogical efforts of Mulliken [49], Ballhausen [50], Cotton [45], and Woodward and Hoffmann [51].

11 N-Electron Wave Functions

We shall now direct our attention towards the description of the electronic structure of atoms and molecules with two or more electrons. As in the previous sections we shall assume that the nuclei are held in fixed positions, and we shall omit specifying these positions.

An electronic bound state of our atom or molecule is then to be described by a normalized N-electron wave function, $\Psi(\vec{r}_i, \sigma_i) \equiv \Psi(\vec{r}_1, \vec{r}_2, \dots, \vec{r}_N; \sigma_1, \sigma_2, \dots, \sigma_N)$, which is an exact or approximate solution to the electronic Schrödinger equation.

$$H\Psi = E\Psi \, . \tag{41}$$

The electronic Hamiltonian, H, may be written as

$$H = H_0 + \sum_{i=1}^{N} h(i) + \frac{1}{2} \frac{e^2}{4\pi\varepsilon_0} \sum_{i=1}^{N} \sum_{j=1}^{N}{}' g(i, j) \, , \tag{42}$$

where H_0 is the nucleus-nucleus repulsion energy, which of course vanishes for an atom, and is a constant for the fixed geometry of a molecule. h(i) is a one-electron operator,

$$h(i) = -\frac{\hbar^2}{2\mu} \nabla_i^2 - \frac{e^2}{4\pi\varepsilon_0} \sum_g \frac{Z_g}{r_{ig}} \, , \tag{43}$$

where g numbers the nuclei, and g(i, j) is a two-electron operator,

$$g(i, j) = \frac{e^2}{4\pi\varepsilon_0} \frac{1}{r_{ij}} \, . \tag{44}$$

The prime on the summation symbol in Eq. (42) indicates that terms for which two indices become equal are to be omitted in the double sum.

Since the Hamiltonian (42) is spin independent we may look for solutions to the Schrödinger equation (41) of the form

$$\Psi(\vec{r}_i, \sigma_i) = \Phi(\vec{r}_i)\, \theta(\sigma_i) \, , \tag{45}$$

where

$$H\Phi = E\Phi \tag{46}$$

and $\theta(\sigma_i)$ is an arbitrary function of the N spin variables. In other words, we may begin by solving the Schrödinger equation without worrying about spin at all, just as in the one-electron case, and then include the spin at the very end. It is, of course, understood that whenever Eq. (46) leads to more than one solution for the same energy, then we may replace the product function on the right hand side of Eq. (45) with any linear combination of product functions corresponding to that energy.

The theme we have developed in the previous sections is that degeneracies are symmetry determined. This is also the case for the solutions of Eq. (46). But now there is a new symmetry group that enters the scene, so let us for the moment assume that the states we consider span one-dimensional representations of the geometrical symmetry group.

The new group is the permutation group S_N, i.e. the group of order N! whose elements permute N objects among each other. It is obvious that the right hand side of Eq. (42) is unchanged by any renumbering of the electrons. Each renumbering corresponds, however, to a permutation of the N electrons, and hence S_N is a symmetry group.

The linear function space corresponding to a given value of E in Eq. (46) will accordingly carry a representation of S_N, and we must expect this representation to be irreducible.

Thus it becomes important to study the group S_N and its representations. We need, however, only take a few steps in that direction before we realize that we may have to exclude some of these representations on physical grounds.

The identity representation is an irreducible representation of any group. This is, of course, a one-dimensional representation, and in the case of S_N it is spanned by a function, $\Phi(\vec{r}_1, \vec{r}_2, \ldots, \vec{r}_N)$, which is invariant under any renumbering of the electrons. There are good reasons to believe that the lowest permissible E value of Eq. (46) goes with such a function. For an atom like carbon, we might try to approximate the function as

$$\Phi = \varphi_{1s}(\vec{r}_1)\,\varphi_{1s}(\vec{r}_2) \ldots \varphi_{1s}(\vec{r}_6) \tag{47}$$

where $\varphi_{1s}(\vec{r})$ is a hydrogen-like atomic orbital of the form A exp $(-\zeta r)$, A being a normalization constant. By varying ζ so as to minimize $\langle\Phi|\, H\, |\Phi\rangle$, the expectation value of the energy, we obtain a reasonable first approximation to the exact Φ, and a reasonable approximation to the lowest E-value.

What we have created in this way is, however, an atom with an energy much below the energy of the real carbon atom. Furthermore, it is a carbon atom with no directional valencies built into it whatsoever. So although it represents a solution of the Schrödinger equation, the function (47) does not correspond to a physically accessible state. Fortunately, this does not imply that the Schrödinger equation is invalid. What it means is, that the Schrödinger equation must be supplemented with a rule that allows us to distinguish between those solutions that correspond to actual physical states, and those that do not.

In 1926, the *ansatz* function (47) would have been rejected immediately, on the ground that it contradicted Bohr's *aufbau* principle [52], according to which no more than two electrons may occupy the same state. But this is of course an empirical argument, because it is tied to an approximate rather than an exact solution of the Schrödinger equation.

Our simple chemical example supplements the spectroscopic examples concerning two- and three-electron atoms that were discussed in great depth by Heisenberg [53, 54] and Wigner [55] in 1926–27. These authors combined the results of their group theoretical analyses with Pauli's exclusion principle [56] in order to associate wave functions with the known spectroscopic states. Again, the procedure was empirical, and for the three-electron atom it lacked uniqueness [55].

A clear selection rule by means of which one could pick the physically allowed solutions of Eq. (46) from the set of its mathematical solutions was thus badly needed from the outset. It was Dirac [57] who found the roots of such a rule. He postulated that elementary particles of the same kind are physically indistinguishable, and he showed that this postulate requires a many-particle wave function to be either symmetrical or antisymmetrical under the interchange of any two particles. He then postulated that electron wave functions must be of the antisymmetric type, because this alone would be in harmony with Pauli's principle.

Dirac's postulate was, however, not complete because he only considered interchange of spatial variables. It was up to Pauli [29], together with his introduction of the spin variable and the spin functions (Section 8), to give the final formulation of the antisymmetry requirement, according to which interchange of two electrons means interchange of spin as well as space variables, so that the integrity of the electron be respected.

Let us again introduce the notation x_i for the combined space and spin coordinates of the i'th electron. The antisymmetry requirement then says that an acceptable Ψ must change sign when two sets of coordinates are interchanged, i.e.

$$\Psi(\dots x_i \dots x_j) = -\Psi(\dots x_j \dots x_i)\,. \tag{48}$$

A simple interchange of two particles is called a transposition, and any permutation $Px_1, Px_2, \dots, Px_N$ of N particles can be generated as a succession of transpositions. Calling the necessary number of transpositions p is follows from Eq. (48) that

$$\Psi(Px_1, Px_2, \dots, Px_N) = (-1)^p\, \Psi(x_1, x_2, \dots, x_N)\,. \tag{49}$$

$(-1)^p$ is called the parity of the permutation.

By defining for each permutation an operator $\hat{P}$ such that

$$\hat{P}\Psi(x_1, x_2, \dots, x_N) = \Psi(Px_1, Px_2, \dots, Px_N) \tag{50}$$

we may write the antisymmetry requirement as

$$\hat{P}\Psi = (-1)^p\, \Psi\,. \tag{51}$$

This relation shows that Ψ must span a one-dimensional representation of S_N, such that the character is +1 for an even permutation (p even) and —1 for an odd permutation (p odd). This representation is called the antisymmetric representation of S_N, and apart from the identity representation it is the only one-dimensional representation of this group [11].

The proper N-electron wave functions corresponding to the Hamiltonian (42) are thus the antisymmetric solutions of Eq. (41). To determine these solutions we may either attack Eq. (41) directly, and we shall consider this approach in the following section, or we may attack Eq. (46) instead, and then apply group theoretical techniques to obtain antisymmetric functions as linear combinations of the product functions (45). This approach will be discussed subsequently. The exclusion of a function like that of Eq. (47) is easily understood in either approach.

12 Slater Determinants

In the previous section we mentioned Wigner's analysis of the three-electron atom [55], based on Eq. (46), the permutation group S_3, and Pauli's exclusion principle. Although his procedure lacked uniqueness it was believed to be a step in the right direction, and Pauli accordingly concluded his monumental article on the spin functions and the antisymmetry requirement with the suggestion, "that it would be interesting to reconsider Wigner's group theoretical analysis under the limitations set by the new physical law".

Pauli's suggestion does indeed lead to a fruitful procedure, as we shall discuss in the next section, but at the time, the methods based on the permutation group were brought into disrepute, because they were relatively complicated. The whole approach was referred to as the "Gruppenpest", the pest of group theory. The situation has been vividly described by Slater [58] who eventually invented a conceptually much more simple procedure of great practical consequence [59]. This was the procedure based on determinantal wave functions, the so-called Slater determinants.

A Slater determinant has the well-known form:

$$D(x_1, x_2, \ldots, x_N) = (N!)^{-1/2} \begin{vmatrix} \psi_1(x_1) & \psi_1(x_2) & \ldots & \psi_1(x_N) \\ \psi_2(x_1) & \psi_2(x_2) & \ldots & \psi_2(x_N) \\ \cdot & \cdot & \cdots & \cdot \\ \cdot & \cdot & \cdots & \cdot \\ \psi_N(x_1) & \psi_N(x_2) & \ldots & \psi_N(x_N) \end{vmatrix} \tag{52}$$

or, in a brief notation (which includes the factor $(N!)^{-1/2}$,

$$D = |\psi_1 \psi_2 \cdots \psi_N| \,. \tag{53}$$

The functions $\psi_1, \psi_2, \ldots, \psi_N$ are spin-orbitals, and it is evident that D is antisymmetric, for an interchange of the i'th and j'th electron corresponds to an interchange of the i'th and j'th column of the determinant.

Now let $\psi_1, \psi_2, \ldots, \psi_i, \ldots$ be a complete set of spin-orbitals, i.e. a set in terms of which any one-electron function may be expanded in the form

$$\psi(x) = \sum_i c_i \psi_i(x) \,. \tag{54}$$

It is then easy to show [60, 61] that any antisymmetric N-electron function, Ψ, may be expanded on the set of Slater determinants constructed from these spin-orbitals

$$\Psi = \sum_{k1} \sum_{<k2<,} \sum_{<kN} C_{k1, k2, \ldots, kN} |\psi_{k1} \psi_{k2} \cdots \psi_{kN}| \,, \tag{55}$$

and it is in this sense that Slater determinants play a fundamental part in the description of many-electron systems. Ψ is inherently antisymmetric and the coefficients $C_{k1, k2, \ldots, kN}$ may be determined by the variation principle.

The description becomes especially simple when the spin-orbitals form an orthonormal set, i.e.

$$\langle \psi_i \mid \psi_j \rangle = \delta_{ij} \, . \tag{56}$$

The determinant (52) is then normalized, and the functions in (55) are mutually orthogonal, so that the determinants $|\psi_{k1}\psi_{k2} \ldots \psi_{kN}|$ themselves form an orthonormal set. The matrix elements of the Hamiltonian (42) are likewise of a particularly simple form in this case [12, 34]. We shall accordingly adopt the condition (56) in what follows.

The simplicity of the above description is obvious, also because it lends itself so easily to an intuitive interpretation. Let us introduce the so-called Hartree product by the definition

$$\Xi = \psi_1(x_1)\,\psi_2(x_2) \ldots \psi_N(x_N) \, . \tag{57}$$

The determinant (52) may then formally be written as

$$D = \hat{A}\Xi \, , \tag{58}$$

where $\hat{A}$ is an antisymmetrization operator,

$$\hat{A} = (N!)^{-1/2} \sum_P (-1)^p \, \hat{P} \, . \tag{59}$$

$\hat{P}$ is a permutation operator of the form (50), and the sum is over the N! permutations of the N electrons.

The Hartree product (57) assigns each of the N electrons to a spin-orbital, and Eq. (58) corrects for the indistinguishability of the electrons. In so doing, it eliminates the Hartree product if it contains the same spin orbital more than once, for in that case the determinant (52) vanishes. This result explains Pauli's exclusion principle and, in the same token, it explains why the wave function (47) cannot describe the ground state of the carbon atom. For an atomic orbital, φ, can only give rise to two independent spin-orbitals, for example

$$\overset{+}{\varphi} = \varphi\alpha \, , \qquad \overset{-}{\varphi} = \varphi\beta \, , \tag{60}$$

and any Hartree product corresponding to (47) will therefore contain the same spin-orbital more than once.

In Sects. 9 and 10 we referred to the SCF (self-consistent field) description of an N-electron atom or molecule, in which each electron is assumed to move in a totally symmetric field which is the sum of the Coulomb field from the nuclei and an averaged field from N − 1 electrons. This description approximates the sum (55) with a single term or a linear combination of just a few terms, the number of terms being chosen such that Ψ becomes a basis function for an irreducible representation of the pertinent symmetry group, and such that it becomes an eigenfunction of the operators S^2 and S_z corresponding to the total spin angular momentum

$$\vec{S} = \vec{s}_1 + \vec{s}_2 + \dots + \vec{s}_N . \tag{61}$$

Formally, we may obtain such a symmetry adapted Ψ by operating on a single determinant with a suitable projection operator, $\hat{Q}$, i.e.

$$\Psi = \hat{Q}D . \tag{62}$$

Thus, Eqs. (58) and (62) tie the SCF function Ψ to the simple Hartree product (57), and thus also to the intuitive ideas that one may associate with such a product (the independent-particle model). This point has, in particular, been strongly emphasized by Löwdin [61, 62].

The actual construction of N-electron SCF functions is amply described elsewhere [11, 12, 34, 44, 49, 63]. We get, for instance, for the ground state of the NH_3 molecule:

$$\Psi = |\overset{+}{\varphi}(1a_1)\,\bar{\varphi}(1a_1)\,\overset{+}{\varphi}(2a_1)\,\bar{\varphi}(2a_1)\,\overset{+}{\varphi}(3a_1)\,\bar{\varphi}(3a_1)\,\overset{+}{\varphi}(1e_x)\,\bar{\varphi}(1e_x)\,\overset{+}{\varphi}(1e_y)\,\bar{\varphi}(1e_y)| \tag{63}$$

with the molecular orbitals as defined in Sect. 10. This wave function is a singlet ($S = 0$), and of A_1 symmetry. Thus, we may denote it $\Psi(^1A_1)$. It is said to arise from the $(1a_1)^2\,(2a_1)^2\,(3a_1)^2\,(1e)^4$ configuration, and it is the only wave function this configuration gives rise to.

The SCF wave function (63) gives a reasonable first description of the ground state of NH_3. A much better wave function may be constructed by the method of configuration interaction (CI), as

$$\Psi(^1A_1) = \sum_{r=0}^{M} C_r \Psi_r(^1A_1) . \tag{64}$$

Ψ_0 is the function (63), and the remaining functions in the sum are $^1A_1$ functions arising from other configurations like for instance, $(1a_1)^2\,(2a_1)^2\,(3a_1)^2\,(1e)^3\,(2e)^1$. With today's computers it is possible to work with very large CI expansions of this type. The coefficients C_r are, of course, variationally determined.

It is obvious that Eq. (64) is nothing but an intelligent version of Eq. (55), with the proper symmetries built in from the outset. In principle, it is possible to skip the construction of symmetry adapted N-electron functions or to merely perform a partial symmetrization. This is quite often done in practical calculations, but it does deserve some remarks. To see what is involved, let us consider a relatively simple example in C_{3v} symmetry, that of the e^2 configuration.

For simplicity, we use the notation φ_1 and φ_2 for the orbitals of e_x and e_y symmetry, respectively. The e^2 configuration gives rise to six Slater determinants, namely,

$$|\overset{+}{\varphi}_1\overset{+}{\varphi}_2| , \quad |\overset{+}{\varphi}_1\overset{-}{\varphi}_1| , \quad |\overset{+}{\varphi}_2\overset{-}{\varphi}_2| , \quad |\overset{+}{\varphi}_1\overset{-}{\varphi}_2| , \quad |\overset{-}{\varphi}_1\overset{+}{\varphi}_2| , \quad |\overset{-}{\varphi}_1\overset{-}{\varphi}_2| , \tag{65}$$

and suitable linear combinations of these span the representations ${}^3A_2$, ${}^1A_1$, and ${}^1E$. The symmetry adapted wave functions are:

$$\begin{cases} \Psi({}^3A_2, M_S = 1) = |\overset{+}{\varphi}_1\overset{+}{\varphi}_2| \\ \Psi({}^3A_2, M_S = 0) = \sqrt{\frac{1}{2}}\,(|\overset{+}{\varphi}_1\overset{-}{\varphi}_2| + |\overset{-}{\varphi}_1\overset{+}{\varphi}_2|) \\ \Psi({}^3A_2, M_S = -1) = |\overset{-}{\varphi}_1\overset{-}{\varphi}_2| \end{cases} \tag{66}$$

$$\Psi({}^1A_1) = \sqrt{\frac{1}{2}}\,(|\overset{+}{\varphi}_1\overset{-}{\varphi}_1| + |\overset{+}{\varphi}_2\overset{-}{\varphi}_2|) \tag{67}$$

$$\begin{cases} \Psi({}^1E_x) = \sqrt{\frac{1}{2}}\,(|\overset{+}{\varphi}_1\overset{-}{\varphi}_2| - |\overset{-}{\varphi}_1\overset{+}{\varphi}_2|) \\ \Psi({}^1E_y) = \sqrt{\frac{1}{2}}\,(|\overset{+}{\varphi}_1\overset{-}{\varphi}_1| - |\overset{+}{\varphi}_2\overset{-}{\varphi}_2|) \end{cases} \tag{68}$$

The ${}^3A_2$ functions are uniquely determined by their spin properties alone, i.e. we need no properties of the C_{3v} group to find them. Of the singlet functions $\Psi({}^1E_x)$ is odd under the σ_a operation, whereas $\Psi({}^1E_y)$ and $\Psi({}^1A_1)$ are even. Accordingly, we may single $\Psi({}^1E_x)$ out by the condition that it be odd under σ_a. The functions $\Psi({}^1A_1)$ and $\Psi({}^1E_y)$ can, however, only be separated by utilizing their transformation properties under one of the C_3 operations. Alternatively, they may be separated by diagonalizing the Hamiltonian of the system.

Although present CI programs are very powerful, they are usually not so advanced that they can exploit the full symmetry of non-commutative point groups. Mostly, only the properties under two-fold operations are taken advantage of. For the above example this has the consequence that if one goes on to improve the $\Psi({}^1A_1)$ function by performing a large CI calculation, then the dimension of the Hamiltonian matrix becomes three times as large as it needs to be. The factor three refers to the fact that the representation spanned by a very large basis, in analogy with the so-called regular representation [12, 13], contains a chosen irreducible representation a number of times proportional to its dimension. Thus, there will be twice as many $\Psi({}^1E_y)$ functions as there are $\Psi({}^1A_1)$ functions.

The construction of symmetry adapted functions as linear combinations of Slater determinants is, in principle, no more difficult for systems with three or more electrons than it is for two-electron systems, and hence we shall not discuss the determinantal method further. Its simplicity is, of course, a direct consequence of this conceptual independence of N.

Let us close the present section with a remark on the total spin angular momentum, as defined by Eq. (61). The Hamiltonian (42) is independent of spin, and the Schrödinger equation (41) does therefore not put any well-defined condition on the spin dependence of its eigenfunctions. Such a condition is, however, imposed by the antisymmetry requirement (51). For this requirement implies that any operator of which Ψ is an eigenfunction must commute with all the permutation operators.

The only N-electron operators that have this property are the totally symmetric ones. Hence the importance of the operator (61), and the necessity of constructing eigenfunctions of S^2 and, say, S_z.

13 Symmetric Group Approach

We shall now consider the problem of constructing approximate solutions to the N-electron Schrödinger equation (41) via Eqs. (45) and (46). This method draws heavily on the representation theory for the permutation group S_N. Mathematicians prefer to call this group the symmetric group of degree N, and since this praxis also has become generally adopted in connection with the problem we now discuss, we shall adhere to it in the following.

In setting up Eq. (45) we allowed θ to be an arbitrary N-electron spin function. Such a function is a linear combination of the 2^N product functions of the type $\eta_1(\sigma_1)\,\eta_2(\sigma_2)\,\dots\,\eta_N(\sigma_N)$, where each $\eta_i(\sigma_i)$ is either $\alpha(\sigma_i)$ or $\beta(\sigma_i)$. So if $\Phi_1, \Phi_2, \dots, \Phi_f$ are the independent solutions of Eq. (46) corresponding to a certain energy E, then the most general solution of Eq. (41) corresponding to that energy may be written

$$\Psi = \sum_{r=1}^{f} \Phi_r \sum_t C_{rt} X_t\,, \tag{69}$$

where X_t is one of the above product functions, and the sum over t therefore includes 2^N terms, with arbitrary coefficients C_{rt}. We assume again that each Φ_r spans a one-dimensional representation of the pertinent point group (the same representation for each Φ_r), and that $\Phi_1, \Phi_2, \dots, \Phi_f$ therefore span an irreducible representation of S_N. Thus we have that

$$\hat{P}\Phi_r = \sum_{s=1}^{f} \Phi_s D_{sr}(P)\,. \tag{70}$$

Let us now impose the antisymmetry condition (51) on Ψ. The coefficients C_{rt} in Eq. (69) will then no longer be arbitrary. It is, in fact, obvious that the f functions

$$\theta_r = \sum_t C_{rt} X_t \tag{71}$$

must also span a representation of S_N, i.e.

$$\hat{P}\theta_r = \sum_{u=1}^{f} \theta_u \Gamma_{ur}(P)\,. \tag{72}$$

We get then that

$$\hat{P}\Psi = \sum_{r=1}^{f} \hat{P}(\Phi_r\theta_r) = \sum_{r=1}^{f}\sum_{s=1}^{f}\sum_{u=1}^{f} \Phi_s\theta_u D_{sr}(P)\,\Gamma_{ur}(P)\,. \tag{73}$$

But the antisymmetry condition requires that

$$\hat{P}\Psi = \sum_{s=1}^{f} (-1)^p \Phi_s\theta_s \,. \tag{74}$$

Hence, we must have that

$$\sum_{r=1}^{f} D_{sr}(P)\, \Gamma_{ur}(P) = (-1)^p\, \delta_{su}\,, \tag{75}$$

or

$$\Gamma(P) = (-1)^p \{\tilde{D}(P)\}^{-1}\,, \tag{76}$$

where the matrix $\tilde{D}(P)$ is the transpose of D(P).

The functions $\Phi_1, \Phi_2, \dots, \Phi_f$ may as usual be so chosen that they form an orthonormal set. The D matrices will then be unitary. Furthermore, it is a property of the symmetric group that all its irreducible representations may be taken to be real [64, 65]. We shall therefore also assume that the functions $\Phi_1, \Phi_2, \dots, \Phi_f$ correspond to real D matrices. A real, unitary matrix is, however, the same thing as an orthogonal matrix, and for such a matrix the transpose is also the inverse. Thus, we may write Eq. (76) as

$$\Gamma(P) = (-1)^p\, D(P)\,. \tag{77}$$

The representations $\Gamma(P)$ and D(P) are said to be the duals of each other.

There are now two possibilities. Either the representation spanned by the 2^N functions X_t contains the dual of the representation spanned by the functions Φ_r, or it does not. In the second case we cannot construct θ_r functions such that the antisymmetry condition (74) becomes fulfilled, and hence the functions Φ_r do not lead to acceptable solutions of the Schrödinger equation (41). In the first case they do.

To investigate the first and favourable case further, we may take advantage of the fact, discussed at the end of the last section, that the acceptable solutions of Eq. (41) must be eigenfunctions of S^2 and, say S_z (with the eigenvalues $S(S+1)\hbar^2$ and $M_S\hbar$, respectively). The X_t functions may obviously be combined to give such eigenfunctions, and it is easy to see that the possible values of S will be N/2, N/2 — 1, ... , (1/2 or 0). Furthermore, a careful analysis [11, 64] shows that the number of linearly independent combinations corresponding to given values of S and M_S is given by the formula

$$f(N, S) = \binom{N}{N/2 - S} - \binom{N}{N/2 - S - 1}. \tag{78}$$

We shall denote an orthonormal set of such N-electron spin functions by

$$\theta_{S,M_S;r} \equiv \theta_{S,M_S;r}(\sigma_1, \sigma_2, \dots, \sigma_N)\,, \qquad r = 1, 2, \dots, f(N, S)\,. \tag{79}$$

It may now be shown [11, 64] that the f(N, S) functions obtained by keeping S and M_S fixed span an irreducible representation of S_N, the representation being the same for each M_S value corresponding to a given value of S. We shall denote it $\Gamma^{(N, S)}$, and have accordingly:

$$\hat{P}\theta_{S, M_S; r} = \sum_{u=1}^{f(N, S)} \theta_{S, M_S; u} \Gamma^{(N, S)}_{u, r}(P). \tag{80}$$

The dual representation is conveniently denoted $D^{(N, S)}$.

Equation (80) is directly comparable with Eq. (72), and the construction of the functions $\theta_{S, M_S; r}$ corresponds, therefore, to the fixation of the C_{rt} coefficients in Eq. (71). We have accordingly solved the problem of determining the condition that the functions $\Phi_1, \Phi_2, \dots, \Phi_f$ must satisfy in order that they determine an acceptable solution to the Schrödinger equation (41). This condition is that the functions span one of the $D^{(N, S)}$ representations of the symmetric group S_N.

To illustrate the conclusions we have arrived at, let us first consider the two-electron problem. The symmetric group approach to this problem is in fact treated in many text books, especially in connection with a discussion of the helium atom and the hydrogen molecule [8, 16–19, 34, 44, 63]. The spin functions define a triplet (S = 1) and a singlet (S = 0). f(2, S) is 1 in both cases. Hence, we suppress the index r in (79) and get the well-known functions

$$\begin{cases} \theta_{11} = \alpha(\sigma_1)\,\alpha(\sigma_2) \\ \theta_{10} = \sqrt{\frac{1}{2}}\,[\alpha(\sigma_1)\,\beta(\sigma_2) + \beta(\sigma_1)\,\alpha(\sigma_2)] \\ \theta_{1-1} = \beta(\sigma_1)\,\beta(\sigma_2), \end{cases} \tag{81}$$

and

$$\theta_{00} = \sqrt{\frac{1}{2}}\,[\alpha(\sigma_1)\,\beta(\sigma_2) - \beta(\sigma_1)\,\alpha(\sigma_2)]. \tag{82}$$

The group S_2 has only two elements, namely, the identity and the operator that interchanges electrons 1 and 2. S_2 is of course isomorphic to the group C_i whose character table we reproduced in Table 4. Its only irreducible representations are the symmetric and the antisymmetric representations. They are the duals of each other.

Each of the triplet functions (81) are symmetric, while the singlet function (82) is antisymmetric. Hence, all solutions of Eq. (46) lead to acceptable states in this case. For a solution of Eq. (46) will either be symmetric, and then it combines with the singlet spin function, or it will be antisymmetric, and then it combines with the triplet spin functions.

In the previous section we presented the symmetry adapted wave functions of the e^2

configuration (C_{3v} symmetry). The functions were listed in Eqs. (66)–(68), and by expanding the determinants we get

$$\Psi(^3A_2, M_S) = \sqrt{\frac{1}{2}}\,[\varphi_1(\vec{r}_1)\,\varphi_2(\vec{r}_2) - \varphi_2(\vec{r}_1)\,\varphi_1(\vec{r}_2)]\,\theta_{1M_S}(\sigma_1, \sigma_2)\,, \qquad (83)$$

$$\Psi(^1A_1) = \sqrt{\frac{1}{2}}\,[\varphi_1(\vec{r}_1)\,\varphi_1(\vec{r}_2) + \varphi_2(\vec{r}_1)\,\varphi_2(\vec{r}_2)]\,\theta_{00}(\sigma_1, \sigma_2)\,, \qquad (84)$$

$$\left\{\begin{aligned} \Psi(^1E_x) &= \sqrt{\frac{1}{2}}\,[\varphi_1(\vec{r}_1)\,\varphi_2(\vec{r}_2) + \varphi_2(\vec{r}_1)\,\varphi_1(\vec{r}_2)]\,\theta_{00}(\sigma_1, \sigma_2) \\ \Psi(^1E_y) &= \sqrt{\frac{1}{2}}\,[\varphi_1(\vec{r}_1)\,\varphi_1(\vec{r}_2) - \varphi_2(\vec{r}_1)\,\varphi_2(\vec{r}_2)]\,\theta_{00}(\sigma_1, \sigma_2)\,. \end{aligned}\right. \qquad (85)$$

The form of these functions is, of course, in agreement with the above prescription for combining spatial functions and spin functions. But, in addition, the ^1E functions illustrate a case of spatial degeneracy.

Spatial degeneracy was excluded in the discussion leading up to Eq. (70). But it is easy to see how it may be incorporated in the general case. Instead of the single set of functions, $\Phi_1, \Phi_2, \ldots, \Phi_f$, we shall now have as many sets as the spatial degeneracy prescribes. If, for instance, the spatial degeneracy is two, as above, then we shall have two sets, $\Phi_1, \Phi_2, \ldots, \Phi_f$, and $\Phi_1', \Phi_2', \ldots, \Phi_f'$, that span the same representation of S_N and may be treated independently under the operations of that group. But Φ_1 and Φ_1' span a two-dimensional representation of the pertinent point group. Φ_2 and Φ_2' span the same representation, etc. This result is a simple consequence of the fact that the elements of the point group commute with all elements of S_N.

Let us now move to the three-electron case and the group S_3. This group is isomorphic to C_{3v} whose representations we listed in Tables 2 and 3. A_1 and A_2 correspond to the symmetric and the antisymmetric representations, respectively. They are the duals of each other. The two-dimensional representation is self-dual.

In writing product functions we shall assume that the arguments always refer to the particles in the order 1, 2, 3. Accordingly, we shall simply write $\alpha\beta\alpha$ instead of $\alpha(\sigma_1)\,\beta(\sigma_2)\,\alpha(\sigma_3)$, etc. The three-electron spin functions of the type (79) may, for instance, be generated from the functions (81) and (82) as follows:

$$\left\{\begin{aligned} \theta_{\frac{3}{2}\frac{3}{2}} &= \theta_{11}\,\alpha && = \alpha\alpha\alpha \\ \theta_{\frac{3}{2}\frac{1}{2}} &= \sqrt{\frac{2}{3}}\,\theta_{10}\alpha + \sqrt{\frac{1}{3}}\,\theta_{11}\beta && = \sqrt{\frac{1}{3}}\,(\beta\alpha\alpha + \alpha\beta\alpha + \alpha\alpha\beta) \\ \theta_{\frac{3}{2}-\frac{1}{2}} &= \sqrt{\frac{2}{3}}\,\theta_{10}\beta + \sqrt{\frac{1}{3}}\,\theta_{1-1}\alpha && = \sqrt{\frac{1}{3}}\,(\alpha\beta\beta + \beta\alpha\beta + \beta\beta\alpha) \\ \theta_{\frac{3}{2}-\frac{3}{2}} &= \theta_{1-1}\beta && = \beta\beta\beta\,, \end{aligned}\right. \qquad (86)$$

$$\begin{cases} \theta_{\frac{1}{2}\frac{1}{2};a} = -\sqrt{\frac{1}{3}}\,\theta_{10}\alpha + \sqrt{\frac{2}{3}}\,\theta_{11}\beta = \sqrt{\frac{1}{6}}\,(-\beta\alpha\alpha - \alpha\beta\alpha + 2\alpha\alpha\beta) \\ \theta_{\frac{1}{2}-\frac{1}{2};a} = \sqrt{\frac{1}{3}}\,\theta_{10}\beta - \sqrt{\frac{2}{3}}\,\theta_{1-1}\alpha = \sqrt{\frac{1}{6}}\,(\alpha\beta\beta + \beta\alpha\beta - 2\beta\beta\alpha) \end{cases} \tag{87}$$

$$\begin{cases} \theta_{\frac{1}{2}\frac{1}{2};b} = \theta_{00}\alpha = \sqrt{\frac{1}{2}}\,(\alpha\beta\alpha - \beta\alpha\alpha) \\ \theta_{\frac{1}{2}-\frac{1}{2};b} = \theta_{00}\beta = \sqrt{\frac{1}{2}}\,(\alpha\beta\beta - \beta\alpha\beta) \end{cases} \tag{88}$$

Each of the functions (86) span the symmetric representation. The functions $\theta_{\frac{1}{2}\frac{1}{2};a}$ and $\theta_{\frac{1}{2}\frac{1}{2};b}$ span the two-dimensional representation of S_3, and so do the functions $\theta_{\frac{1}{2}-\frac{1}{2};a}$ and $\theta_{\frac{1}{2}-\frac{1}{2};b}$. We note that the antisymmetric representation is absent. Hence, a symmetric solution of Eq. (46) cannot lead to an acceptable solution of Eq. (41). This situation is general. There is no antisymmetric spin function for $N > 2$, and hence a spatial function like (47) cannot produce an acceptable Ψ.

We have now carried the discussion of the symmetric group approach far enough that its theoretical basis should be clear. The spatial functions $\Phi_1, \Phi_2, \ldots, \Phi_f$, that combine with a set of spin functions, $\theta_1, \theta_2, \ldots, \theta_f$, to form the antisymmetric combination

$$\Psi = \sum_{r=1}^{f} \Phi_r \theta_r \tag{89}$$

may be constructed from orbital products, or more general trial functions may be used. When orbital products are used, as in the two-electron example above, we may introduce configuration interaction as in the method based on Slater determinants, and the two approaches become directly comparable.

For further discussion of the basis of the symmetric group approach we refer to the comprehensive treatment by Pauncz [64], and for a thorough discussion of its practical implementation we refer to a review article by Duch and Karwowski [66].

We shall now round our discussion of N-electron systems off with a few remarks on the so-called unitary group approach.

14 Unitary Group Approach

Like the symmetric group, the unitary group belongs to the so-called classical groups discussed by Weyl in his fundamental exposition from 1938 [67]. In his preface Weyl writes: "The task may be characterised precisely as follows: with respect to the assigned group of linear transformations in the underlying vector space, to decompose the space of tensors of given rank into its irreducible invariant subspaces." Let us see what this quotation has to do with our subject.

The method of configuration interaction (CI), whether it is formulated by means of Slater determinants or in the symmetric group approach, has as its building blocks one or more sets of one-electron functions, $u_1, u_2, \dots, u_n$, which may be spatial orbitals, spin functions, or spin-orbitals. We assume that the set is orthonormal, i.e.

$$\langle u_i \mid u_j \rangle = \delta_{ij} \,. \tag{90}$$

The n functions define an n-dimensional vector space, Ω_n, in the sense discussed in Sect. 5. The functions $u_1, u_2, \dots, u_n$ define a basis in Ω_n, but in analogy with Eq. (4) we may define a new basis,

$$u'_r = \sum_{s=1}^{n} u_s S_{sr} \,, \qquad r = 1, 2, \dots, n \tag{91}$$

where the matrix S in non-singular.

It is now easy to show, that if we also require the new basis to be orthonormal, i.e.

$$\langle u'_i \mid u'_j \rangle = \delta_{ij} \,, \tag{92}$$

then S must be unitary. Any unitary matrix will thus define a change of basis such that the scalar product (90) is conserved.

The set of all $n \times n$ unitary matrices forms a group. This is the unitary group, denoted U(n). It is a compact Lie group with n^2 real parameters (cf. Sect. 8). The elements of the corresponding Lie algebra, u(n), may be expressed in terms of the operators $|u_r\rangle \langle u_s|$, which are defined by the relation

$$|u_r\rangle \langle u_s \mid u_t\rangle = |u_r\rangle \, \delta_{st} \,, \tag{93}$$

where $|u_r\rangle$ is the Dirac notation for the function u_r. Thus, such operators play a fundamental role in the theory of U(n).

The construction of N-electron wave functions from the functions $u_1, u_2, \dots, u_n$ involves the formation of product functions of the type $u_{r1}(1)\, u_{r2}(2) \dots u_{rN}(N)$, where the particle coordinates simply have been denoted 1, 2, ..., N. All such products form an Nth rank tensor space. This is also a vector space, and it carries a unitary representation of U(n). This representation is defined by the matrices which describe the way the product functions are transformed under the transformations (91). The representation is in general reducible, and the reduction of the representation is exactly the task Weyl refers to in his preface.

Let us, as a simple example, consider the case $n = 2$. We have then the two functions u_1 and u_2. If we also let $N = 2$, then the second rank tensor space will be spanned by the functions

$$u_1(1)\, u_1(2) \,, \qquad u_2(1)\, u_2(2) \,, \qquad u_1(1)\, u_2(2) \,, \qquad u_2(1)\, u_1(2) \,. \tag{94}$$

It is now not too difficult to show that this space can be decomposed into two invariant subspaces, of dimensions 3 and 1, respectively. The two subspaces are given by the basis functions

$$\begin{cases} u_1(1)\, u_1(2) \\ \sqrt{\frac{1}{2}}\,[u_1(1)\, u_2(2) + u_2(1)\, u_1(2)] \\ u_2(1)\, u_2(2) \end{cases} \tag{95}$$

and

$$\sqrt{\frac{1}{2}}\,[u_1(1)\, u_2(2) - u_2(1)\, u_1(2)] \tag{96}$$

The first subspace spans a three-dimensional irreducible representation of U(2), the second subspace spans a one-dimensional representation.

If we now identify u_1 and u_2 with the α and β spin functions, then the functions (95) become the triplet functions (81), and the function (96) becomes the singlet function (82).

If, on the other hand, we identify u_1 and u_2 with the e-functions φ_1 and φ_2, then we see that the space spanned by the functions (95) is the same as the space spanned by the spatial parts of the $^1A_1$ and ^1E functions, (84) and (85). The function (96) is the same as the spatial part of the $^3A_2$ function given by Eq. (83).

The fact that the $^1A_1$ and ^1E functions come together to form a representation of U(2) shows that we cannot diagonalize the Hamiltonian matrix by means of U(2). The form of the functions (95) and (96) points, however, towards a connection between the second rank tensor representations of U(2) and the representations of S_2, for the functions (95) are symmetric, whereas the function (96) is antisymmetric.

The indicated connection between the Nth rank tensor representations of U(n) and the representations of S_N is an important reason for introducing the unitary group in the present context. Another reason is that the Hamiltonian may be expressed in terms of operators of the form

$$E_{rs} = \sum_{i=1}^{N} |u_r(i)\rangle \langle u_s(i)|\,. \tag{97}$$

It is, however, important to note that the Hamiltonian does not commute with these operators. U(n) is not a symmetry group of the Hamiltonian.

The meaning of U(n) is that it describes the geometry of the underlying vector space Ω_n. This geometry is given by Eq. (90), and U(n) is its invariance group. The importance of expressing a geometry by its invariance group was the kernel of Klein's celebrated Erlangen Program (formulated in 1872, reprinted in [68]).

The unitary group plays a very important role in today's large scale CI calculations. Here, we have merely indicated that U(n) may have some significance for the N-electron problem, because a serious discussion would force us to introduce many new constructs. Hence, we must refer the reader to the literature. Fortunately, the pioneers of the application to many-electron theory have discussed the unitary group approach in a multi-author book [69], and a recent text book by Matsen and Paunez [70] gives a thorough discussion of the approach, with all important references. We also refer the reader to an interesting and pedagogic overview by Sutcliffe [71], expecially because it discusses the application of group theory to CI methods in general, and hence extends the discussion we have given in Sects. 11–14. Sutcliffe also discusses the problem to which we alluded in Sect. 12, namely, that present CI programs, in spite of their strength, usually are unable to exploit the full symmetry of non-commutative point groups. This is of course a lack which should be rectified, and some extensive studies have in fact recently been directed at that aim [72–74].

15 Concluding Remarks

Our essay on symmetry in molecules has now come to a natural end, although we have merely scratched the surface of the topic. Thus, we have not discussed the vibration and rotation of molecules at all. Nor have we dwelled on the application of group theory in spectroscopy. It has been necessary to severely confine the discussion in order to be reasonably exact. The author's background has made it natural to put emphasis on the description of electronic structures, but the principles that lie behind the application of group theory to other aspects of molecular theory are essentially the same as those presented here. It is accordingly my hope that the present paper may serve as a general first introduction to the broad subject called symmetry in molecules.

Acknowledgements. I thank the Department of Chemistry at the University of Chicago for hospitality whilst this paper was written. The work was supported by the Danish Natural Science Research Council and Julie Damms Studiefond.

16 References

1. Weyl H (1952) Symmetry. Princeton University Press, Princeton, New Jersey
2. Hargittai I (ed) (1986) Symmetry. Unifying human understanding. Pergamon, New York
3. Moore FJ (1918) A history of chemistry. McGraw-Hill, New York
4. Benfey OT (ed) (1963) Classics in the theory of chemical combination. Dover, New York
5. Kauffman GB (1968) Classics in coordination chemistry. Dover, New York
6. Phillips FC (1946) An introduction to crystallography. Longmans, London
7. Hahn T (ed) (1983) International tables for crystallography. Volume A. Space-group symmetry. Published for the International Union of Crystallography. Reidel, Dortrecht
8. Landau LD, Lifshitz EM (1958) Quantum mechanics. Pergamon, London
9. Atkins PW, Child MS, Phillips CSG (1970) Tables for group theory. Oxford University Press, Oxford
10. Lomont JS (1959) Applications of finite groups. Academic, New York
11. Wigner EP (1959) Group theory and its application to the quantum mechanics of atomic spectra. Academic, New York

12. Tinkham M (1964) Group theory and quantum mechanics. McGraw-Hill, New York
13. Hamermesh M (1962) Group theory and its application to physical problems. Addison-Wesley, Reading, MA
14. Dahl JP (1972) The independent-particle model. Polyteknisk Forlag, Lyngby, Denmark
15. Schrödinger E (1926) Ann. d. Physik 79: 361, 489; 80: 437; 81: 409
16. Schiff LI (1968) Quantum mechanics, 3rd ed, McGraw-Hill, New York
17. Eyring H, Walter J, Kimball GE (1944) Quantum chemistry. Wiley, New York
18. Atkins PW (1983) Molecular quantum mechanics, 2nd ed, Oxford University Press, Oxford
19. Berry RS, Rice SA, Ross J (1980) Physical chemistry. Wiley, New York
20. Cornwell JF (1984) Group theory in physics. Academic, London
21. Runge C (1919) Vektoranalysis. S. Hirzel, Leipzig
22. Lenz W (1924) Z. Physik 24: 197
23. Pauli W (1926) Z. Physik 36: 336
24. Fock V (1935) Z. Physik 98: 145
25. Bargmann V (1936) Z. Physik 99: 576
26. Bander M, Itzykson C (1966) Rev. Mod. Phys. 38: 330
27. Dahl JP (1968) Physics Letters 27A: 62
28. Uhlenbeck GE, Goudsmit S (1925) Naturwiss. 13: 953
29. Pauli W (1927) Z. Phys. 43: 601
30. Rauch H, Zeilinger A, Badurek G, Wilfing A, Bauspiess W, Bonse U (1975) Phys. Lett. 54A: 425
31. Werner SA, Colella R, Overhauser AW, Eagen CF (1975) Phys. Rev. Lett. 35: 1053
32. Dahl JP (1977) Kgl. Danske Vid. Selsk. Mat. Fys. Medd. 39 no. 12
33. Dirac PAM (1928) Proc. Roy. Soc. A117: 610
34. Slater JC (1960) Quantum theory of atomic structure, Vol. 1. McGraw-Hill, New York
35. Born M, Oppenheimer R (1927) Ann. d. Phys. 84: 457
36. Born M (1951) Gött. Nachr. math. phys. Kl. no. 6
37. Born M, Huang K (1954) Dynamical theory of crystal lattices. Oxford University Press, Oxford
38. Ballhausen CJ, Hansen AE (1972) Ann. Rev. Phys. Chem. 23: 15
39. Berry RS (1980) in: Woolley RG (ed) Quantum dynamics of molecules. Plenum, New York. p 143
40. Ezra GS (1982) Symmetry properties of molecules. Springer, Berlin Heidelberg New York (Lecture Notes in Chemistry, vol 28)
41. Berry RS (1987) in: Avery J, Dahl JP, Hansen AE (eds) Understanding molecular properties. Reidel, Dortrecht, p 425
42. Sutcliffe BT, Tennyson J (1987) in: Avery J, Dahl JP, Hansen AE (eds) Understanding molecular properties. Reidel, Dortrecht, p 449
43. Dahl JP, Feng X (1984) Theor. Chim. Acta 66: 261
44. McWeeny R, Sutcliffe BT (1969) Methods of molecular quantum mechanics. Academic, London
45. Cotton FA (1963) Chemical applications of group theory. Wiley-Interscience, New York
46. Bethe H (1929) Ann. Physik 5: 133
47. Dambus T (1980) Linear Algebra Appl. 32: 125
48. Altmann SL (1986) Rotations, quarternions, and double groups. Clarendon, Oxford
49. Mulliken RS (1933) Phys. Rev. 43: 279
50. Ballhausen CJ (1962) Introduction to ligand field theory. McGraw-Hill, New York
51. Woodward RB, Hoffmann R (1970) The conservation of orbital symmetry. Verlag Chemie, Weinheim
52. Bohr N (1922) Z. Physik 9: 1
53. Heisenberg W (1926) Z. Physik 38: 411; 39: 499
54. Heisenberg W (1927) Z. Physik 41: 239
55. Wigner E (1926) Z. Physik 40: 492, 883
56. Pauli W (1925) Z. Physik 31: 765
57. Dirac PAM (1926) Proc. Roy. Soc. (London) A112: 661
58. Slater JC (1975) Solid-state and molecular theory: A scientific biography. Wiley, New York
59. Slater JC (1929) Phys. Rev. 34: 1293
60. Boys SF (1950) Proc. Roy. Soc. A200: 542
61. Löwdin P-O (1960) Rev. Mod. Phys. 32: 328

62. Löwdin P-O (1969) Adv. Chem. Phys. 14: 283
63. Slater JC (1963) Quantum theory of molecules and solids. McGraw-Hill, New York
64. Pauncz R (1979) Spin eigenfunctions. Construction and use. Plenum, New York
65. Kotani M, Amemiya A, Ishiguro E, Kimura T (1963) Tables of molecular integrals. Maruzen, Tokyo
66. Duch W, Karwowski J (1985) Computer Physics Report 2: 93
67. Weyl H (1938) The classical groups. Princeton University Press, Princeton, NJ
68. Coxeter HSM (1977) Mathematical Intelligencer 0: 22
69. Hinze J (ed) (1981) The unitary group. Springer, Berlin Heidelberg New York (Lecture Notes in Chemistry, vol 22)
70. Matsen FA, Pauncz R (1986) The unitary group in quantum chemistry. Elsevier, Amsterdam (Studies in physical and theoretical chemistry, vol 44)
71. Sutcliffe BT (1983) in: Diercksen GHF, Wilson S (eds) Computational molecular physics. Reidel, Dortrecht (NATO ASI Series C, vol 113)
72. Chen JQ, Wang F, Gao MJ, Yu ZR (1980) Scientia Sinica 9: 1116
73. Rettrup S, Sarma CR, Dahl JP (1982) Int. J. Quantum Chem. 22: 127
74. Gao MJ, Chen JQ, Paldus J (1987) Int. J. Quantum Chem. 32: 133

Chirality of Molecular Structures — Basic Principles and Their Consequences

Laurence D. Barron

Chemistry Department, The University, Glasgow G12 8QQ, U.K.

This article reviews the concept of molecular chirality from the standpoint of modern physics. The application of the fundamental symmetry operations of space inversion and time reversal together are shown to lead to a more precise definition of a chiral object than that usually employed. It follows that, although spatial enantiomorphism is sufficient to guarantee chirality in a stationary object, enantiomorphous systems are not necessarily chiral when motion is involved, leading to the concept of true and false chirality associated with time-invariant and time non-invariant enantiomorphism, respectively. Although only a truly chiral influence can induce absolute asymmetric synthesis in a reaction that has reached true thermodynamic equilibrium, it is shown how false chirality can suffice in a reaction under kinetic control due to a breakdown of microscopic reversibility and hence of detailed balancing. The discussion then turns to a consideration of symmetry violation and how it differs from symmetry breaking. This provides considerable insight into molecular chirality and exposes productive analogies between the quantum states of a chiral molecule and those of various elementary particles. As well as facilitating a sound understanding of the structure and properties of chiral molecules and of the factors involved in their synthesis and transformations, it is hoped that this article will encourage greater awareness of the value of concepts from modern elementary particle and condensed matter physics in theoretical chemistry.

1 Introduction

The idea that molecules could be chiral, or handed, first arose in the early years of the last century following the discovery of natural optical activity. Optical activity was first observed by Arago in 1811 in the form of colours in sunlight that had passed along the optic axis of a quartz crystal placed between crossed polarizers. Subsequent experiments by Biot established that the colours originated in a rotation of the plane of polarization of linearly polarized light (optical rotation), with different rotations for light of different wavelengths (optical rotatory dispersion). He also discovered a second form of quartz which rotated the plane of polarization in the opposite direction. Then in 1815 Biot observed optical rotation in certain organic liquids such as turpentine, which indicated that the optical activity of fluids must reside in the individual molecules and may be observed even when the molecules are arranged in a random fashion; that of quartz is a property of the crystal structure since molten quartz is not optically active.

In 1824, following his discovery of circularly polarized light, Fresnel was able to describe optical rotation in terms of different refractive indices for the coherent right and left circularly polarized components of equal amplitude into which linearly polarized light can be resolved. This immediately provided an insight into the symmetry requirements for an optically active molecule or crystal. In the words of Fresnel [1]:

There are certain refracting media, such as quartz in the direction of its axis, turpentine, essence of lemon, etc., which have the property of not transmitting with the same velocity circular vibrations from right to left and those from left to right. This may result from a peculiar constitution of the refracting medium or of its molecules, which produces a difference between the directions right to left and left to right; such, for instance, would be a helicoidal arrangement of the molecules of the medium, which would present inverse properties accordingly as these helices were dextrogyrate or laevogyrate.

Pasteur's celebrated resolution, in 1848, of a sample of the optically inactive (racemic) version of tartaric acid into equal numbers of molecules giving equal and opposite optical rotations in solution confirmed Fresnel's insight that optically active molecules must be handed. This realization that the molecules which exhibit optical rotation have an essentially helical structure meant that from early on molecules were being thought about in three dimensions, leading eventually to the concept of tetrahedral valencies for the carbon atom and the subject of stereochemistry.

In 1846, Faraday demonstrated conclusively the intimate connection between electromagnetism and light by observing magnetic optical activity in the form of optical rotation induced in a rod of lead borate glass placed between the poles of an electromagnet. In fact a Faraday rotation is found when light is transmitted through any crystal or fluid in the direction of a static magnetic field, the sense of rotation being reversed on reversing the direction of either the light beam or the magnetic field. The Faraday effect has been a source of much confusion in chirality studies; but we shall see that a careful analysis of the different symmetry aspects of natural and magnetic optical activity gives helpful insight into the phenomenon of molecular chirality.

A finite cylindrical helix provides a good example of figures exhibiting what Pasteur [2] called *dissymmetry* if they possess structural forms which "differ only as an image in a mirror differs from the object which produces it", the two distinct forms being called enantiomers (from the Greek *enantios morphe*, opposite shape). Dissymmetric figures are not necessarily asymmetric, meaning devoid of all symmetry elements, since they may possess one or more proper rotation axes (the finite cylindrical helix has a twofold rotation axis through the mid-point of the coil, perpendicular to the long helix axis). However, dissymmetry excludes improper rotation axes; that is centres of inversion, reflection planes and rotation-reflection axes.

Pasteur's extension of the concept of dissymmetry to other aspects of the physical world [3, 4] has had a considerable influence on attempts to induce absolute asymmetric synthesis and on theories of the origin of optical activity in nature. For example, he thought that a magnetic field, since it can induce optical rotation, generates the same type of dissymmetry as that possessed by an optically active molecule; similarly the combination of a rotation with a linear motion.

Lord Kelvin, Professor of Natural Philosophy in the University of Glasgow, coined the word *chiral* to describe a geometrical figure "if its image in a plane mirror, ideally realized, cannot be brought to coincide with itself", and this word has now replaced dissymmetric in the literature of stereochemistry [5]. However, the two words are not strictly synonymous in a broader context. Dissymmetry means the absence of certain symmetry elements; improper rotation axes in Pasteur's usage. Chirality is a more positive concept in that it refers to the possession of the attribute of handedness which, as discussed later, has a physical content: in molecular physics this is the ability to support time-even pseudoscalar observables; in elementary particle physics chirality is defined as the eigenvalue of the Dirac matrix operator γ_5 [6].

Although dissymmetry is sufficient to guarantee chirality in a stationary object such as a finite helix, dissymmetric systems are not necessarily chiral when motion is involved. I have introduced the concept of "true" and "false" chirality to draw attention to this distinction [7–9] but do not intend that this should become standard nomenclature; rather, I suggest that the word "chiral" be reserved in future for systems that I call here truly chiral. It will be appreciated from what follows that true and false chirality correspond to time-invariant and time-noninvariant enantiomorphism, respectively. We shall see that the combination of a rotation with a linear motion does indeed generate true chirality, but that a magnetic field does not (in fact it is not even false chirality). Examples of dissymmetric systems that show false chirality are a stationary rotating cone, and collinear electric and magnetic fields.

Jumping to our own era, the triumph of theoretical physics in unifying the weak and electromagnetic forces into a single "electroweak" force has provided a new perspective on chirality. Since the weak and electromagnetic forces have turned out to be different aspects of the same, but more fundamental, unified force, the absolute parity violation associated with the weak force is now thought to infiltrate to a tiny extent into all electromagnetic phenomena so that free atoms, for example, show tiny optical rotations, and a tiny energy difference exists between the enantiomers of a chiral molecule. Indeed, we shall see that the distinction between true and false chirality hinges on the symmetry operations that interconvert enantiomers, and that parity violation provides a cornerstone for the identification of true chirality. It is intriguing that parity violation has now provided a scientific basis for the general cosmic dissymmetry that Pasteur sensed over a century ago [10, 11].

This article is the latest in a series [9, 12] that attempts to provide a description of molecular chirality rooted in the principles of modern physics in order to facilitate a proper understanding of the structure and properties of chiral molecules, and of the factors involved in their synthesis and transormations.

2 Basic Symmetry Principles

Symmetry arguments are of course central to considerations of molecular chirality, and they take up much of this article. As well as conventional point group symmetry, the fundamental symmetries of space inversion, time reversal and even charge conjugation have something to say about optical activity and chirality at all levels: the experiments that show up optical activity observables, the objects generating these observables and the nature of the quantum states that these objects must be able to support.

2.1 Non-observables and Symmetry Operations

The spatial symmetry aspects of molecules are a central feature of modern chemistry. An object is said to possess a particular spatial symmetry if, after subjecting it to a symmetry operation such as inversion, reflection or rotation with respect to a corresponding geometrical symmetry element within the object, it looks the same as it did before. In particular, symmetry operations constituting the point group of a molecule leave one point invariant.

Less familiar in chemistry, but central to modern physics, is the remarkable existence of symmetries in the laws which determine the operation of the physical world. According to Lee [13], all symmetry principles originate in the assumption that, in microscopic processes, it is impossible to observe certain basic quantities called non-observables. This implies invariance of physical laws under an associated transformation and usually generates a conservation law or selection rule. The non-observables of interest here are absolute chirality (absolute right- or left-handedness), absolute direction of time flow (from past to future or future to past), and absolute sign of electric charge.

The transformation associated with absolute chirality is *space inversion* represented by the parity operator $\hat{P}$ which, in the active convention, inverts the system through the origin of an arbitary set of space-fixed axes so that each constituent particle i at some point $\boldsymbol{r}_i$ is moved to $-\boldsymbol{r}_i$. This is equivalent to a reflection of the system in any plane containing the coordinate origin, followed by a rotation through 180° about an axis perpendicular to the reflection plane. Most physical laws (but not those describing processes such as β-decay which involve the weak interaction) are unchanged by space inversion: in other words the equations representing the physical laws are unchanged if the space coordinates (x, y, z) are replaced everywhere by $(-x, -y, -z)$, and the corresponding processes are said to conserve parity.

The transformation associated with absolute direction of time flow is *time reversal*, represented classically by the operator $\hat{T}$, and this has the effect of reversing the motions of all the particles in the system. In fact it is better to think of $\hat{T}$ as motion reversal since this does not have the same mysterious connotations as travelling backwards in time. If replacing the time coordinate (t) by $(-t)$ everywhere in equations describing physical laws leaves those equations unchanged, the physical processes represented by those laws are said to conserve time reversal invariance, or to have reversality. One must not confuse reversality with the thermodynamic concept of reversibility: a process involving a macroscopic system will have reversality provided each constituent microscopic process has reversality, even though the time-reversed macroscopic process is most unlikely. For example, the mechanical shuffling of a pack of cards has reversality although thermodynamics would classify it as an irreversible process.

The transformation associated with absolute sign of electric charge is *charge conjugation*, represented by the operator $\hat{C}$ which interconverts corresponding particles and antiparticles (if an elementary particle carries an electric charge, the corresponding antiparticle carries the opposite charge). Although this exotic operation might appear to have no relevance to chemistry, it is shown later to have conceptual significance in studies of molecular chirality.

An important consequence of the existence of symmetries in the laws which determine the operation of the physical world is that, if a *complete* experiment is subjected to an associated symmetry operation such as space inversion or time reversal, the resulting experiment should, in principle, be realizable [14, 15]. A detailed consideration of the natural and magnetic optical rotation experiments shows that they do indeed conserve parity and reversality [12, 16]; and such arguments can also be used to predict or discount possible new effects (such as an electric analogue of the Faraday effect) without recourse to mathematical theories [12, 16–18]. The same conclusions are also obtained using a distinct but complementary approach based on photon selection rules [19, 20].

2.2 Symmetry Classification of Physical Quantities

A central concept is the behaviour of a physical quantity (an observable) under a symmetry operation. Physical quantities are first classified as *scalars*, *vectors* or *tensors* depending on their directional properties: a scalar such as temperature has magnitude but no associated direction; a vector such as velocity has magnitude and one associated direction; and a tensor such as electric polarizability has magnitudes associated with two or more directions. Scalars, vectors and tensors are then further classified according to their behaviour under $\hat{P}$ and $\hat{T}$.

Vectors such as position $\boldsymbol{r}$, velocity $\boldsymbol{v}$ and linear momentum $\boldsymbol{p}$ which change sign under the inversion operation $\hat{P}$ are called *polar* or true vectors. A vector such as angular momentum $\boldsymbol{L} = \boldsymbol{r} \times \boldsymbol{p}$ whose sign is not changed by $\hat{P}$ is called an *axial* or pseudo vector: $\boldsymbol{L}$ is defined relative to the sense of rotation by a right-hand rule, and $\hat{P}$ does not change the sense of rotation. A *pseudoscalar* quantity is a number with no directional properties but which changes sign under $\hat{P}$. A pseudoscalar is generated by taking the scalar product of a polar and an axial vector.

Physical entities are classified as *time-even* or *time-odd* depending on whether they are invariant or change sign under the time reversal operation $\hat{T}$. This behaviour is usually immediately obvious from a consideration of the motions of the constituent particles. Of course, many physical entities do not involve motion and so are time-even, examples being the energy scalar W, position vector $\boldsymbol{r}$, electric field vector $\boldsymbol{E}$ and electric dipole moment vector $\boldsymbol{\mu}$. Many other entities do involve motion and are time-odd: for example the velocity vector $\boldsymbol{v}$, linear momentum vector $\boldsymbol{p}$, angular momentum vector $\boldsymbol{L}$, magnetic field vector $\boldsymbol{B}$ and magnetic dipole moment vector $\boldsymbol{m}$. The elusive magnetic monopole, which has never been observed, transforms as a time-odd pseudoscalar.

Pseudosclar quantities are of central importance in the discussion of molecular chirality because, as shown in Sect. 3.1, the natural optical activity phenomena supported by chiral molecules are characterized by time-even pseudoscalar observables (e.g. optical rotation angle, rotational strength, Raman circular intensity difference, etc.). It is instructive at this point to introduce Neumann's principle [21, 22], which states that any type of symmetry exhibited by the symmetry group of a system is possessed by every physical property of the system. In fact it is Curie's re-statement of this principle, rather than the original, that is most helpful here [23]: "dissymmetry makes the phenomenon". Thus no dissymmetry can manifest itself in a physical property which does not already exist in the system. In this instance the pseudoscalar natural optical activity observables change sign under space inversion, which parallels the interconversion under space inversion of the distinguishable enantiomeric chiral molecules which make up the distinguishable enantiomeric bulk isotropic samples.

2.3 Symmetry in Quantum Mechanics

The application of symmetry operations in quantum mechanics introduces some important new features not present in the classical discussion.

Parity

The starting point for parity considerations is the invariance of the conventional (i.e. parity-conserving) Hamiltonian for a closed system of interacting particles to an inversion of the coordinates of all the particles. $\hat{P}$ is now interpreted as a linear unitary Hermitian operator that changes the sign of the space coordinates in the Hamiltonian and the wavefunction. Consider first the wavefunction:

$$\hat{P}\psi(\boldsymbol{r}) = \psi(-\boldsymbol{r}) . \tag{1}$$

If $\psi(\boldsymbol{r})$ happens to be an eigenfunction of $\hat{P}$ we can write

$$\hat{P}\psi(\boldsymbol{r}) = p\psi(\boldsymbol{r}) . \tag{2}$$

The eigenvalues p are found by realizing that a double application amounts to the identity so that

$$\hat{P}^2\psi(\boldsymbol{r}) = p^2\psi(\boldsymbol{r}) = \psi(\boldsymbol{r}) , \tag{3}$$

from which we obtain

$$p^2 = 1 , \qquad p = \pm 1 . \tag{4}$$

Thus even (+) and odd (—) parity wavefunctions are defined according as they are invariant or simply change sign under $\hat{P}$:

$$\hat{P}\psi(+) = \psi(+) , \qquad \hat{P}\psi(-) = -\psi(-) . \tag{5}$$

Turning now to the Hamiltonian, its invariance under space inversion means we can write

$$\hat{P}\hat{H}\hat{P}^{-1} = \hat{H} , \quad \text{or} \quad [\hat{P}, \hat{H}] \equiv \hat{P}\hat{H} - \hat{H}\hat{P} = 0 . \tag{6}$$

Since $\hat{P}$ does not depend explicitly on time and commutes with the Hamiltonian, we can say that, if the state of a closed system has a definite parity, that parity is conserved (the law of conservation of parity) [24].

If two eigenfunctions $\psi(+)$ and $\psi(-)$ of opposite parity have energy eigenvalues that are degenerate, or nearly so, the system can exist in states of mixed parity with wavefunctions

$$\psi_1 = \frac{1}{\sqrt{2}} [\psi(+) + \psi(-)] , \tag{7a}$$

$$\psi_2 = \frac{1}{\sqrt{2}} [\psi(+) - \psi(-)] . \tag{7b}$$

Clearly these two mixed parity states are interconverted by $\hat{P}$:

$$\hat{P}\psi_1 = \psi_2 , \qquad \hat{P}\psi_2 = \psi_1 . \tag{8}$$

It follows from (6) that a central property of definite parity states is that they are true stationary states with constant energy $W(+)$ or $W(-)$, i.e.

$$\psi(\pm) = \psi^{(0)}(\pm)\, e^{-iW(\pm)t/\hbar} , \tag{9}$$

but mixed parity states are not. We shall see in Section 6 below that mixed parity states can become quasi-stationary when $W(+) \approx W(-)$, and true stationary states if $\hat{H}$ contains a parity-violating term.

All observables can be classified as having even or odd parity depending on whether they are invariant or change sign under space inversion. Even and odd parity operators $\hat{A}(+)$ and $\hat{A}(-)$ associated with these observables are therefore defined by

$$\hat{P}\hat{A}(+)\hat{P}^{-1} = \hat{A}(+) , \qquad \hat{P}\hat{A}(-)\hat{P}^{-1} = -\hat{A}(-) . \tag{10}$$

Since integrals taken over all space are only non-zero for totally symmetric integrands, the expectation values of these operators in a mixed parity state such as (7a) reduce to

$$\langle\psi_1|\hat{A}(+)|\psi_1\rangle = \frac{1}{2}[\langle\psi(+)|\hat{A}(+)|\psi(+)\rangle + \langle\psi(-)|\hat{A}(+)|\psi(-)\rangle], \tag{11a}$$

$$\langle\psi_1|\hat{A}(-)|\psi_1\rangle = \frac{1}{2}[\langle\psi(+)|\hat{A}(-)|\psi(-)\rangle + \langle\psi(-)|\hat{A}(-)|\psi(+)\rangle], \tag{11b}$$

from which it follows that the expectation value of any odd parity observable vanishes in any state of definite parity, i.e. a state for which either $\psi(+)$ or $\psi(-)$ is zero. This means that measurements on a system in a state of definite parity can reveal only observables with even parity, examples being electric charge, angular momentum, magnetic dipole moment, electric quadrupole moment, etc.; whereas measurements on a system in a state of mixed parity can reveal, in addition, observables with odd parity, examples being magnetic monopole, linear momentum, electric dipole moment, etc. [12, 25]. The optical rotatory parameter, being pseudoscalar, has odd parity: this leads to the important deduction that resolved chiral molecules exist in mixed parity quantum states, the detailed nature of which will be elaborated later.

Time Reversal

The classical operator $\hat{T}$ introduced above does not translate directly into a satisfactory quantum-mechanical operator. Instead, the operator

$$\hat{\Theta} = \hat{T}\hat{K} \tag{12}$$

where $\hat{T}$ again represents the transformation $t \rightarrow -t$ and $\hat{K}$ is the operator of complex conjugation, is taken as the time reversal operator in quantum mechanics [12, 25, 26]. Although it is possible to classify time-even and time-odd Hermitian operators and their associated observables according to whether they are invariant or change sign under time reversal, it is not possible to classify a quantum state as being even or odd under time reversal because $\hat{\Theta}$, unlike $\hat{P}$, does not have eigenvalues. (On the other hand, the operator $\hat{\Theta}^2$ does have eigenvalues, these being $+1$ for an even-electron system and -1 for an odd-electron system).

A simple illustration is the effect of $\hat{\Theta}$ on a general atomic state $|J, M\rangle$ where both orbital and spin angular momenta can contribute to the total electronic angular momentum characterized by the usual quantum numbers J and M. Using a particular phase convention it is found that [12]

$$\hat{\Theta}\,|J, M\rangle = (-1)^{J-M+q}\,|J, -M\rangle\,, \tag{13}$$

where q is the sum of the individual orbital quantum numbers of all the electrons in the atom. Thus time reversal has generated a new quantum state, orthogonal to the original, corresponding to a reversal of the sense of the total angular momentum of the atom. Since they are interconverted by time reversal, such states can be loosely regarded as having "mixed reversality", analogous to the mixed parity states (7), even though associated states of definite reversality do not exist: such states can support time-odd observables [12, 25]. Notice, however, that states $|J, M\rangle$ do have definite parity since they are eigenstates of $\hat{P}$:

$$\hat{P}\,|J, M\rangle = (-1)^q\,|J, M\rangle\,. \tag{14}$$

This follows from the behaviour under space inversion of the spherical harmonics, and the standard convention that the "intrinsic parity" of an electron spin state is +1.

Charge Conjugation

A discussion of the effect of the charge conjugation operator $\hat{C}$ in quantum mechanics requires a formulation in terms of relativistic quantum field theory [13, 27]. For the purposes of this article, all we need to appreciate is that a charged particle is not in an eigenstate of $\hat{C}$ [27].

3 True Chirality

A central point in this article is that optical activity is not necessarily the hallmark of chirality, and that a proper symmetry classification of the corresponding observables leads to a more precise definition of a chiral object. The distinction between natural and magnetic optical activity, which is often a source of confusion in the literature of both chemistry and physics, provides a good example.

3.1 Natural and Magnetic Optical Activity

It was recognized in the last century that the symmetry aspects of natural and magnetic optical activity are quite different. For example, in his Baltimore Lectures delivered in 1884, Lord Kelvin (then Sir William Thomson) said the following [5]:

The magnetic rotation has neither left-handed nor right-handed quality (that is to say, no chirality). This was perfectly understood by Faraday, and made clear in his writings, yet even to the present day we frequently find the chiral rotation and the magnetic rotation of the plane of polarized light classed together in a manner against which Faraday's original description of his discovery of the magnetic polarization contains ample warning.

This viewpoint was reinforced by Zocher and Török [28], who discussed the space-time symmetry aspects of natural and magnetic optical activity from a general classical viewpoint and recognized that quite different asymmetries are involved.

The symmetry classification of the natural and magnetic optical activity observables is obtained by comparing the results of optical rotation measurements before and after subjecting the sample plus any applied field to space inversion and time reversal (this is a different procedure to that mentioned at the end of Sect. 2.1 in which the complete experiment, including the probe light beam, is subjected to symmetry operations in order to demonstrate conservation of parity and reversality).

Consider first the natural optical rotation experiment. Under space inversion, an isotropic collection of chiral molecules is replaced by a collection of the enantiomeric molecules, and an observer with a linearly polarized probe light beam will measure equal and opposite optical rotation angles before and after the inversion. This indicates that the observable has odd parity, and it is easy to deduce that it is a pseudoscalar (rather than, say, a polar vector) because it is invariant with respect to any proper rotation in space of the complete sample. Under time reversal, an isotropic collection of chiral molecules is unchanged, so the optical rotation is unchanged. Thus the natural optical rotation observable is a time-even pseudosclar.

Now consider the Faraday effect, where optical rotation is induced in an isotropic collection of achiral molecules by a static uniform magnetic field parallel to the light beam. Under space inversion, the molecules and magnetic field direction are unchanged, so the same magnetic optical rotation will be observed. This indicates that the observable has even parity, and we can further deduce that it is an axial vector (rather than a scalar) by noticing that a proper rotation of the complete sample, including the magnetic field, through π about any axis perpendicular to the field reverses the relative directions of the magnetic field and the probe beam and so changes the sign of the observable. Under time reversal, the collection of molecules (even if they are individually paramagnetic) can be regarded as unchanged provided it is isotropic in the absence of the field, but again the relative directions of the magnetic field and the probe light beam are reversed and so the optical rotation changes sign. Thus the magnetic optical rotation observable is a time-odd axial vector.

These conclusions are reinforced by a more fundamental approach in which operators are defined whose expectation values generate the optical activity observables [12, 29]. It is found that the natural optical rotation observable is generated by a time-even odd-parity operator, and the magnetic optical rotation observable by a time-odd even-parity operator. Another approach is to look at the associated molecular property tensors: it is found that all the contributions to natural optical rotation are generated by time-even tensors, and that all the contributions to magnetic optical rotation are generated by time-odd tensors [30, 31].

This tells us that the nature of the quantum states of molecules that can support natural optical rotation is quite different from that of the quantum states that can support magnetic optical rotation. From the discussion in Sect. 2.3 it is clear that the former must have, among other things, mixed parity and the latter mixed reversality. The former is associated with spatial dissymmetry and corresponds to true chirality; whereas the latter originates in a different types of dissymmetry associated with lack of time reversal invariance (in previous articles I have called this an example of false chirality but now prefer to reserve the term for systems exhibiting time-noninvariant enantiomorphism). From Pasteur onwards these two types of optical activity have often been confused: indeed Pasteur was probably the target of Lord Kelvin's rebuke. Recent examples include the assertion that an achiral molecule in a pure rotational state can be regarded as chiral [32, 33], and that there exists a chiral discrimination in the intermolecular forces between co- and counter-rotating pairs of such molecules [34]. It is certainly correct to call an achiral molecule in a pure rotational quantum state $|J, M\rangle$ (or $|J, K, M\rangle$ for a symmetric top) optically active, since it will induce optical rotation in a light beam travelling parallel to the space-fixed quantization axis. An equal and opposite optical rotation is induced by $|J, -M\rangle$. But this is equivalent to magnetic optical activity since a magnetic field, or some other time-odd influence such as a rotation of the bulk sample [35], is required to lift the degeneracy of $|J, M\rangle$ and $|J, -M\rangle$.

3.2 A New Definition of Chirality

It should now be clear that the hallmark of a chiral system is that it can support time-even pseudoscalar observables. This leads to the following definition that enables chirality to be distinguished from other types of dissymmetry [7–9, 12, 29]:

True chirality is exhibited by systems that exist in two distinct enantiomeric states that are interconverted by space inversion, but not by time reversal combined with any proper spatial rotation.

This means that the enantiomorphism shown by truly chiral systems is time-invariant. Enantiomorphism that is time-noninvariant has different characteristics that I call false chirality in order to emphasise the distinction.

It is easy to see that stationary objects, such as a finite helix, that are chiral according to the traditional stereochemical definition, are accomondated by the first part of this definition: space inversion is a more fundamental operation than the mirror reflection traditionally invoked but provides an equivalent result; and the second part of the definition is irrelevant for a stationary object. However, the full definition is required to identify more subtle sources of chirality in which motion is an essential ingredient. A few examples will make this clear.

3.3 Translating Spinning Cones, Spheres and Elementary Particles

Consider a cone spinning about its symmetry axis. This system certainly supports enantiomorphism because the space-inverted version is not superposable on the original (Fig. 1a), so it might be thought that a spinning cone is a chiral object. However, according to the definition above, it is false chirality because time reversal followed by a rotation $\hat{R}_\pi$ through 180° about an axis perpendicular to the symmetry axis generates the same object as space inversion (Fig. 1a). But if the spinning cone is also translating along the axis of spin, time reversal followed by a 180° rotation now generates a different system to that generated by space inversion (Fig. 1b). Thus a *translating* spinning cone exhibits true chirality.

In fact the translating spinning object does not need to be a cone. A sphere translating along the axis of spin also shows true chirality. This can be appreciated by looking at just the pattern of arrows in Fig. 1b and ignoring the cone.

The molecular equivalent of a stationary spinning cone is a symmetric top in a rotational quantum state $|J, K, M\rangle$. The parity operation transforms $|J, K, M\rangle$

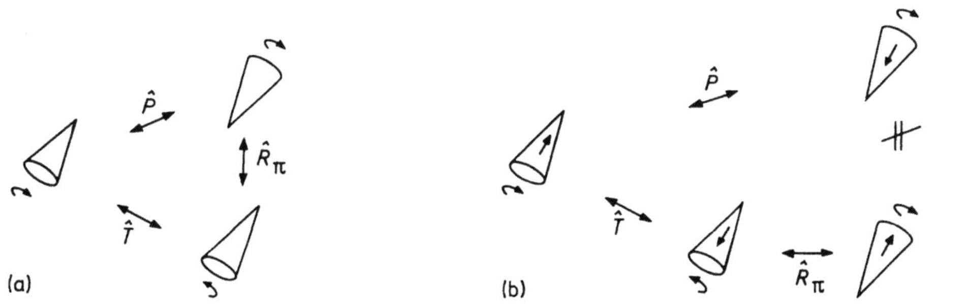

Fig. 1 a, b. The effect of $\hat{P}$, $\hat{T}$ and $\hat{R}_\pi$ on (a) a stationary spinning cone, and (b) a translating spinning cone

into $|J, -K, M\rangle$, which therefore has mixed parity [36], so that these two states correspond to the two non-superposable cones in Fig. 1a. And just as the two cones can be interconverted by time reversal followed by a rotation through 180° about an axis perpendicular to the symmetry axis, so this sequence of operations interconverts $|J, K, M\rangle$ and $|J, -K, M\rangle$.

This shows that mixed parity in a molecular quantum state is not necessarily sufficient to generate chirality, despite the fact that it can result in two enantiomeric objects. Although mixed parity is a necessary condition for any odd-parity observable, further characteristics are required for different types of odd-parity observable. In this instance the mixed parity characteristic of the rotational quantum state $|J, K, M\rangle$ results in a symmetric top with $K \neq 0$ showing a space-fixed electric dipole moment (an odd-parity observable transforming as a polar vector) and hence a first order Stark effect provided the top is dipolar to start with; but of course in order for the symmetric top to be dipolar there is the additional requirement of mixed parity internal (vibrational-electronic) quantum states associated with a molecular framework of C_n or C_{nv} symmetry. On the other hand natural optical rotation in isotropic samples is an odd-parity observable transforming as a pseudosclar and so requires mixed parity vibrational-electronic quantum states associated with a chiral molecular framework (symmetry C_n, D_n, O, T or I), but there is no requirement for mixed parity rotational states. The origin of these mixed parity internal states is described in Sect. 6.

The protons in a circularly polarized light beam propagating as a plane wave are in spin angular momentum eigenstates characterized by a spin quantum number $s = 1$, with quantum numbers $m_s = +1$ and -1 corresponding to projections of the spin angular momentum vector parallel and antiparallel, respectively, to the propagation direction. The absence of states with $m_s = 0$ is connected with the fact that photons, being massless, have no rest frame and so always move with the velocity of light (the usual $2j + 1$ projections for a general angular momentum vector are defined in the rest frame) [27]. In the usual convention, the electric vector of a right-circularly polarized light beam rotates in a clockwise sense when viewed towards the source of the beam, so right- and left-circularly polarized photons have spin angular momentum projections $-\hbar$ and $+\hbar$, respectively, along the propagation direction. Considerations analogous to those above for a translating spinning sphere then show that a circularly polarized photon exhibits true chirality.

The case of a spinning electron ($s = 1/2$, $m_s = \pm 1/2$) is rather different to that of a circularly polarized photon because an electron has rest mass. From the foregoing, it is clear that, whereas a stationary spinning electron is not a chiral object, an electron translating with its spin projection parallel or antiparallel to the propagation direction exhibits true chirality, with opposite spin projections corresponding to opposite handedness. Indeed, beams of spin-polarized electrons impinging on targets composed of chiral molecules are expected to exhibit effects analogous to the polarization effects in the light beams used as probes in conventional optical activity phenomena [37–39]. One such effect has recently been observed: an asymmetry in the attenuation of beams of right- and left-handed spin-polarized 5 eV electrons on passing through camphor vapour [40]. A central aspect of such experiments is that, all other things being equal, the magnitudes of the optical activity observables should increase with increasing electron velocity because electron chirality is velocity-dependent. This is emphasized by a mechanism proposed for asymmetric decomposition of enantiomeric chiral molecules by longitudinally spin-polarized electrons that is a function of v/c [41], and by the discussion in Sect. 7 of the relativistic aspects of chirality and the associated velocity-dependence of the amplitude of parity violation in the weak interaction. The fact that such a large effect was observed in the experiment just mentioned with electron beams of such low energy is probably due to the fact that camphor has a broad resonance associated with $\pi^* \leftarrow n$ carbonyl transitions at about this energy.

Similar experiments have been proposed for beams of spin-polarized neutrons [42–45], which are also spin $-1/2$ particles with mass.

3.4 Electric, Magnetic and Gravitational Fields

It is clear that neither a static uniform electric field $\boldsymbol{E}$ (a time-even polar vector) nor a static uniform magnetic field $\boldsymbol{B}$ (a time-odd axial vector) constitutes a chiral system; likewise time-dependent uniform electric and magnetic fields. Furthermore, contrary to a suggestion first made by Curie [23], no combination of a uniform electric and a uniform magnetic field (static or time-dependent) can constitute a chiral system. Collinear electric and magnetic fields do indeed generate enantiomorphism, but it is time-noninvariant and so corresponds to false chirality. Thus parallel and antiparallel arrangements are intercoverted by space inversion and are not superposable:

But they are also interconverted by time reversal combined with a rotation through 180°:

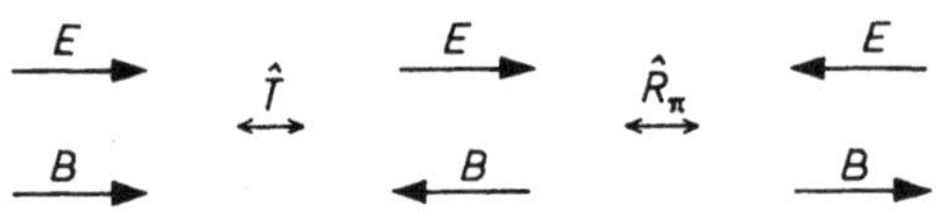

Zöcher and Török [28] also recognized the flaw in Curie's suggestion: they called the collinear arrangement of electric and magnetic fields a time-asymmetric enantiomorphism, and said that it does not permit a time-symmetric optical activity.

In fact the basic requirement for two collinear vectorial influences to generate chirality is that one transforms as a polar vector and the other as an axial vector, with both either time-even or time-odd. The second case is exemplified by the rotating translating cone or sphere discussed above, and by *magneto-chiral* phenomena such as a birefringence and a dichroism induced in a chiral sample by a uniform magnetic field parallel to the propagation vector $\boldsymbol{k}$ of a light beam of arbitrary polarization [46–49]. Thus parallel and antiparallel arrangements of $\boldsymbol{B}$ and $\boldsymbol{k}$ are true chiral enantiomers because they cannot be interconverted by time reversal since $\boldsymbol{k}$, unlike $\boldsymbol{E}$, is time-odd. The magneto-chiral observable transforms as a time-odd polar vector [48].

Analogous to collinear electric and magnetic fields is the case of a rapidly rotating vessel with the axis of rotation perpendicular to the earth's surface [50]. Here we have the time-odd axial angular momentum vector of the spinning vessel either parallel or antiparallel to the earth's gravitational field, itself a time-even polar vector. The physical influence here therefore exhibits false chirality.

4 Absolute Asymmetric Synthesis

The use of an external physical influence to produce an enantiomeric excess in what would otherwise be a racemic product of a prochiral chemical reaction is known as an absolute asymmetric synthesis. The subject still attracts much interest and controversy [8, 9, 51], and although it concerns molecular synthesis and transformation rather than structure, it is worth considering here because some rather subtle and unexpected parallels with elementary particle physics emerge.

No problems arise with the use of truly chiral influences such as circularly polarized photons and spin-polarized electrons. But opinion is still divided on the rôle of falsely chiral influences such as collinear electric and magnetic fields and spinning vessels.

4.1 Falsely Chiral Influences

It should first be realized that, in considering absolute asymmetric synthesis, a fundamental distinction must be made between reactions that have been allowed to reach thermodynamic equilibrium (thermodynamic control) and reactions that have not attained equilibrium (kinetic control).

It is generally accepted that, because a chiral molecule M and its enantiomer $\overline{\mathrm{M}}$ are isoenergetic in the presence of collinear electric and magnetic fields, or a spinning vessel with its axis perpendicular to the earth's surface (neglecting the very small differences due to parity violation discussed later), such falsely chiral influences cannot induce absolute asymmetric synthesis in a reaction mixture which is isotropic in the absence of the influence and which has been allowed to reach thermodynamic equilibrium [52–56]. This can be seen from a consideration of the invariance properties of the Hamiltonian in the presence of the influence, or from the application of Neumann's principle.

The situation is less straightforward for reactions under kinetic control. Some insight can be obtained from the fact that Neumann's principle cannot be applied in space-time to a system in which the entropy is changing [21], which is certainly the case for a reaction mixture away from equilibrium. This indicates that false chirality might suffice, for the intrinsically preferred direction of time associated with the changing entropy destroys the time reversal symmetry of the reacting system so that the time-noninvariant enantiomeric influences remain distinct [9, 57].

A general argument based on the principal of detailed balancing appears to negate this last conclusion [55], and was used to reinforce the criticisms of the claims for absolute asymetric synthesis in collinear electric and magnetic fields [58] and in spinning vessels [59]. However, as discussed in the next section, detailed balancing itself might not hold.

4.2 The Breakdown of Microscopic Reversibility: Enantiomeric Detailed Balancing

Recently it has been suggested that conventional detailed balancing, and the associated kinetic principles, might not be valid for reactions involving chiral molecules in a time-noninvariant enantiomorphous influence [60]. This suggestion was inspired by a remark of Lifshitz and Pitaevskii [61] that, for a system comprising chiral molecules of just one enantiomer, detailed balancing in the literal sense does not obtain because space inversion as well as time reversal is applied to each microscopic process, so that a completely different system is generated which cannot be compared with the original in order to deduce new information as to its properties. This is seen most clearly from the quantum-mechanical description of the microscopic processes [26]. The amplitude for a transition from some initial linear momentum state $\boldsymbol{p}$ to some final state $\boldsymbol{p}'$ is written $\langle \boldsymbol{p}' | \hat{\mathscr{T}} | \boldsymbol{p} \rangle$, where $\hat{\mathscr{T}}$ is the operator responsible for the transition. If $\hat{\mathscr{T}}$ involves purely electromagnetic interactions it will be invariant under both parity and time reversal, which enables us to write

$$\langle \boldsymbol{p}' | \hat{\mathscr{T}} | \boldsymbol{p} \rangle \overset{\text{under } \hat{T}}{=} \langle -\boldsymbol{p} | \hat{\mathscr{T}} | -\boldsymbol{p}' \rangle \overset{\text{under } \hat{P}}{=} \langle \bar{\boldsymbol{p}} | \hat{\mathscr{T}} | \bar{\boldsymbol{p}}' \rangle , \tag{15}$$

where we have allowed the particle to be chiral, the bar denoting the $\hat{P}$-enantiomer. The first equality in (15), obtained from time reversal alone, is the basis of the conventional principle of microscopic reversibility and, when averaged over the complete system of reacting particles at equilibrium, of the principal of detailed balancing [62]. The second equality, obtained by applying space inversion to the time-reversed transition amplitude, describes the inverse process involving the *enantiomeric* particles. Conventional detailed balancing is usually adequate for the kinetic analysis of reactions, even those involving chiral molecules, because conventional microscopic reversibility expressed by the first equality is usually valid. However, in the presence of a time-noninvariant enantiomorphous influence such as collinear electric and magnetic fields, time reversal alone is not a symmetry operation since a different influence is generated: space inversion must also be applied in order to recover the original relative orientations of $\boldsymbol{E}$ and $\boldsymbol{B}$. The first equality is therefore no longer valid, and we must base any kinetic analysis on the relationship

$$\langle p' | \hat{\mathscr{T}} | p \rangle \overset{\hat{T}\hat{P}}{=} \langle \bar{p} | \hat{\mathscr{T}} | \bar{p}' \rangle \, . \tag{16}$$

Lifshitz and Pitaevskii considered a system containing just one enantiomer. But if the system contains *equal numbers* of enantiomeric molecules at equilibrium, it does appear to be possible to obtain new information using arguments based on (16). A new principal of enantiomeric detailed balancing can be invoked in which the statistical average of all the microscopic processes involving one enantiomer can be balanced by the average of the inverse processes, in the sense of (16), involving the enantiomeric molecules [8, 57].

Consider a unimolecular process in which a prochiral molecule R generates a chiral molecule M or its enantiomer $\bar{\mathrm{M}}$:

$$\mathrm{M} \underset{k_b}{\overset{k_f}{\rightleftharpoons}} \mathrm{R} \underset{\bar{k}_f}{\overset{\bar{k}_b}{\rightleftharpoons}} \bar{\mathrm{M}} \, . \tag{17}$$

In refs. 8 and 57, enantiomeric detailed balancing was applied to the separate reactions in this scheme (i.e. $[\mathrm{R}]k_f = [\bar{\mathrm{M}}]\bar{k}_b$ and $[\mathrm{R}]\bar{k}_f = [\mathrm{M}]k_b$) to show that a time-noninvariant enantiomorphous influence allows a difference in rate constants for enantiomeric processes, i.e. $k_f \neq \bar{k}_f$ and $k_b \neq \bar{k}_b$. However, the analysis was incomplete because difficulties arise from the condition that the concentrations [M] and $[\bar{\mathrm{M}}]$ of the two enantiomers must be equal at thermodynamic equilibrium. These difficulties were subsequently resolved by realizing that the scheme (17) represents just one racemization pathway: by considering a reaction quadrangle which allows at least one alternative interconversion pathway between the enantiomers, it was shown that the thermodynamic and kinetic requirements can be reconciled [60]. The reason that the conditions $[\mathrm{M}] = [\bar{\mathrm{M}}]$, $k_f \neq \bar{k}_f$ and $k_b \neq \bar{k}_b$ can hold simultaneously is that, at *true* thermodynamic equilibrium, the different enantiomeric excesses associated with each separate racemization pathway sum to zero.

The conrotatory ring closure of a substituted butadiene to produce a cyclobutene was used as a simple example to show how collinear electric and magnetic fields can bring about a difference in potential energy profiles, and hence rate constants, for enantiometric reactions [60]. This example exposes the origin of the breakdown of microscopic reversibility in the situation under discussion: the transient magnetic dipole moment in the transition state, and hence the interaction with the magnetic field, is velocity-dependent. In the presence of a magnetic field, time reversal invariance and hence microscopic reversibility can usually be recovered if the moving charged particles generating the magnetic field are also reversed [63]. This prescription fails here because the system is only invariant under $\hat{T}\hat{P}$, not $\hat{T}$ alone, so microscopic reversibility is only recovered in the time-reversed *enantiomeric* system.

4.3 Unitarity and Thermodynamic Equilibrium

The unitarity of the scattering matrix [27], together with $\hat{T}\hat{P}$ invariance in the context of a collection of interconverting chiral enantiomers M and $\bar{\mathrm{M}}$, can be used to generalize this analysis [60].

Let $\mathscr{T}_{ji}$ be the amplitude for a transition from a state i to a state j. The requirement of unitarity for the scattering matrix (which corresponds to the fact that the sum of the transition probabilities from a given initial state to all final states is unity) leads to the following relationship for the transition amplitude [27]:

$$\sum_j \mathscr{T}_{ij}\mathscr{T}_{ij}^* = \sum_j \mathscr{T}_{ji}^*\mathscr{T}_{ji} \tag{18}$$

where the sum over j includes all states j and $\bar{j}$ of both enantiomers. It is important to realize that Hermiticity is not invoked in proving this result: $\mathscr{T}_{ij}$ is only Hermitian in a first approximation [27]. If $\hat{T}\hat{P}$ invariance holds, we can use (16) to write

$$\mathscr{T}_{ji} = \mathscr{T}_{\bar{i}\bar{j}} \tag{19}$$

and this, together with (18), gives

$$\sum_j |\mathscr{T}_{ji}|^2 = \sum_j |\mathscr{T}_{\bar{i}\bar{j}}|^2 = \sum_j |\mathscr{T}_{ij}|^2 . \tag{20}$$

If the system is racemic and in thermal equilibrium, equivalent enantiomeric states are equally populated. The second equality in (20) then shows that transitions out of these states must produce molecules in a given state i and the enantiomeric molecules in the equivalent state $\bar{i}$ in equal numbers. Thus no excess of one enantiomer over the other can develop at thermal equilibrium even when the presence of a time-noninvariant enantiomorphous influence destroys the equality between rates for specific enantiomeric transitions, i.e. when

$$|\mathscr{T}_{ji}|^2 \neq |\mathscr{T}_{\bar{j}\bar{i}}|^2 . \tag{21}$$

Notice that, had we allowed $\mathscr{T}_{ji}$ to be Hermitian, rates for specific enantiomeric transitions *would* be equal. From (18) and (19) we can also write

$$\sum_j |\mathscr{T}_{ji}|^2 = \sum_j |\mathscr{T}_{j\bar{i}}|^2 , \tag{22}$$

which shows that the total transition rates out of equivalent enantiomeric states are equal. Since in thermal equilibrium no excess of i over $\bar{i}$ may develop, this implies that any initial pre-existing excess tends to be dimished. This argument is similar to that used to demonstrate the existence of equal numbers of particles and antiparticles at thermodynamic equilibrium, despite $\hat{C}\hat{P}$ violation, in the big bang model of the early universe [64].

Thus if an absolute asymmetric synthesis starting with a pure prochiral reagent R could indeed be induced by collinear electric and magnetic fields, say, the enantiomeric excess will ultimately disappear if sufficient time is allowed for the establishment of true thermodynamic equilibrium in which all possible racemization pathways have separately equilibrated. Similarly, the influence could not generate an enantiomeric excess in a racemic mixture. Thus no fundamental thermodynamic principles are violated by the suggestion that a time-noninvariant enantiomorphous influence can, via a breakdown in microscopic reversibility, generate a difference in rate constants for enantiomeric processes.

This analysis is important for chemical physics generally because it reinforces the conclusion that it is unitarity, rather than microscopic reversibility, that is necessary for the validity of Boltzmann's H-theorem [64, 65].

5 Symmetry Violation

A symmetry violation (often called "non-conservation") shows up one of the non-observables discussed in Sect. 2.1 above. Although the study of symmetry violation belongs to the realm of physics rather than of chemistry, a consideration of symmetry violation, and how it differs from symmetry breaking, provides considerable insight into the phenomenon of molecular chirality.

5.1 The Fall of Parity

Before 1957 it had been accepted as self evident that handedness is not built into the world at any level. Thus if two objects exist as non-superposable mirror images of each other, such as the two enantiomers of a chiral molecule, it did not seem reasonable that nature should prefer one over the other. Any difference between enantiomeric systems was thought to be confined to the sign of odd-parity observables: the mirror image of any complete experiment involving one enantiomer should be realizable, with any odd parity observable (such as optical rotation angle) changing sign but retaining *precisely* the same magnitude. Then in 1956 Lee and Yang [66] pointed out that, unlike the electromagnetic and strong interactions, there was no evidence for parity conservation in processes involving the weak interaction. Of the experiments they suggested, that executed by Wu et al. [67] in 1957 is the most famous.

The Wu experiment studied the β-decay process

$$^{60}\mathrm{Co} \rightarrow {}^{60}\mathrm{Ni}^{+} + \mathrm{e}^{-} + \tilde{\nu}_{\mathrm{e}}$$

in which, essentially, a neutron n has decayed via the weak interaction into a proton p, an electron e^{-}, and an electron antineutrino $\tilde{\nu}_{\mathrm{e}}$. The nuclear spin magnetic moment $\boldsymbol{I}$ of each $^{60}\mathrm{Co}$ nucleus in the sample was aligned with an external magnetic field $\boldsymbol{B}$, and the angular distribution of electrons measured. It was found that the electrons are emitted preferentially in the direction *antiparallel* to that of the magnetic field (Fig. 2a). In accordance with the discussion in Sect. 2.2, $\boldsymbol{B}$ and $\boldsymbol{I}$ are axial vectors and so do not change sign under space inversion, whereas the electron propagation vector $\boldsymbol{k}$ does because it is a polar vector. Thus in the space-inverted experiment the electrons are emitted *parallel* to the magnetic field (Fig. 2b). It is only possible to reconcile Figs. 2a and 2b with parity conservation if there was no preferred direction for electron emission (an isotropic distribution), or if the electrons were emitted preferentially in the plane perpendicular to $\boldsymbol{B}$. The observation of (a) alone provides

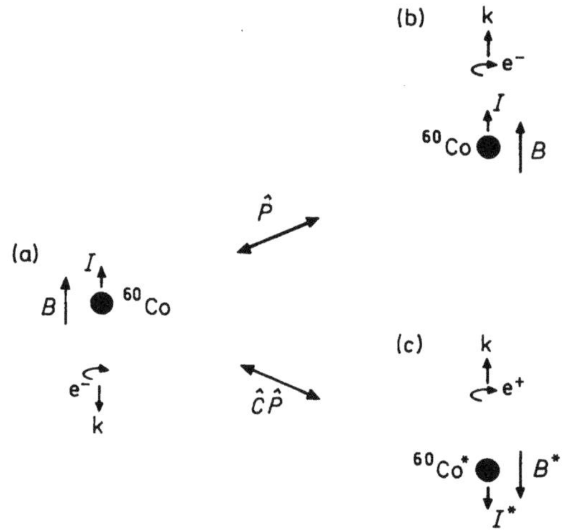

Fig. 2a–c. Parity violation in β-decay. Only experiment (**a**) is found; the space-inverted version (**b**) cannot be realized. Symmetry is recovered in experiment (**c**), which obtains from (**a**) by invoking charge conjugation simultaneously with space inversion (Co* is anti-Co, and $\boldsymbol{B}^*$ and $\boldsymbol{I}^*$ are reversed relative to $\boldsymbol{B}$ and $\boldsymbol{I}$ because the charges of the moving source particles have reversed)

unequivocal evidence for parity violation. Another important aspect of β-decay is that the emitted electrons have a "left-handed" longitudinal spin polarization, being accompanied by "right-handed" antineutrinos. The corresponding antiparticles emitted in β-decays, namely positrons and neutrinos, have the opposite handedness. (The projection of the spin angular momentum s of a particle along its direction of motion is called the helicity, $\lambda = \boldsymbol{s} \cdot \boldsymbol{p}/|\boldsymbol{p}|$. Spin $^1/_2$ particles can have $\lambda = \pm{}^1/_2\hbar$, the positive and negative states being called right- and left-handed; but this corresponds to the opposite sence of circularity to that used in the usual definition of right- and left-circularly polarized light).

In fact symmetry is recovered by invoking charge conjugation simultaneously with space inversion: the missing experiment is to be found in the antiworld! Thus it can be seen from Fig. 2c that the combined operation of $\hat{C}\hat{P}$ interconverts two equivalent experiments for which nature appears to have no preference (assuming $\hat{T}$ is not violated). This result implies that P-violation is accompanied here by $\hat{C}$-violation since absolute charge is distinguished: the charge that we call negative is carried by the electrons, which are emitted with a left-handed spin polarization.

Notice that the Wu experiment provides a good example of true chirality, as defined in Sect. 3.2. The two experiments (a) and (b) in Fig. 2 are enantiomeric with respect to space inversion, but cannot be interconverted by time reversal combined with any proper spatial rotation.

5.2 Parity Violation in Atoms and Molecules: The Weak Neutral Current

Since the electromagnetic interaction is formulated in terms of an exchange of virtual photons, it was natural to postulate the existence of a particle, denoted W, that mediated the weak interaction. Like the photon, the W is a boson; but unlike the photon, which is neutral, the W must be charged (W^+ or W^-) since β-decay, for example, involves an exchange of charge between particles. A second difference is that, whereas photons have zero mass, the Ws are massive (this follows from the Yukawa-Wick argument that the range of a force is inversely proportional to the mass of the exchanged quantum: the electromagnetic and weak interactions have infinite and very short ranges, respectively).

Following the Wu experiment, the original Fermi theory of the weak interaction [68] was upgraded in order to take account of parity violation. This was achieved by reformulating the theory in such a way that the interaction takes the form of a left-handed pseudoscalar. However, a number of technical problems remained, which were finally overcome in the 1960s in the celebrated work of Weinberg [69], Salam [70] and Glashow [71], which unified the weak and electromagnetic interactions into a single electroweak interaction. The conceptual basis of the theory rests on two pillars: gauge invariance and spontaneous symmetry breaking [72, 73], but the details are beyond the scope of this article. In addition to accommodating the massless photon and the two massive charged W^+ and W^- particles, a new massive neutral particle called Z° (the neutral intermediate vector boson) was predicted which can generate a whole new range of *neutral current* phenomena, including parity-violating effects in atoms and molecules. The theory provides a simple relation between the weak and electromagnetic coupling constants ($g \sin \theta_w = e$, where g and e are the weak and electromagnetic unit charges, and θ_w is the Weinberg angle), and also gives the masses of the W^+, W^- and Z°. In one of the most important experiments of all time, these three particles were detected in 1983 at CERN in proton-antiproton scattering experiments [74].

This weak neutral current generates parity-violating interactions between electrons, and between electrons and nucleons. The latter leads to the following electron-nucleus contact interaction in atoms and molecules (in a. u. where $\hbar = e = m_e = 1$) [75, 76]:

$$\hat{V}_{eN}^{PV} = \frac{G\alpha}{4\sqrt{2}} Q_W \{\boldsymbol{\sigma}_e \cdot \boldsymbol{p}_e, \varrho_N(\boldsymbol{r}_e)\}_+ , \tag{23 a}$$

where G is the Fermi weak coupling constant, α is the fine structure constant, $\boldsymbol{\sigma}_e$ and $\boldsymbol{p}_e$ are the Pauli spin operator and linear momentum operator of the electron, $\varrho_N(\boldsymbol{r}_e)$ is a normalised nuclear density function and

$$Q_W = Z(1 - 4\sin^2\theta_W) - N \tag{23 b}$$

is an effective weak charge which depends on the proton and neutron numbers Z and N. $\{\ \}_+$ denotes an anticommutator. The electron-electron interaction is usually neglected, so (23) is taken as the parity-violating term to be added to the Hamiltonian of an atom or molecule. Since $\boldsymbol{\sigma}_e$ and $\boldsymbol{p}_e$ are axial and polar vectors, respectively, and all other factors are scalars, $\hat{V}_{eN}^{PV}$ transforms as a pseudoscalar, as required, and so can mix even and odd parity electronic states at the nucleus.

Manifestations of parity violation in atoms have now been observed in the form of optical activity phenomena such as tiny optical rotations in vapours of heavy metals [77, 78].

Chiral molecules support a unique manifestation of parity violation in the form of a lifting of the exact degeneracy of the energy levels of mirror-image enantiomers [76, 79–83]. Being pseudoscalars, the parity-violating weak neutral current terms in the molecular Hamiltonian are odd under space inversion:

$$\hat{P}\hat{V}^{\mathrm{PV}}\hat{P}^{-1} = -\hat{V}^{\mathrm{PV}} . \tag{24}$$

As discussed in Sect. 6, the enantiomeric quantum states ψ_{L} and ψ_{R} of a chiral molecule are examples of the mixed parity states (7) and so are interconverted by $\hat{P}$. It then follows that $\hat{V}^{\mathrm{PV}}$ shifts the energies of the enantiomeric states in opposite directions:

$$\begin{aligned}\langle\psi_{\mathrm{L}}|\,\hat{V}^{\mathrm{PV}}\,|\psi_{\mathrm{L}}\rangle &= \langle\hat{P}\psi_{\mathrm{R}}|\,\hat{V}^{\mathrm{PV}}\,|\hat{P}\psi_{\mathrm{R}}\rangle = \langle\psi_{\mathrm{R}}|\,\hat{P}^{\dagger}\hat{V}\hat{P}\,|\psi_{\mathrm{R}}\rangle \\ &= -\langle\psi_{\mathrm{R}}|\,\hat{V}^{\mathrm{PV}}\,|\psi_{\mathrm{R}}\rangle = \varepsilon .\end{aligned} \tag{25}$$

Attempts to calculate ε are faced with the following difficulty. The electronic coordinate part of $\hat{V}^{\mathrm{PV}}_{\mathrm{eN}}$ in (23) is linear in $\boldsymbol{p}_{\mathrm{e}}$ and is therefore pure imaginary. Since, in the absence of external magnetic fields, the molecular wavefunction can always be chosen to be real, $\hat{V}^{\mathrm{PV}}_{\mathrm{eN}}$ has zero expectation values. Also, the presence of $\boldsymbol{\sigma}_{\mathrm{e}}$ means that only matrix elements between different spin states survive. Consequently, it is necessary to invoke a magnetic perturbation of the wavefunction involving spin, the favourite candidate being spin-orbit coupling [76, 81, 82]. This leads to a tractable method for detailed quantum-chemical calculations of parity-violating energy differences between enantiomers, giving values of the order 10^{-20} a. u. [76, 84, 85]. (The atomic unit of energy, the Hartree, is equivalent to 27.2 eV or to 4.36×10^{-18} J). There are also slight structural differences between enantiomers [86]. It is intriguing that, in all the cases treated so far, the L-amino acids and the D-sugars, which dominate the biochemistry of living organisms, are found to be the more stable enantiomers [87, 88].

5.3 Violation of Time Reversal and the $\hat{C}\hat{P}\hat{T}$ Theorem

Violation of time reversal was observed in the famous experiment of Cronin, Fitch et al. in 1964 [89]. The measurements involved decay modes of the neutral K-meson [90, 91]; but despite intensive effort since then, no other system has shown the effect. As Cronin has said [90], nature has provided us with just one extraordinarily sensitive system to convey a cryptic message that has still to be deciphered.

Although unequivocal, the effects are very small; certainly nothing like the parity-violating effects in weak processes, which can sometimes be absolute. In fact $\hat{T}$ violation itself is not observed directly: rather the observations show $\hat{C}\hat{P}$ violation, from which $\hat{T}$ violation is implied by the celebrated $\hat{C}\hat{P}\hat{T}$ theorem. The $\hat{C}\hat{P}\hat{T}$ theorem is derived from general considerations within relativistic quantum field theory [13, 27], and states that the Hamiltonian is invariant to the combined operations of $\hat{C}\hat{P}\hat{T}$ even if it is not invariant to one or more of those operations.

One manifestation of $\hat{C}\hat{P}$ violation is the following decay rate asymmetry of the long-lived neutral K-meson, K_L^o [13, 73]:

$$a = \frac{\text{Rate}\,(K_L^o \rightarrow \pi^- e_r^+ \nu_l)}{\text{Rate}\,(K_L^o \rightarrow \pi^+ e_l^- \tilde{\nu}_r)} \approx 1.00648 \,. \tag{26}$$

As the formula indicates, K_L^o can decay into either positive pions π^+, left helical electrons e_l^- and right helical antineutrions $\tilde{\nu}_r$; or into negative antipions π^-, right helical positrons e_r^+ and left helical neutrinos ν_l. Since these two sets of decay products are interconverted by $\hat{C}\hat{P}$, this decay rate asymmetry indicates that $\hat{C}\hat{P}$ is violated. If we naively represent this decay process in the form of "chemical equilibria" as in (17),

$$\pi^+ + e_l^- + \tilde{\nu}_r \underset{k_b}{\overset{k_f}{\rightleftharpoons}} K_L^0 \underset{\bar{k}_f}{\overset{\bar{k}_b}{\rightleftharpoons}} \pi^- + e_r^+ + \nu_l \,, \tag{27}$$

a parallel is established with absolute asymmetric synthesis associated with a breakdown of microscopic reversibility discussed in Sect. 4.2 since in both cases $k_f \neq \bar{k}_f$. Thus the K^o and the two sets of decay products are the equivalents, with respect to $\hat{C}\hat{P}$, of R, M and $\bar{M}$ with respect to $\hat{P}$. We can therefore conceptualize the decay rate asymmetry here as arising from a breakdown in microscopic reversibility due to a time-noninvariant $\hat{C}\hat{P}$ enantiomorphism in the forces of nature [60] (the $\hat{C}\hat{P}\hat{T}$ theorem guarantees that the two distinct $\hat{C}\hat{P}$ enantiomorphous influences are interconverted by $\hat{T}$). The analogy is completed by the fact that, as mentioned in Sect. 4.3, the asymmetries cancel out over all possible channels at true thermodynamic equilibrium.

Another manifestation of $\hat{C}\hat{P}$ violation arises that should be mentioned, but first more needs to be said about the K^o system. Four distinct states are displayed: particle and antiparticle states $|K^o\rangle$ and $|\tilde{K}^o\rangle$, and two combined states

$$|K_1^o\rangle = \frac{1}{\sqrt{2}}(|K^o\rangle + |\tilde{K}^o\rangle)\,, \tag{28a}$$

$$|K_2^o\rangle = \frac{1}{\sqrt{2}}(|K^o\rangle - |\tilde{K}^o\rangle)\,, \tag{28b}$$

which have different energies because of coupling between $|K^o\rangle$ and $|\tilde{K}^o\rangle$ via the weak force. Since the particle and antiparticle states are interconverted by $\hat{C}\hat{P}$,

$$\hat{C}\hat{P}|K^o\rangle = |\tilde{K}^o\rangle, \qquad \hat{C}\hat{P}|\tilde{K}^o\rangle = |K^o\rangle\,, \tag{28c}$$

we can appreciate that the combined states are even and odd eigenstates of $\hat{C}\hat{P}$:

$$\hat{C}\hat{P}|K_1^o\rangle = |K_1^o\rangle, \qquad \hat{C}\hat{P}|K_2^o\rangle = -|K_2^o\rangle\,. \tag{28d}$$

However, the $\hat{C}\hat{P}$ eigenstates are not pure: for example $|K_2^o\rangle$, which is odd with respect to $\hat{C}\hat{P}$, is occasionally observed to decay into products which are even with respect to $\hat{C}\hat{P}$. This implies that the Hamiltonian contains a small $\hat{C}\hat{P}$-violating term which mixes $|K_1^o\rangle$ and $|K_2^o\rangle$. In fact the long-lived state $|K_L^o\rangle$ introduced above is actually $|K_2^o\rangle$ with a small admixture of $|K_1^o\rangle$; and there is an associated short-lived state $|K_S^o\rangle$ that is $|K_1^o\rangle$ with a small admixture of $|K_2^o\rangle$. One example is the decay into the pair of oppositely charged pions $\pi^+\pi^-$: the state of zero angular momentum of this pair is even with respect to $\hat{C}\hat{P}$ so only the decay of $|K_1^o\rangle$ should yield such a product state. What is found is a very small but significant amplitude for K_L^o to decay into $\pi^+\pi^-$:

$$\eta_{+-} = \frac{\text{amplitude}\,(K_L^o \to \pi^+\pi^-)}{\text{amplitude}\,(K_S^o \to \pi^+\pi^-)} \approx 2.274 \times 10^{-3}\,. \tag{29}$$

The $\hat{C}\hat{P}$-violating parameters a and η_{+-} extracted from the two at first sight rather different experiments summarized by (26) and (29), together with a third parameter η_{00} derived from a similar experiment to (29), are in fact all related, and the results of all the experiments involving $\hat{C}\hat{P}$ violation can be summarized by a single complex number [90].

6 The Mixed Parity of a Chiral Molecule

It was shown above (Sects. 2.3 and 3.1) that, since a chiral molecule can support pseudoscalar observables, it must exist in mixed parity internal quantum states (vibrational-electronic) associated with a chiral molecular framework. We now explore the nature of these quantum states, and investigate the consequences of a small parity-violating term in the Hamiltonian.

6.1 The Double Well Model

The origin of these mixed parity states is well known. It is best appreciated by considering vibrational wavefunctions associated with the "inversion" mode ν_2 of a molecule such as NH_3 which is said to invert between the two equivalent configurations shown in Fig. 3 [92, 93], although this motion does not in fact correspond to an inversion through the centre of mass. If the planar configuration were the most stable, the adiabatic potential energy function would have the parabolic form shown on the left with simple harmonic vibrational levels equally spaced. If a potential hill is raised gradually in the middle, the two pyramidal configurations become the most stable and the energy levels approach each other in pairs. For an infinitely high potential hill, the pairs of energy levels are exactly degenerate, as shown on the right. The rise of the central potential hill modifies the wavefunctions as shown, but does not destroy their parity. The even and odd parity wavefunctions $\psi(+)$ and $\psi(-)$ describe stationary states in all circumstances. On the other hand, the wavefunctions ψ_L and ψ_R, corresponding to the system in its lowest state of oscillation and localized completely in the left and right wells, respectively, are not true stationary states. They

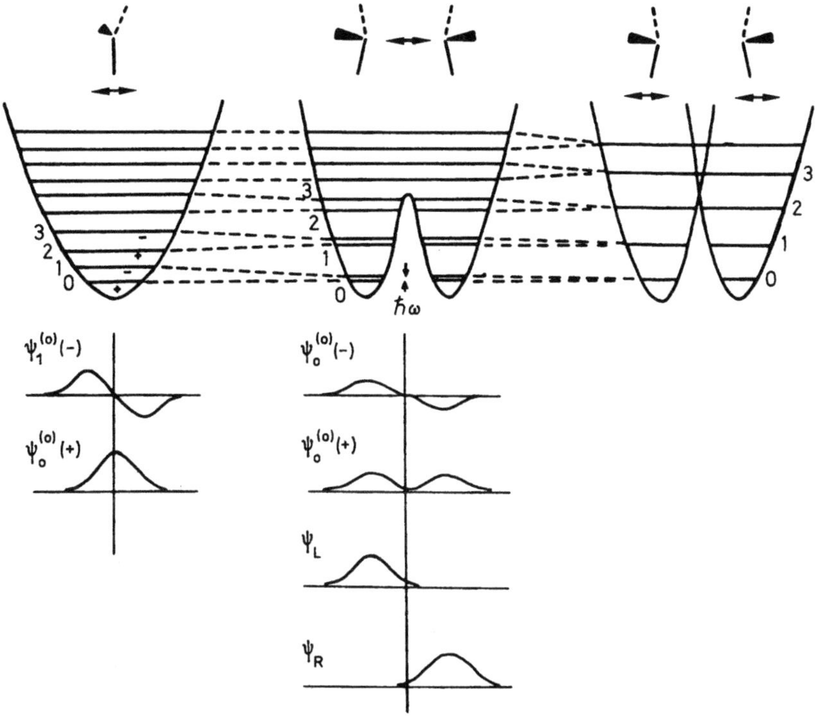

Fig. 3. The vibrational states of a molecule that can invert between two equivalent configurations. $\psi^{(o)}(+)$ and $\psi^{(o)}(-)$ are the amplitudes of the definite parity stationary states with energy $W(+)$ and $W(-)$, and ψ_L and ψ_R are the two mixed parity non-stationary states at $t = 0$ and $t = \pi/\omega$, where $\hbar\omega$ is the tunnelling splitting

are obtained from the following combinations of the even and odd parity wavefunctions

$$\psi_L = \frac{1}{\sqrt{2}} [\psi^{(0)}(+) + \psi^{(0)}(-)], \tag{30a}$$

$$\psi_R = \frac{1}{\sqrt{2}} [\psi^{(0)}(+) - \psi^{(0)}(-)], \tag{30b}$$

which are explicit examples of the general mixed parity wavefunctions (7).

The wavefunctions (30) are in fact specializations of the general time-dependent wavefunction for a degenerate two-state system (see Sect. 6.2). To be precise, let us assume that the system is in the left well at $t = 0$. Then at a later time we have [26]

$$\begin{aligned}\psi(t) &= \frac{1}{\sqrt{2}} [\psi^{(0)}(+)\, e^{-iW(+)t/\hbar} + \psi^{(0)}(-)\, e^{-iW(-)t/\hbar}] \\ &= \frac{1}{\sqrt{2}} [\psi^{(0)}(+) + \psi^{(0)}(-)\, e^{-i\omega t}]\, e^{-iW(+)t/\hbar}, \end{aligned} \tag{31}$$

where $\hbar\omega = W(-) - W(+)$ is the energy separation of the two opposite parity states, which in this context is interpreted as a splitting arising from tunnelling through the potential energy barrier separating the two wells. Thus at $t = 0$ (31) reduces to (30a) corresponding to the molecule being found in the left well, as required; and at $t = \pi/\omega$ (31) reduces to (30b) corresponding to the right well. The angular frequency ω is interpreted as the frequency of a complete inversion cycle. The tunnelling splitting $\hbar\omega$ is determined by the height and width of the barrier, and is zero if the barrier is infinite.

It is emphasized that a splitting of the energy levels will occur for any normal mode of vibration in which the height of the pyramid changes [93]. In NH_3, for example, the height of the pyramid changes somewhat in each of the four normal modes, although the "inversion" mode ν_2 discussed above shows the greatest change and the corresponding energy levels the greatest splitting.

One source of confusion in this model is that the parity of the vibrational wavefunctions is defined with respect to a *reflection* σ across the plane of the nuclei [92],

$$\sigma\psi_{\text{v}} = (-1)^{\text{v}}\,\psi_{\text{v}}, \tag{32}$$

where v is the vibrational quantum number (the normal vibrational coordinate for ν_2 changes sign under σ); whereas the basic definition of the parity operation is an *inversion* with respect to space-fixed axes. In the conventional treatment of inverting non-planar symmetric tops [92, 94], the rotational wavefunction of a planar symmetric top such as BF_3 is multiplied by the time-dependent wavefunction (31) corresponding to the "inversion" vibration. The parity operation corresponds to an inversion of all the particle positions (nuclei and electrons), and is achieved by rotating the complete BF_3 molecule through π about the threefold axis, followed by a reflection across the plane containing the nuclei. Since the rotation is an external affair, it affects only the rotational wavefunctions and is used to classify their parity. The reflection is a purely internal affair, so the parity of the vibrational-electronic parts of the quantum state is determined by their behaviour under reflection across the plane of the nuclei. This sort of consideration has been placed on a more sophisticated footing by the use of permutation-inversion groups to specify the parity of the complete wavefunction of a general non-rigid molecule in the gas phase [95, 96].

Since an analogous potential energy diagram can be drawn for any chiral molecule with a high barrier separating left and right wells which now correspond to the two enantiomeric states, we now have a model for the source of the mixed parity internal (vibrational-electronic) states of a resolved enantiomer. The horizontal axis might represent the position of an atom above a plane containing three different atoms, the torsion coordinate of a chiral biphenyl, or some more complicated collective coordinate of the molecule. If such a state is prepared, but the tunnelling splitting is finite, its energy will be indefinite because it is a superposition of two opposite parity states of different energy. The splitting of the two definite parity states, and hence the uncertainty in the energy of an enantiomer, is inversely proportional to the left-right conversion time π/ω (this is an explicit example of the general result that the width of an energy level corresponding to a quasi-stationary state with average lifetime T is $\Delta W = \hbar/T$ [97]).

A crucial point is therefore the relation between the time scale of the optical activity measurement and the lifetime of the resolved enantiomer. A manifestation of the uncertainty principle appears to arise here, which I have stated loosely as follows [12, 98]: if, for the duration of the measurement, there is complete certainty about the enantiomer, there is complete uncertainty about the parity of its quantum state; whereas if there is complete uncertainty about the enantiomer, there is complete certainty about the parity of its quantum state. Thus experimental resolution of the definite parity states of tartaric acid, say, an enantiomer of which has a lifetime probably greater than the age of the universe, is impossible unless the duration of the experiment is virtually infinite; whereas for a non-resolvable chiral molecule such as H_2O_2, spectroscopic transitions between states of definite parity are observed routinely.

6.2 Two-state Systems and Parity Violation

We have just seen how the mixed parity states of a resolved chiral molecule can be pictured in terms of a double well potential. This aspect can be developed further by considering the quantum mechanics of a degenerate two-state system in order to gain insight into the apparent paradox of the stability of optical enantiomers, which was recognized at the beginning of the quantum era since the existence of optical enantiomers was difficult to reconcile with basic quantum mechanics. In the words of Hund [99].

If a molecule admits two different nuclear configurations being the mirror images of each other, then the stationary states do not correspond to a motion around one of these two equilibrium configurations. Rather, each stationary state is composed of left-handed and right-handed configurations in equal shares . . . The fact that the right-handed or left-handed configuration of a molecule is not a quantum state (eigenstate of the Hamiltonian) might sppear to contradict the existence of optical isomers.

Similarly Rosenfeld [100]:

A system (state) with sharp energy is optically inactive.

And Born and Jordan [101]:

Since each molecule consists of point charges interacting via Coulomb's law, the energy function (Hamiltonian) is always invariant with respect to space inversion. Consequently there could not exist any optically active molecules, which contradicts experience.

These translated quotations are taken from a review by Pfeifer [102]. These points have been emphasized recently by Woolley as a part of a re-evaluation of the whole concept of molecular structure [103–106].

Hund's resolution of the paradox involves arguments of the type given in the previous section, namely that typical chiral molecules have such large barriers to inversion that the lifetime of a prepared enantiomer is virtually infinite. Recently, Hund's approach has been brought up to date by injecting a small parity-violating term into the Hamiltonian, which can result in the two enantiomeric states becoming the true stationary states [12, 82, 83].

For a general two-state system in the orthonormal basis (ψ_1, ψ_2), not necessarily degenerate, the exact energy eigenvalues and eigenfunctions corresponding to the true stationary states are [15, 107]

$$W_{\pm} = \frac{1}{2}(H_{11} + H_{22}) \pm \frac{1}{2}[(H_{11} - H_{22})^2 + 4\,|H_{12}|^2]^{1/2}, \tag{33a}$$

$$\psi_{+}^{(0)} = \cos\theta\, e^{-i\varphi/2}\, \psi_1 + \sin\theta\, e^{i\varphi/2}\, \psi_2\,, \tag{33b}$$

$$\psi_{-}^{(0)} = -\sin\theta\, e^{-i\varphi/2}\, \psi_1 + \cos\theta\, e^{i\varphi/2}\, \psi_2\,, \tag{33c}$$

where

$$\tan 2\theta = \frac{2\,|H_{12}|}{(H_{11} - H_{22})} \quad \text{with} \quad 0 \leqq 2\theta < \pi\,, \tag{33d}$$

$$H_{21} = |H_{21}|\, e^{i\varphi}\,. \tag{33e}$$

The superscripts (o) denote the amplitudes of the corresponding time-dependent wavefunctions, and $H_{ab} = \langle\psi_a^{(o)}|\,\hat{H}\,|\psi_b^{(o)}\rangle$ are matrix elements of the total Hamiltonian of the system. The subscripts $\pm$ here denote higher and lower energy levels, not the parity.

Thus ψ_1 and ψ_2 are not the eigenstates (stationary states) of the Hamiltonian of the system and so couple with each other through H_{21}; whereas the stationary states ψ_+ and ψ_- do not. So if a two-state system is prepared in a non-stationary state ψ_1 or ψ_2, it might appear falsely to be influenced by a time-dependent perturbation lacking some fundamental symmetry of the internal Hamiltonian of the system. In general, ψ_1 and ψ_2 will be interconverted by a particular symmetry operation of the Hamiltonian, whereas $\psi_+^{(o)}$ and $\psi_-^{(o)}$ will transform according to one or other of the irreducible representations of the symmetry group comprising the identity and the operation in question.

By restricting attention to the ground and first excited state of the normal mode of vibration that interconverts the enantiomers in Fig. 3, we can identify ψ_1 and ψ_2 with ψ_R and ψ_L. If the small parity-violating terms in the Hamiltonian are neglected, the Hamiltonian has inversion symmetry, and since $\hat{P}\psi_R = \psi_L$ and $\hat{P}\psi_L = \psi_R$, the enantiomeric state ψ_R and ψ_L are degenerate. The stationary state amplitudes (33b and c) now specialize to

$$\psi_{+}^{(0)} = \frac{1}{\sqrt{2}}\, e^{-i\varphi/2}\, (\psi_R + e^{i\varphi}\, \psi_L)\,, \tag{34a}$$

$$\psi_{-}^{(0)} = \frac{1}{\sqrt{2}}\, e^{-i\varphi/2}\, (-\psi_R + e^{i\varphi}\, \psi_L)\,, \tag{34b}$$

and so transform according to one or other of the irreducible representations of the inversion group comprising $\hat{P}$ plus the identity. Which has even and which has odd parity depends on the choice of φ: for example if $\varphi = \pi$ (so that H_{LR} is real and negative), $\psi_+^{(o)}$ is odd and $\psi_-^{(o)}$ is even. The separation of the two stationary states is simply twice the coupling energy of the two enantiomeric states,

$$W_+ - W_- = 2\,|\langle\psi_L|\,\hat{H}\,|\psi_R\rangle| = 2\delta\,, \tag{35}$$

and is interpreted as a splitting caused by tunnelling through the potential energy barrier separating the two enantiomers (Fig. 3).

We now allow the Hamiltonian to contain a small parity-violating term $\hat{V}^{PV}$ such as the electron-nucleus weak neutral current interaction (23a). According to (25) this shifts the energies of the two enantiomeric states in opposite directions by an amount ε. The enantiomeric states are now no longer degenerate, so using the general two-state results (33) we have

$$W_+ - W_- = 2(\varepsilon^2 + \delta^2)^{1/2}\,, \tag{36a}$$

$$\tan 2\theta = \delta/\varepsilon\,. \tag{36b}$$

The general time-dependent wavefunction is given by the sum of each stationary state amplitude multiplied by its exponential time factor:

$$\psi(t) = \frac{1}{\sqrt{2}}\, e^{-i\varphi/2}\, \{(\cos\theta\psi_R + \sin\theta\, e^{i\varphi}\,\psi_L)\, e^{-iW_+t/\hbar} + (-\sin\theta\psi_R + \cos\theta\, e^{i\varphi}\,\psi_L)\, e^{-iW_-t/\hbar}\}\,. \tag{37}$$

This only has a simple interpretation in the two limits of $\varepsilon = 0$ and $\delta = 0$. When $\varepsilon = 0$ (zero parity violation)

$$\psi(t) = \frac{1}{2}\, e^{-i\varphi/2}\, \{(\psi_R + e^{i\varphi}\,\psi_L)\, e^{-i\delta t/\hbar} + (-\psi_R + e^{i\varphi}\,\psi_L)\, e^{i\delta t/\hbar}\}\, e^{-i(W_+ + W_-)t/2\hbar} \tag{38}$$

which reduces, within a phase factor, to (31) (do not confuse the notation $\psi_\pm$ and $W_\pm$ for higher- and lower-energy states with $\psi(\pm)$ and $W(\pm)$ for even- and odd-parity states). Thus at $t = 0$ the system is entirely in ψ_L and at $t = \pi\hbar/2\delta$ it is entirely in ψ_R; the system oscillates between ψ_L and ψ_R with $2\delta/\hbar$ being the frequency of a complete inversion cycle and $\psi_+ = \psi(-)$ and $\psi_- = \psi(+)$ the stationary states. But when $\delta = 0$ (zero tunnelling splitting)

$$\psi(t) = \frac{1}{\sqrt{2}}\, e^{-i\varphi/2}\, \{\psi_R\, e^{-i\varepsilon t/\hbar} + e^{i\varphi}\,\psi_L\, e^{i\varepsilon t/\hbar}\}\, e^{-i(W_+ + W_-)t/2\hbar}\,, \tag{39}$$

so at $t = 0$ the system is entirely in ψ_+ and at $t = \pi\hbar/2\varepsilon$ it is entirely in ψ_-: now the system oscillates between ψ_+ and ψ_- with $2\varepsilon/\hbar$ being the frequency of a complete cycle and ψ_R and ψ_L the stationary states.

The time-depenendence of the optical activity observable depends on the nature of the state in which the molecule is prepared initially. Consider first the molecule prepared in a handed state ψ_L or ψ_R, which means that at $t = 0$ the state is given by (30a) or (30b), respectively. At some later time t the corresponding states will be

$$\psi_L(t) = e^{-i\varphi/2} (\sin \theta\psi_+^{(0)} e^{-iW_+t/\hbar} + \cos \theta\psi_-^{(0)} e^{-iW_-t/\hbar}), \quad (40a)$$

$$\psi_R(t)' = e^{i\varphi/2} (\cos \theta\psi_+^{(0)} e^{-iW_+t/\hbar} - \sin \theta\psi_-^{(0)} e^{-iW_-t/\hbar}), \quad (40b)$$

which are obtained by inverting (33b and c) and multiplying each stationary state amplitude by its exponential time factor. Thus for a molecule prepared in ψ_L, the time-dependence of the optical rotation angle is given by [12, 82, 83]

$$\alpha(t) = \alpha_L \left\{ \frac{\varepsilon^2 + \delta^2 \cos[2(\delta^2 + \varepsilon^2)^{1/2} t/\hbar]}{\delta^2 + \varepsilon^2} \right\}, \quad (41)$$

where α_L is the optical rotation angle of the left-handed enantiomer. So if $\varepsilon \neq 0$, the optical rotation oscillates asymmetrically (but if $\varepsilon = 0$ it oscillates between equal and opposite values associated with the two enantiomers). Taking the time average, we find

$$\frac{\bar{\alpha}}{\alpha_{max}} = \frac{\varepsilon^2}{\delta^2 + \varepsilon^2}. \quad (42)$$

Thus parity violation causes a shift away from zero of $\bar{\alpha}$. This is the basis of an experiment suggested by Harris and Stodolsky [82] to detect the parity-violating energy shift between enantiomers.

But if the molecule is prepared in one of the stationary states

$$\psi_+(t) = (\cos \theta\, e^{-i\varphi/2} \psi_R + \sin \theta\, e^{i\varphi/2} \psi_L)\, e^{-iW_+t/\hbar}, \quad (43a)$$

$$\psi_-(t) = (-\sin \theta\, e^{-i\varphi/2} \psi_R + \cos \theta\, e^{i\varphi/2} \psi_L)\, e^{-iW_-t/\hbar}, \quad (43b)$$

the optical rotation will be given by

$$\alpha_+(t) = -\alpha_-(t) = -\alpha_L \frac{\varepsilon}{(\varepsilon^2 + \delta^2)^{1/2}}. \quad (44)$$

Thus if $\varepsilon = 0$, the optical rotation will be zero, as required, since the stationary states will have definite parity; but if $\varepsilon \neq 0$, the stationary states will acquire equal and opposite parity-violating optical rotation that does not change with time.

It is clear from (37) and (33) that, as $\delta/\varepsilon \to 0$, ψ_L and ψ_R become the true stationary stares. In fact for typical chiral molecules, δ corresponds to tunnelling times of the order of millions of years: Harris and Stodolsky [82] have estimated ε to correspond to times of the order of seconds to days, so at low temperature and in a vacuum, a prepared enantiomer will retain its handedness essentially for ever. So the ultimate answer to the paradox of the stability of optical enantiomers might lie in the weak interactions.

6.3 Parity Violation and Parity Breaking

The existence of parity-violating phenomena is interpreted quantum-mechanically by saying that the Hamiltonian has a lower symmetry than previously thought (since the weak interaction potential is a pseudoscalar). This means that $\hat{P}$ and $\hat{H}$ no longer commute, so the corresponding conservation law no longer holds. Such symmetry *violation* must be clearly distinguished from symmetry *breaking*: current usage in the physics literature applies the latter term to describe the situation arising when a system displays a lower symmetry than expected from its Hamiltonian [108]. Natural optical activity is therefore a phenomenon arising from parity breaking since, as we have just seen, a resolved chiral molecule displays a lower symmetry than its associated Hamiltonian: if the small parity-violating term in the Hamiltonian is neglected, the symmetry operation that the Hamiltonian possesses but the chiral molecule lacks is inversion, and it is this inversion operation that interconverts the two enantiomeric parity-broken states.

The conventional view, expressed in Sect. 6.1, is that parity violation plays no part in the stabilization of chiral molecules. The optical activity is assumed to remain observable only so long as the observation time is short compared with the interconversion time between enantiomers, which is proportional to the inverse of the tunnelling splitting. Such parity-breaking optical activity therefore averages to zero over a sufficiently long observation time. Thus the statement that [109] "... processes involving pseudoscalar quantities will not obey the law of parity" betrays a common misconception: the law of parity is saved in systems displaying broken parity because their pseudoscalar properties average to zero over a sufficiently long observation period or, equivalently, the space-inverted experiment is realizable. In either interpretation absolute chirality is not observable.

A related misconception has arisen in connection with the correlation of entities that have opposite parities, such as the coupling of rotational and translational motion in collections of chiral molecules. Thus it has been stated that [110] "... in a system such that the Hamiltonian has inversion symmetry, properties of different parity are totally uncorrelated for all time"; also "... if the system contains optically active molecules, the Hamiltonian does not have even parity and none of these theorems apply". Any Hamiltonian involving only electromagnetic interactions always has even parity: it is the broken parity states of the chiral molecules that mediate the coupling of opposite parity entities, and any associated pseudoscalar properties will therefore average to zero over a sufficiently long period of time.

This brings us to a crucial distinction between parity-breaking and parity-violating natural optical activity phenomena: the former are time-dependent and average to zero; the latter are constant in time (recall the stationary states acquiring time-independent optical activity in the previous section when $\varepsilon \neq 0$). Hence if a small chiral molecule could be isolated sufficiently from the environment, a parity-violating element is indicated if the optical activity remains observable for longer than the expected interconversion time [111].

6.4 Symmetry Breaking in Isolated Molecules and in Condensed Media

Symmetry breaking has attracted much attention in recent years in both elementary particle and condensed matter physics, but with rather different emphasis on the various aspects. Anderson has written at length on broken symmetry in condensed matter, with some valuable asides on the molecular aspects [112, 113].

Ferromagnetism provides an important example. The Hamiltonian for an iron crystal is invariant under spatial rotations. But the ground state is not invariant: it distinguishes a specific direction, the direction of magnetization. A non-zero magnetization in zero applied field also breaks time reversal symmetry. When the temperature is raised above the Curie point, the magnetization disappears and the rotational and time reversal symmetries become manifest. The term spontaneous symmetry breaking that is often encountered is borrowed from the term spontaneous magnetization. Notice that the rotational symmetry is still present in the ferromagnetic phase in that the sense of magnetization is arbitrary; but this would be hidden from an observer living inside the crystal.

Temperature is a central feature here, because behaviour reflecting the full symmetry of the Hamiltonian can be recovered at sufficiently high temperature. Molecules behave rather differently from macroscopic systems in that there are no sharp transitions between symmetric and asymmetric states in molecules [113]. For example, in a molecule described by the double well model, thermal agitation will cause the "inversion" transition to take place (unless the barrier is effectively infinite) so that there is no absolute "one-sidedness" at any temperature: in other words its broken parity characteristics decrease continuously with increasing temperature. In condensed media, on the other hand, large number of particles can cooperate to produce sudden rather than gradual thermal transitions between symmetric and asymmetric states of the complete macroscopic sample. In a ferroelectric crystal below its phase transition temperature, for example, the field associated with each molecular dipole (or the dipole associated with each unit cell) acts to hold others in the same direction: the reversal of any one will not reverse the total but simply represents a local fluctuation. At the phase transition temperature, a sufficient number of dipoles are reversed so that the system suddenly transforms into a state where the sign of each occurs with equal probability. Thus a distinction must be made between broken symmetry in a macroscopic system and in the individual microscopic constituents, although in many cases, including the example of ferroelectricity considered here, broken symmetry in the microscopic constituents (so that they are dipolar) is a prerequisite for the macroscopic broken symmetry state associated with ferroelectricity. This distinction between microscopic and macroscopic broken symmetry can sometimes become rather blurred, as in superconductivity [113].

There has been much discussion concerning the rôle of the environment in the problem of the stabilization of enantiomeric molecules and of molecular chirality generally [102–106, 111, 114–117]. While most would agree that the environment must have some influence, Woolley has argued that environment is everything so that "optical activity has to be understood in a macroscopic context as a loss of inversion symmetry of the whole material medium, and that chirality is not a property that can be related to isolated molecules (the rotational strength for an isolated molecule vanishes identically if one ignores the weak neutral current)" [105]. Here an isolated molecule means a closed system of electrons and nuclei that evolve in time in vacuum. The essential ingredient, according to Woolley, is that molecules are coupled to an "environment" that is described by a quantum field theory [118]: This "environment" may be given a traditional representation as a "reaction field" [116], or all other molecules in the substance apart from a reference molecule [106], but it may also be the quantized electromagnetic field [102] or other boson fields (e.g. phonons) [119]. In my view, Woolley is suggesting that the dynamical mechanism causing optical activity is qualitatively similar to that producing broken symmetry in condensed matter; moreover that it corresponds to the special situation where the in condensed marter; moreover that it corresponds to the special situation where the individual molecules and that of the bulk sample) is lost. Although I am not yet entirely convinced about this particular viewpoint [120, 121], it certainly merits careful consideration.

7 Chirality and Relativity

In Sect. 3.3 it was demonstrated that a spinning cone or sphere translating along the axis of spin possesses true chirality. This is an interesting concept because it exposes a link between chirality and special relativity. Suppose a particle is moving away from an observer with a right-handed helicity. If the observer accelerates to a sufficiently high velocity that he starts to catch up with the particle, it will now appear to be moving towards the observer and so takes on a left-handed helicity. In its rest frame the helicity of the particle is undefined and its chirality vanishes. Only for massless particles such as photons and neutrinos is the chirality conserved since they always move at the velocity of light in any reference frame.

Indeed, this relativistic aspect of chirality is a central feature of modern elementary particle theory, especially in relation to the weak interaction where the parity-violating aspects are velocity-dependent. The interaction of electrons with neutrinos provides a good illustration: neutrinos are quintessential chiral objects since only left-handed neutrinos and right-handed antineutrinos exist [13, 27, 72, 73]. Consider first the extreme case of electrons moving close to the velocity of light. Only left-handed relativistic electrons interact with left-handed neutrinos via the weak force; right-handed relativistic electrons do not interact at all with neutrinos. But right-handed relativistic positrons interact with right-handed antineutrinos. For non-relativistic electron momenta, the weak interaction still violates parity, but the amplitude of the violation is reduced to order v/c [73]. This is used to explain the strange fact that the $\pi^- \rightarrow e^- \tilde{\nu}_e$ decay is a factor of 10^4 rarer than the $\pi^- \rightarrow \mu^- \tilde{\nu}_\mu$ decay, even though the available energy is much larger in the first decay [73]. In the rest frame of the pion, the lepton (electron or muon) and the antineutrino are emitted in opposite directions so that their linear momenta cancel. Also, since the pion is spinless, the lepton must have a righthanded helicity in order to cancel the right-handed helicity of the antineutrino. Thus both decays would be forbidden if e and μ had the velocity c because the associated maximal parity violation dictates that both be pure left-handed. However, on account of its much greater mass, the muon is emitted much more slowly than the electron, so there is a much greater amplitude for it to be emitted with a right-handed helicity.

It should be mentioned that the discussion in the previous paragraph applies only to charge-changing weak processes, mediated by W^+ or W^- particles. Weak neutral current processes, mediated by Z^0 particles, are rather different since, even in the relativistic limit, both left- and right-handed electrons participate, but with slightly different amplitudes [72, 73].

So far in this article, the term "chirality" has been used in its qualitative chemical sense. In elementary particle physics, "chirality" is given a precise quantitative meaning: it is the eigenvalue of the Dirac matrix operator $\hat{\gamma}_5$, with values of $+1$ and -1 corresponding to right- and left-handed leptons. But only massless leptons (such as neutrinos), which always move at the velocity of light, are in eigenstates of $\hat{\gamma}_5$ and so have precise chirality. Leptons with mass (such as electrons) always move more slowly than c and so do not have well-defined chirality. Indeed, the very existence of mass is associated with "chiral symmetry breaking" [13, 122]. On the other hand, helicity (defined in Sect. 5.1) can be defined for both massless and massive particles, but only for the former is it completely invariant to the frame of the observer. For massless particles the helicity is actually equivalent to the chirality (for an antiparticle the helicity and chirality have the opposite sign). The interesting suggestion has been made recently that, if the physical problem singles out a preferred spatial origin, such as the source of an electromagnetic field, then chirality becomes sharply defined even for a particle with mass [123]: this could have important consequences for for the foundations of atomic and molecular physics.

There appears to be another, very different, connection between chirality and relativity. We saw in Sect. 6.3 that parity-breaking and parity-violating optical activity are distinguished by the fact that the first is time-dependent while the second is independent of time. Because a clock on a moving object slows down relative to a stationary observer, a molecule exhibiting spontaneous parity-breaking optical activity will become increasingly stable with increasing velocity relative to a stationary observer, and as it approaches the speed of light it will become infinitely stable. Thus spontaneous parity-breaking optical activity in a chiral object moving at the speed of light becomes indistinguishable from parity-violating optical activity.

8 True Enantiomers and Parity Violation

The conceptual value of parity violation in the discussion of molecular chirality now emerges, because only the space-inverted enantiomers of truly chiral systems show a parity-violating energy difference [7]. Space-inverted enantiomers of systems showing false chirality, such as a stationary rotating cone, or co-linear electric and magnetic fields, are strictly degenerate. This follows from the fact that, although the parity-violating weak neutral current Hamiltonian (23a) is odd under space inversion, it is invariant under both time reversal and any proper spatial rotation: since the last two operations together interconvert the two space-inverted enantiomers of a system displaying false chirality, it follows from a development analogous to (25) that the energy difference is zero.

Since the space-inverted enantiomers of a truly chiral object are not strictly degenerate, they are not true enantiomers (since the concept of enantiomer implies the exact opposite). So where is the true enantiomer of a chiral molecule to be found? In the antiworld of course! The molecule with the opposite absolute configuration but composed of antiparticles will have exactly the same energy as the original [12, 29]. This follows from the $\hat{C}\hat{P}\hat{T}$ theorem and the assumption that $\hat{T}$ is not violated. So true enantiomers are interconverted by $\hat{C}\hat{P}$. Since $\hat{P}$ violation automatically implies $\hat{C}$ violation here, it also follows that there is a small energy difference between a chiral molecule in the real world and the corresponding chiral molecule with the same absolute configuration in the antiworld.

This more general definition of the enantiomers of truly chiral objects is consistent with the chirality that free atoms display on account of parity violation [75, 77, 78, 83]. The weak neutral current generates only one type of chiral atom in the real world: the conventional enantiomer of a chiral atom obtained by space inversion alone does not exist. Clearly, the enantiomer of a chiral atom is generated by the combined $\hat{C}\hat{P}$ operation. Thus the corresponding atom composed of antiparticles will of necessity have the opposite "absolute configuration" and will show an opposite sense of optical rotation.

The space-inverted enantiomers of objects such as translating spinning cones or spheres that only exhibit chirality on account of their motion also show parity-violating differences. One manifestation is that, as mentioned in Sect. 7, left- and right-handed particles (or antiparticles) have different weak interactions. Again, true enantiomers are interconverted by $\hat{C}\hat{P}$: for example, a left-handed electron and a right-handed positron are interconverted by $\hat{C}\hat{P}$ (the fact that the left-handed electron and the resulting right-handed positron are moving in opposite directions can be corrected by invoking an operation $\hat{P}_\pi$, in place of $\hat{P}$, in which space inversion is followed by an appropriate spatial rotation through 180° [73]).

9 Concluding Remarks

This article has discussed the basic principles and consequences of the chirality of molecular structures from the standpoint of modern physics. The analysis has important lessons for the theory of chemical bonding and molecular structure generally. For example, we have seen how the pursuit of analogies between the quantum states of a chiral molecule and those of various elementary particles reinforces Heisenberg's perception of a kinship between molecules and elementary particles [124, 125]. This insight should encourage theoretical chemists to keep abreast of developments in elementary particle physics in order to introduce concepts that could form the basis of a new quantum chemistry.

Elementary particle physicists can also learn something from chemistry through such analogies. Thus Bohm et al. [126] have extended the physical ideas and mathematical methods of molecular and nuclear physics into the relativistic domain in order to develop a model of collective motions of extended relativistic objects to calculate the mass spectrum and radiative transitions of hadrons (particles which undergo strong interactions — baryons and mesons). Also Wigner [127] has likened the four distinct states $|K^o\rangle$, $|\bar{K}^o\rangle$, $|K_1^o\rangle$ and $|K_2^o\rangle$ of the neutral K-meson to the four possible states ψ_L, ψ_R, $\psi(+)$ and $\psi(-)$ of a chiral molecule: just as $|K^o\rangle$ and $|\bar{K}^o\rangle$ are interconverted by $\hat{C}\hat{P}$ and $|K_1^o\rangle$ and $|K_2^o\rangle$ are even and odd eigenstates with respect to $\hat{C}\hat{P}$, so ψ_L and ψ_R are interconverted by $\hat{P}$ and $\psi(+)$ and $\psi(-)$ have even and odd parity. However, Wigner's analogy falters when we introduce $\hat{C}\hat{P}$ violation into the K^o system and $\hat{P}$ violation into the chiral molecule. Although $\hat{C}\hat{P}$ violation mixes $|K_1^o\rangle$ and $|K_2^o\rangle$ just as $\hat{P}$ violation mixes $\psi(+)$ and $\psi(-)$, there does not appear to be a $\hat{C}\hat{P}$ analogue of the lifting of the degeneracy of ψ_L and ψ_R through $\hat{P}$ violation because one of the consequences of the $\hat{C}\hat{P}\hat{T}$ theorem is that a particle and its associated antiparticle have the same mass and lifetime [13]. A better molecular analogy would be with the four states of a chiral molecule in a time-noninvariant $\hat{P}$-enantiomorphous influence such as collinear electric and magnetic fields. Indeed, a detailed quantum-mechanical analysis of a chemical reaction system such as the butadiene-cyclobutene interconversion in collinear electric and magnetic fields, in which expressions for analogues of the $\hat{C}\hat{P}$-violating parameters mentioned in Section 5.3 were derived, might help to remove some of the mystery surrounding $\hat{C}\hat{P}$ violation.

Acknowledgement. I thank Dr. R. G. Woolley for helpful criticisms.

10 References

1. Fresnel A (1824) Bull. Soc. Philomath. p 147
2. Pasteur L (1848) Ann. Chim. 24: 457
3. Pasteur L (1884) Rev. Scientifique 7: 2
4. Mason SF (1982) Molecular optical activity and the chiral discriminations. Cambridge University Press, Cambridge
5. Lord Kelvin (1904) Baltimore lectures. C. J. Clay, London
6. Sakurai JJ (1964) Invariance principles and elementary particles. Princeton University Press, Princeton
7. Barron LD (1986) Chem. Phys. Lett. 123: 423
8. Barron LD (1986) J. Am. Chem. Soc. 108: 5539
9. Barron LD (1986) Chem. Soc. Rev. 17: 189
10. Haldane JBS (1960) Nature (London) 185: 87
11. Mason SF (1987) Nouveau Journal de Chimie 10: 739
12. Barron LD (1982) Molecular light scattering and optical activity. Cambridge University Press, Cambridge
13. Lee TD (1984) Particle physics and introduction to field theory. Harwood, Chur
14. Wigner EP (1927) Z. Phys. 43: 624
15. Feynman RP, Leighton RB, Sands M (1964) The Feynman lectures on physics. Addison-Wesley, Reading, MA
16. Barron LD (1972) Nature (London) 238: 17
17. Figueiredo IMB, Raab RE (1980) Proc. Roy. Soc. A369: 501
18. Graham C (1980) Proc. Roy. Soc. A369: 517
19. Stedman GE (1983) Am. J. Phys. 51: 753
20. Stedman GE (1985) Adv. Phys. 34: 513
21. Birss RR (1966) Symmetry and magnetism. North-Holland, Amsterdam
22. Shubnikov AV, Koptsik VA (1974) Symmetry in science and art. Plenum, New York
23. Curie P (1884) J. Phys. (Paris) (3) 3: 393

24. Landau LD, Lifshitz EM (1977) Quantum mechanics. Pergamon, Oxford
25. Kaempffer FA (1965) Concepts in quantum mechanics. Academic, New York
26. Sakurai JJ (1985) Modern quantum mechanics. Benjamin/Cummings, Menlo Park, CA
27. Berestetskii VB, Lifshitz EM, Pitaevskii LP (1982) Quantum electrodynamics. Pergamon, Oxford
28. Zocher H, Török C (1953) Proc. Natl. Acad. Sci. USA 39: 681
29. Barron LD (1981) Mol. Phys. 43: 1395
30. Buckingham AD, Graham C, Raab RE (1971) Chem. Phys. Lett. 8: 622
31. Buckingham AD (1979) Philos. Trans. Roy. Soc. A293: 239
32. Atkins PW, Gomes JANF (1976) Chem. Phys. Lett. 39: 519
33. Mislow K, Bickart P (1976/77) Isr. J. Chem. 15: 1
34. Atkins PW (1980) Chem. Phys. Lett. 74: 358
35. Jones RV (1976) Proc. Roy. Soc. A349: 423
36. Bohm A (1979) Quantum mechanics. Springer, Berlin Heidelberg NewYork
37. Beerlage MJM, Farago PS, Van der Wiel MJ (1981) J. Phys. B 14: 3245
38. Farago PS (1981) J. Phys. B 14: L743
39. Rich A, Van House J, Hegstrom RA (1982) Phys. Rev. Lett. 48: 1341
40. Campbell DM, Farago PS (1985) Nature (London) 318: 52
41. Zel'dovich Ya B, Saakyan DB (1980) Sov. Phys. JETP 51: 1118
42. Kabir PK, Karl G, Obryk E (1974) Phys. Rev. D 10: 1471
43. Cox JN, Richardson FS (1977) J. Chem. Phys. 67: 5702
44. Harris RA, Stodolsky L (1979) J. Chem. Phys. 70: 2789
45. Gazdy B, Ladik J (1982) Chem. Phys. Lett. 91: 158
46. Baranova NB, Zel'dovich B Ya (1979) Mol. Phys. 38: 1085
47. Wagnière G, Meier A (1982) Chem. Phys. Lett. 93: 78
48. Barron LD, Vrbancich J (1984) Mol. Phys. 51: 715
49. Wagnière G (1984) Z. Naturforsch. Teil A 39: 254
50. Dougherty RC (1980) J. Am. Chem. Soc. 102: 380
51. Mason SF (1983) Int. Rev. Phys. Chem. 3: 217
52. Rhodes W, Dougherty RC (1978) J. Am. Chem. Soc. 100: 6247
53. de Gennes PG (1970) A.R. Hebd. Seances Acad. Sci. Ser. B 270: 891
54. Mead CA, Moscowitz A, Wynberg H, Meuwese F (1977) Tetrahedron Lett. p 1063
55. Mead CA, Moscowitz A (1980) J. Am. Chem. Soc. 102: 7301
56. Peres A (1980) J. Am. Chem. Soc. 702: 7390
57. Barron LD (1987) Biosystems 20: 7
58. Gerike P (1975) Naturwissenschaften 62: 38
59. Edwards D, Cooper K, Dougherty RC (1980) J. Am. Chem. Soc. 102: 381
60. Barron LD (1987) Chem. Phys. Lett. 135: 1
61. Lifshitz EM, Pitaevskii LP (1981) Physical kinetics. Pergamon, Oxford, p 6
62. Tolman RC (1938) The principles of statistical mechanics. Oxford University Press, Oxford
63. Onsager L (1931) Phys. Rev. 37: 405
64. Kolb EW, Wolfram S (1980) Nucl. Phys. B 172: 224
65. Aharony A (1973) in: Gal-Or B (ed) Modern developments in thermodynamics. Wiley, New York, p 95
66. Lee TD, Yang CN (1956) Phys. Rev. 104: 254
67. Wu CS, Ambler E, Hayward RW, Hoppes DD, Hudson RP (1957) Phys. Rev. 105: 1413
68. Fermi E (1934) Z. Phys. 88: 161
69. Weinberg S (1980) Rev. Mod. Phys. 52: 515
70. Salam A (1980) Rev. Mod. Phys. 52: 525
71. Glashow SE (1980) Rev. Mod. Phys. 52: 539
72. Aitchison IJR, Hey AJ (1982) Gauge theories in particle physics. Adam Hilger, Bristol
73. Gottfried K, Weisskopf VF (1984) Concepts of particle physics vol. 1. Clarendon, Oxford
74. Rubbia C (1985) Rev. Mod. Phys. 57: 699
75. Bouchiat MA, Bouchiat C (1974) J. Phys. (Paris) 35: 899
76. Hegstrom RA, Rein DW, Sanders PGH (1980) J. Chem. Phys. 73: 2329
77. Fortson EN, Lewis LL (1984) Phys. Rep. 113: 289
78. Bouchiat MA, Pottier L (1986) Science 234: 1203

79. Rein DW (1974) J. Mol. Evol. 4: 15
80. Letokhov V (1975) Phys. Lett. 53A: 275
81. Zel'dovich BYa, Saakyan DB, Sobel'man II (1977) Sov. Phys. JETP Lett. 25: 95
82. Harris RA, Stodolsky L (1978) Phys. Lett. 78B: 313
83. Harris RA (1980) in: Woolley RG (ed) Quantum dynamics of molecules. Plenum, New York, p 357
84. Mason SF, Tranter GE (1984) Mol. Phys. 53: 1091
85. Mason SF, Tranter GE (1985) Proc. Roy. Soc. A397: 45
86. Tranter GE (1985) Chem. Phys. Lett. 121: 339
87. Tranter GE (1986) J. Theor. Biol. 119: 467
88. Tranter GE (1987) Chem. Phys. Lett. 135: 279
89. Christenson JH, Cronin JW, Fitch VL, Turlay R (1964) Phys. Rev. Lett. 13: 138
90. Cronin JW (1981) Rev. Mod. Phys. 53: 373
91. Fitch VL (1981) Rev. Mod. Phys. 53: 367
92. Townes CH, Schawlow AL (1955) Microwave spectroscopy. McGraw-Hill, New York
93. Herzberg G (1945) Infrared and Raman spectra of polyatomic molecules. Van Nostrand, New York
94. Allen HC, Cross PC (1963) Molecular vib-rotors. Wiley, New York
95. Oka T (1973) J. Mol. Spectrosc. 48: 503
96. Bunker PR (1979) Molecular symmetry and spectroscopy. Academic, New York
97. Davydov AS (1976) Quantum mechanics. Pergamon, Oxford
98. Barron LD (1979) J. Am. Chem. Soc. 101: 269
99. Hund, F (1927) Z. Phys. 43: 805
100. Rosenfeld L (1928) Z. Phys. 52: 161
101. Born M, Jordan P (1930) Elementare quantenmechanik. Springer, Berlin Heidelberg NewYork
102. Pfeifer P (1980) Chiral molecules — a superselection rule induced by the radiation field. Thesis, Swiss Federal Institute of Technology, Zurich
103. Woolley RG (1975) Adv. Phys. 25: 27
104. Woolley RG (1978) J. Am. Chem. Soc. 100: 1073
105. Woolley RG (1980) Isr. J. Chem. 19: 30
106. Woolley RG (1982) Struct. Bonding 52: 1
107. Cohen-Tannoudji C, Diu B, Laloë F (1977) Quantum mechanics. Wiley, New York
108. Michel L (1980) Rev. Mod. Phys. 52: 617
109. Ulbricht TLV (1959) Q. Rev. Chem. Soc. 13: 48
110. Berne BJ, Pecora R (1976) Dynamic light sattering. Wiley, New York
111. Harris RA, Stodolsky L (1981) J. Chem. Phys. 74: 2145
112. Anderson PW (1972) Science 177: 393
113. Anderson PW (1983) Basic notions of condensed matter physics. Benjamin/Cummings, Menlo Park, CA
114. Simonius W (1978) Phys. Rev. Lett. 40: 980
115. Joos E, Zeh HD (1985) Z. Phys. B 59: 223
116. Claverie P, Jona-Lasinio G (1986) Phys. Rev. A 33: 2265
117. Leggett AJ, Chakravarty S, Dorsey AT, Fisher MPA, Garg Q, Zwerger W (1987) Rev. Mod. Phys. 59: 1
118. Woolley RG (in press) Molecular organization and engineering, volume 1
119. Davies EB (1979) Comm. Math. Phys. 64: 191; ibid (1980) 75: 263
120. Barron LD (1981) Chem. Phys. Lett. 79: 392
121. Woolley RG (1981) Chem. Phys. Lett. 79: 395
122. Okun LB (1985) Particle physics. Harwood, Chur
123. Biedenharn LC, Horwitz LP (1984) Found. Phys. 14: 953
124. Heisenberg W (1966) Introduction to the unified field theory of elementary particles. Wiley, New York, p 6
125. Heisenberg W (1976) Physics Today. vol 29 no. 3 p 32
126. Bohm A, Boya LJ, Kielanowski P, Kmiecik M, Loewe M, Magnollay P (in press) International journal of modern physics A
127. Wigner EP (1965) Sci. Am. 213: No. 6, 28

Interplay of Experiment and Theory in Determining Molecular Geometries A. The Experiments

James E. Boggs

Department of Chemistry, The University of Texas, Austin, Texas 78712, U.S.A.

1 Introduction

This chapter and the following one are not intended as monographs for specialists, but rather are directed toward those seeking an overview of the experimental and theoretical methods available for determining molecular structure as a probe of the nature of chemical bonding. Since the modern trend in structural chemistry is in the direction of correlatiing and integrating all possible approaches to obtain the desired information, the present chapter on experimental methods may be of some value to the theoretician and the following one on theory may be of help to experimentalists. No rigorous derivation of the methods will be attempted, but an effort will be made to point out certain limitations of the various techniques and to offer some precautions that the reader of papers in the diverse fields may wish to keep in mind.

The most basic information that a chemist can derive about a molecule is its geometric structure. The concept of structure has evolved along with the science of chemistry from the earliest efforts to characterize a compound as being made up of simpler chemical elements, through research directed toward determining the number of atoms of each kind bound together into a molecule, to discovery of the importance of the order of attachment of the atoms within the molecule, to concepts of the three-dimensional nature of molecules and the development of sophisticated physical methods for characterizing that three-dimensional structure.

In addition to the location of nuclear positions, structural chemistry is also interested in the electronic structure of molecules. Experimental probes of electronic structure are generally much less direct and specific than are the diagnostic tools available for determining nuclear position. Interestingly enough, a very accurate and detailed knowledge of the nuclear geometry may provide one of the best sources of information about molecular electronic structure. The small variations in electronic structure that give each chemical substance its chemical individuality, that are responsible, for example, for the physical, chemical, and biological differences between methyl and ethyl alcohols, are reflected in small but interpretable shifts in nuclear positions. Very precise structural determinations, then, are capable of providing the information required to understand, interpret, and predict the properties and reactions of individual chemical substances. This, after all, is the fundamental goal of chemistry.

Molecules are not static objects, but are always undergoing complex vibrations and, often, internal conformational rearrangements, even at room temperature. Methods for the determination of molecular structure must therefore pay close attention to the vibrational and conformational behavior of the molecules under study. This motion can complicate the studies, but it is also an important source of additional structural information.

The variations in electronic structure that are responsible for the characteristic properties and reactions of different chemically related substances are typically reflected in energy differences on the order of a few kilojoules or, at most, of some tens of kilojoules per mole. Such energy differences can be correlated with differences in bond lengths of not more than ten picometers or variations in bond angles of less that 10 degrees. An experimental determination of structure, then, to be useful as a source of information on the most fundamental questions of chemistry must be accurate to a small fraction of this uncertainty, certainly to some tenths of a picometer or tenths of a degree. Such accuracy can be achieved, if sufficient care is taken, by a number of experimental and theoretical methods as described in this and the following chapter.

Most chemicals, as commonly encountered, are in the solid or liquid phase or in solution. For such systems, it is necessary to be concerned not only with the internal dynamics of an individual molecule but also with its interactions with its neighbors. Intermolecular interaction energies in condensed phases are sufficiently high that the resulting distortions of the nuclear positions can mask the structural variations being investigated as a source of information on the details of chemical bonding within the molecule. Thus one may distinguish two separate types of structural investigations with distinct purposes. For the highest accuracy of information about the molecule itself and its structural response solely to internal chemical bonding forces, experiments on gas-phase samples are usually required. For information on the structure of the molecule in the environment in which it commonly exists in natural systems, the experiment should be done in the phases or solvents of interest. This chapter is primarily concerned with structural investigations to obtain information on chemical bonding and therefore emphasizes gas-phase methods. The theoretical methods described in the following chapter also yield information on the isolated molecule directly.

2 Spectroscopic Methods

It is not the intent of this chapter to discuss experimental details of the methods of structural chemistry beyond the minimum level needed to understand the unique contributions and limitations of each technique, but rather to emphasize the procedures used to obtain molecular information from the experimental results. In addition to specialized monographs, there are a number of introductory textbooks that cover many aspects of molecular spectroscopy [1–5].

2.1 Microwave Spectroscopy

The microwave spectral region is, in general, characterized by discrete, sharp, rotational transitions, the frequencies of which can easily be measured to an accuracy of a few parts in 10^6. Unfortunately, this figure is not carried over to a comparable relative accuracy in the molecular parameters for polyatomic molecules because of numerous problems in analysis and interpretation as sketched briefly below. In spite of the very large amount of effort required to analyze a spectrum fully, there are numerous characteristics that have made this experimental method one of the most powerful sources of detailed information about molecular structure. Some of the advantages and handicaps are listed in Table 1. A recent compilation [6] listed 56 laboratories currently engaged in structural studies by means of microwave spectro-

Table 1. Characteristics of Microwave Spectroscopy

Advantages	Disadvantages
Extremely high resolution.	Sample must be in gas phase.
All components of a mixture give separate spectra	Molecule must have a permanent dipole moment.
Every conformer present gives an independent spectrum.	Limited structural information available from the spectrum of a single isotopic species.
Each vibrational state of a molecule gives an independent spectrum.	Much effort often required to assign spectrum.
Fine structure gives information on low-barrier motions in molecules.	Limited to relatively small molecules with few modes of internal motion.
Fine structure gives information on nuclear quadrupole coupling, dipole moment, etc.	
High redundancy in information content of spectrum makes errors rare.	

scopy. Several general textbooks on the method are available [7–10], none of them very recent.

The molecular information commonly obtained after a microwave spectrum is fully assigned consists of, at most, three moments of inertia for each vibrational state, conformer, and isotopic species observed, a number of centrifugal distortion constants, the three directional components of the dipole moment, possibly one or two nuclear quadrupole coupling tensors, and rough estimates of the energy differences between vibrational states or molecular conformers as obtained from relative intensities. Other information may be available in certain cases or obtained from special experiments.

2.1.1 Molecular Conformation

Many molecules exist in a variety of forms due to multiple minima in low-frequency internal motions. These forms may be equivalent, as resulting from the ammonia inversion motion or in the three equivalent forms of ethane reached by successive rotation by 120° around the C—C bond. With greater asymmetry in the molecule, however, the forms may be non-equivalent and, depending on the height of the barrier separating them, be considered as conformers or as geometric isomers. For our purposes, geometric isomers, separated by energy barriers sufficiently high to permit their physical separation under normal laboratory conditions, can be considered as separate molecules.

An example with relatively low potential barriers separating multiple minima is shown in Fig. 1 [11]. The most stable form of CH_3CH_2COF has the carbonyl group *syn* to the methyl group. Two other, equivalent minima are found by rotation of 120 and 240° around the CH_2—COF bond. The heights of the barriers are not sufficient to permit separation of the forms as stable substances under normal conditions, but they are high enough to sustain separate vibrational energy levels in each minimum. Although not shown, this molecule also has another mode of internal rotation, around the CH_3—CH_2 bond, which leads to additional conformational possibilities. A review of conformational equilibria is give in Ref. 12.

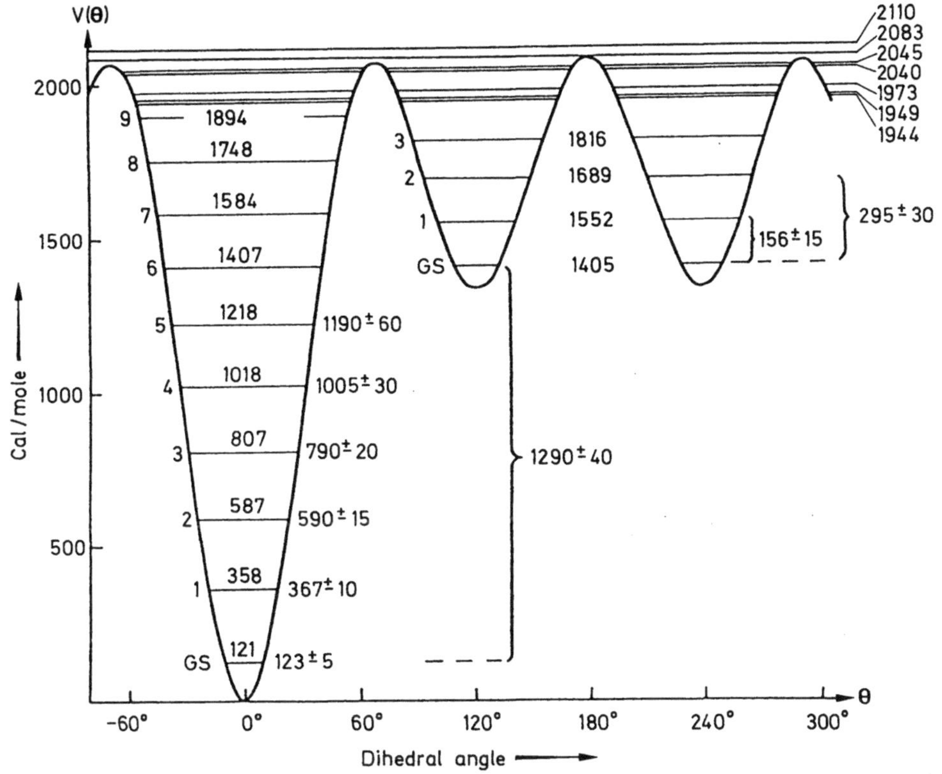

Fig. 1. Potential energy function and vibrational energy levels for propionyl fluoride as a function of the angle of rotation around the CH_3CH_2—COF bond. The *syn* conformer (at 0°) is more stable than the two equivalent *gauche* conformers. The observed microwave spectrum arises within the manifolds of rotational energy levels based on each of the vibrational levels shown. Taken from Ref. 11

The microwave spectrum of a molecule such as CH_3CH_2COF is a composite of the spectra resulting from transitions within the rotational manifold of each of the vibrational states shown, while transitions which involve a change in the vibrational quantum number typically occur in the far infrared region. The intensity of the microwave spectra from different conformers falls off according to the Boltzmann distribution, so that this method is most useful in cases where the different conformational minima are not too widely different in energy. In favorable cases, all of the information can be obtained for each conformer that could be obtained for completely separate molecules plus an estimate from intensity measurements of the energy difference between the states.

A special situation arises when an energy barrier is low and narrow. In such a case, quantum-mechanical tunneling through the barrier results in a splitting of the observed transitions which can lead to a very accurate measure of the height of the barrier above the energy level involved.

Another special situation in which microwave spectroscopy can give very accurate information is in establishing the planarity or linearity of a molecule. Except for inertial defects (see the general Ref. 7–9), a linear molecule has two moments of inertia equal to zero, while for a planar molecule the sum of two moments on inertia equals the third. These relationships provide a very delicate test without the need to evaluate any specific bond distances or angles.

2.1.2 r_o Structures

At most, the microwave spectrum of a single form of a molecule can provide three independent moments of inertia (two for a planar molecule or symmetric top, one for a linear molecule). Only for a diatomic molecule is this information alone sufficient to determine the complete geometric structure of the molecule. In this case,

$$I = m_1 m_2 r^2/(m_1 + m_2) , \tag{1}$$

so that the internuclear distance r can be obtained from the single spectroscopic moment of inertia I, assuming the atomic masses m_1 and m_2 are known. Since the molecule must be vibrating even in the vibrational ground state, both I and r represent average values over their ranges during the vibrational motion. These are designated I_o and r_o for the ground state, or I_v and r_v for a general vibrational state v. While all experiments measure vibrational averages of the geometric parameters, different experimental methods perform different kinds of averaging, so it is necessary to be more explicit. The reported r_0 distance from microwave spectroscopy is actually $\langle I/r^2 \rangle^{-1/2}$, the averaging being performed over the zero-point vibrational motion.

For a non-linear triatomic molecule, there are three geometric parameters to be determined but only two independent moments of intertia from experiment. The situation becomes rapidly worse for larger molecules. If it could be assumed that the internuclear distances were the same for all isotopic species of a molecule, only the masses differing, observation of the spectrum of a second isotopic species would give additional independent moments of inertia. The experiment could then be repeated with additional molecular isotopic forms until a sufficient number of moments of inertia were obtained to evaluate all the r_0 values for a molecule of any desired size. Aside from the practical difficulties that some elements do not have available isotopic modifications and that a very large number of independent spectra must be observed for any but the smallest structures, there is the limitation that the fundamental assumption is not strictly correct. The minimum in the vibrational potential well, and hence the equilibrium internuclear distance r_e, is nearly invariant to isotopic substitution, but the mass difference causes a given vibrational energy level to be

Table 2. r_0 Structures of OCS[a]

Isotopic Species Used	C—O	C—S
$^{16}O^{12}C^{32}S$ and $^{16}O^{12}C^{34}S$	1.1647	1.5576
$^{16}O^{12}C^{32}S$ and $^{16}O^{13}C^{32}S$	1.1629	1.5591
$^{16}O^{12}C^{34}S$ and $^{16}O^{13}C^{34}S$	1.1625	1.5594
$^{16}O^{13}C^{32}S$ and $^{18}O^{12}C^{32}S$	1.1552	1.5653

[a] Ref. 9, p. 527.

located differently within that well so that any average over the vibrational state, *e.g.* r_o, is different. The magnitude of the errors introduced is indicated by the data in Table 2, which shows bond distances in OCS as computed independently from microwave spectra of different pairs of isotopic species.

It often happens that only certain aspects of the geometry of a molecule are of chemical interest — questions, for example, as to whether a ring is planar or whether the location of a hydrogen atom indicates intramolecular hydrogen bonding. In these common cases, very accurate bond lengths are not needed and r_o structures may solve the problem. Indeed, it is often possible to assume that many of the structural parameters of a molecule are the same as those of known related molecules, so that data from a limited number of isotopic species can produce a quasi-r_o structure whose validity must depend entirely on the correctness of the structural assumptions.

2.1.3 r_s Structures

Kraitchman [13] has outlined a method for determining the coordinates of one atom in a molecule by concentrating on the changes in the moments of inertia when isotopic substitution is made for that atom. Full details can be found in the general references [7–10]. It is, for example, possible to determine the distance between a pair of atoms A and B that may be of particular interest from the moments of inertia of the normal species, the species with isotopic substitution at A, and the species with isotopic substitution at B. Such partial r_s structures can answer many chemical questions without the necessity for working out the full structure of the complete molecule. The method fails, of course, when applied to such cases as bonds to fluorine atoms where there are no isotopic modifications with which to work.

Kraitchman's equations are rigorous only for imaginary rigid molecules, but Costain [14] has shown that vibrational effects tend to cancel in comparisons between isotopically substituted molecules so that r_s substitution structures are more internally consistent than are r_o structures. For a diatomic molecule the r_s internuclear distance is approximately the mean of the r_o and r_e lenths. This is not true for polyatomic molecules however. An r_s bond distance does not have a rigorous physical definition, only an operational one in terms of its method of derivation from the changes in the moments of inertia. Consequently, there is no real way to tell in advance what the error in such a bond distance will be, and in some cases, particularly with hydrogen atoms [15], it can be very large. This source of error is not generally included in error estimates given in papers on microwave spectroscopy even though it may be several times (or even up to ten times!) larger than the predictable uncertainties quoted. With this note of caution, the r_s structure can generally be taken as a practical experimental approximation for the r_e structure which, unfortunately, cannot be measured directly.

2.1.4 The Approach to r_e

The vibrational potential function of a diatomic molecule can be approximated by the Morse function

$$U(r) = D[1 - e^{-\alpha(r - re)}]^2 \qquad (2)$$

where D is the dissociation energy measured from the bottom of the well and r_e is the equilibrium internuclear distance. For such a molecule, and ignoring centrifugal distortion, the rotational transition frequencies in a given vibrational state are directly related to B_v, which is given by

$$B_v = B_e - \alpha_e(v + {}^1/_2) + \gamma_e(v + {}^1/_2)^2 + \ldots \quad (3)$$

where

$$B_e = h/8\pi^2\mu r_e^2 \quad (4)$$

and μ is the reduced mass of the molecule [9]. The constants α and γ are usually obtained from the observed spectrum by measurement of the frequences of a series of lines for a given rotational transition in successively higher vibrational states. Other expressions for the vibrational energy function can be used and higher order anharmonic terms can also be included. These have been evaluated for certain diatomic molecules [16, 17].

For a polyatomic molecule, there are three rotational constants, A_v, B_v, and C_v, each containing a correction term with an α coefficient for every vibrational mode of the molecule. Corrections for centrifugal distortion and for Coriolis interactions are also necessary if accurate r_e parameters are to be obtained [9]. Such complete analyses have been carried out only for relatively few small molecules.

Unlike substitution (r_s) structures, the equilibrium structure has a clear physical definition. Another well-defined physical quantity is the average structure ($\langle r \rangle$ or r_z) in the ground vibrational state. For a harmonic oscillator, $\langle r \rangle = r_e$. Differences between the two, usually small, reflect the degree of anharmonicity of the vibrational mode. It has been shown [18] that the average bond distance can be obtained from the effective distance, r_v, using only the mean-square harmonic vibrational amplitude. The average structure in a given vibrational state is particularly important since it can be converted into the type of structure measured in electron diffraction experiments by averaging over the populations of the vibrational states occupied at the temperature of the experiment.

An idea of the magnitude of the corrections being considered can be obtained from the results in Table 3, showing the values obtained for various structures of two

Table 3. Structures of SO_2 and OF_2 in the ground vibrational state according to various structural definitions[a]

Type of Structure	SO_2		OF_2	
	r	θ	r	θ
Equilibrium, r_e	143.08	119°19′	140.53	103°4′
Effective, r_o	143.36	119°25′	140.87	103°19′
Substitution, r_s	143.12	119°30′	—	—
Average, $\langle r \rangle$	143.49	119°21′	141.24	103°10′

[a] Refs. 19, 20.

bent, triatomic molecules, SO_2 and OF_2. In other cases where information is available, it is also found that the differences are on the order of some tenths of a picometer for distances and tenths of a degree in angles. While these differences are small, it must be remembered that structural determinations are frequently done in order to obtain information about variations in bonding between related molecules, reflecting, for example, different degrees of interactions with substituent groups. Such use of structural variations as a probe of the subtleties of intramolecular bonding requires accuracy of high order.

2.1.5 Special Techniques

The application of microwave spectroscopy to species having vapor pressures below a few microns at room temperature has been a very difficult problem. A few very early attempts to build high-temperature spectrometers were successful [21], but their use proved so difficult that the methods were never exploited very fully, and they were restricted, of course, to substances which did not decompose at the temperatures required to make them sufficiently volatile. An interesting recent approach involves entraining the sample in a supersonic jet. This procedure has produced good quality spectra of such molecules as uracil, thymine, and adenine [22].

The spectroscopy of transient species is an active field. Relatively unstable materials with half-lives on the order of seconds to minutes can be produced by thermal decomposition, chemical reaction, or electrical discharge. The gas stream containing the species of interest can then be flowed through the spectrometer in a time too short to allow excessive decomposition. A few studies have also been done on ions formed in electric discharges.

Aside from molecular geometric structures, microwave spectroscopy provides a highly accurate method for determining the dipole moments of molecules, both the total moment and the directional components. Quadrupole coupling constants and electric field gradients at nuclei can provide additional information on chemical bonding.

2.2 Vibrational Spectroscopy

The absorption of infrared radiation and the Raman scattering of light by a substance are among the oldest techniques used for studying molecular structure. The two experimental techniques are complementary in the information they supply, and contributions from Raman spectroscopy have become increasingly important in recent years with the availability of spectrometers using laser light sources that are both intense and highly monochromatic.

The spectrum in routine infrared spectroscopy normally consists of unresolved vibration-rotation bands. The resolution is not very high compared with the spacing between bands. Nevertheless, it is sometimes possible to analyze complex spectra to the extent of identifying a score or more of fundamental vibrational modes. Even though bond distances and angles cannot come from the measured center frequencies of unresolved bands, it is often possible to obtain other valuable structural information. There are many experimental techniques to assist in band identification which are covered in standard texts. Aside from the classic works by Herzberg [23], an excellent modern treatise is the one by Califano [24].

2.2.1 Vibrational force fields

Within the Born-Oppenheimer approximation, the vibrational motion of a polyatomic molecule takes place on a potential energy surface which can be described by the equation

$$E = E_{ref} + \sum_i g_i q_i + \frac{1}{2} \sum_i \sum_j F_{ij} q_i q_j + \frac{1}{6} \sum_i \sum_j \sum_k F_{ijk} q_i q_j q_k + \dots \quad (5)$$

where the q's represent a complete set of 3N-6 vibrational coordinates and the g's are the forces on the nuclei, which are zero in any minimum energy conformation. The F_{ij} coefficients are the harmonic force constants if $i = j$ and coupling constants if they are not. Similarly, F_{ijk} is a cubic force constant or coupling constant, and so on to higher order terms. All of the information about chemical bonding that can be found from the molecular vibrational motions is contained in the various F constants in this equation.

Evaluation of the constants in Eq. (4) from experimental information requires drastic simplification. First, the harmonic oscillator approximation is usually assumed, which corresponds to terminating the equation after the term involving the double summation. Within this approximation, the object is to evaluate the harmonic constants F_{ij}, which are conveniently represented in a matrix form, $\boldsymbol{F}$, with the diagonal elements F_{ii} being the usual harmonic force constants. To illustrate the magnitude of the task, a totally non-symmetric molecule containing, say, 20 atoms has 3N-6 = 54 modes of vibration which produce an $\boldsymbol{F}$ matrix with $1 + 2 + \dots + 54 = 1{,}485$ force or coupling constants. Since such a molecule has only 54 fundamental vibrational frequencies, it is clear that the desired information, even with the harmonic oscillator approximation, is severely underdetermined. The problem is somewhat simplied if molecular symmetry makes it possible to factor the $\boldsymbol{F}$ matrix into symmetry blocks, and the use of supplementary data from isotopic species, centrifugal distortion constants, and other experimental methods makes the problem completely soluble within the harmonic oscillator approximation for considerably smaller molecules containing only a few atoms.

For molecules beyond the very small size range where experimental data can provide a complete quadratic $\boldsymbol{F}$ matrix, still further approximations can be used. Depending on the coordinate system used to describe the molecular vibrations, the $\boldsymbol{F}$ matrix may be roughly diagonal, i.e. the coupling between vibrational modes may be small. Assuming this is so, and one can never be sure that the assumption is correct, the off-diagonal coupling constants may be completely ignored and the vibrational behavior described in terms of harmonic force constants, F_{ii}, only. If it is suspected that some of the coupling constants might be important, a few of those could also be included in the fitting procedure if sufficient experimental information is available. It is also possible to make the approximation that some given type of vibration ought to be similar to that in a different, known molecule, and the corresponding force constant simply set at the value it has in the reference species.

If an empirical harmonic force field, F, is obtained by fitting experimental data, even though there might be sufficient experimental data to evaluate all of the parameters, it is still necessarily an effective force field rather than one made up of the true harmonic constants. This follows from the fact that molecular vibrations are really anharmonic, and the fitting procedure takes up the contributions of all the anharmonic terms in providing the best fit to the truncated form of Eq. (4) which is being used.

The force field matrix, F, can be expressed in different coordinate systems, and transformational procedures between them are described in many textbooks. It is often convenient to use symmetry coordinates in which, within the harmonic oscillator approximation, the molecular vibrations are separable into normal modes. To interpret the vibrational behavior in terms of molecular bonding properties, however, it is useful to convert F into an internal coordinate system in which the coordinates are individual bond stretches, bends, etc.

The determination of anharmonic force constants, F_{ijk}, from experiment has been limited to very small molecules, generally not larger than triatomic. Where such information is available, it can provide useful insight into the nature of the chemical bond. As described in chapter 7, this is an area in which computational methods can be of great assistance in the study of molecular vibrations.

2.2.2 High-Resolution Infrared Spectroscopy

For very small molecules in the gas phase, it is sometimes possible to resolve a vibration-rotation band so that the individual rotational structure can be observed. Such a spectrum provides information comparable to that obtained from microwave spectroscopy, including the rotational constants from which moments of inertia can be calculated and the centrifugal distortion constants. The resolved vibrational spectrum provides the particular advantage that information is also available to give corresponding constants for excited vibrational states. As with microwave spectroscopy, the separate bond distances and angles can be determined provided a sufficient number of independent moments of inertia can be measured.

Vibrational laser spectroscopy provides a relatively new and powerful tool for obtaining structural information. Diode lasers have high monochromaticity and sufficient tunability to give very accurate measurements of resolved vibration-rotation bands. The group at the Institute for Molecular Sciences in Okazaki, Japan, working under Professor Hirota, has made especially impressive progress in this area [25]. The technique is applicable to ions, radicals, and other unstable species and has been very valuable for such systems which are difficult to study by other methods, Unfortunately, the analysis is limited to very small molecules.

2.2.3 Conformational Analysis

As illustrated in Fig. 1, periodic internal motions in a molecule can lead to the existence of alternative conformations, providing the minima are of sufficient depth to support at least one vibrational state. Microwave spectroscopy is a valuable tool for investigating the various conformations separately if the difference in energy between the minima is small enough (below about 10 kJ mol^{-1}) to provide appreciable populations at room temperature. Vibrational spectroscopy is an alternative and complementary source of information and has the advantage that it is easier to use at higher temperatures where the populations may be more favorable for study of the less stable conformational forms.

Vibrational transitions between torsional states within one conformational minimum typically occur in the far infrared region below a few hundred cm^{-1}. Since these torsional states are so closely spaced, many of them are normally occupied even at room temperature and overtone bands are often clearly observed. Furthermore the periodic vibrational potential function deviates strongly from harmonic oscillator shape, so that the overtone bands are frequently well separated from the fundamental. In practice, since the shape of the potential well is unknown in advance, it is sometimes a difficult task to assign the rich variety of torsional bands that are observed. When it is done correctly, however, such experiments provide a probe of the shape of the potential function that can give a full description of the conformational alternatives.

A series of studies on the ring puckering motion in halogenated cyclobutanes [26–28] has demonstrated the caution that must be exercised in using far infrared spectroscopy alone to investigate conformational potential curves. The observed, rather extensive series of bands that was observed for the torsional vibration of fluorocyclobutane could be fit extremely well, to high precision, by either of two very different sorts of potential functions. One had two rather deep minima for equatorial and axial substitutions while the other had essentially one minimum for the halogen in the equatorial position with only a high-energy shallow minimum or flattening of the potential curve at the axial position. The infrared data alone could not distinguish between the two conflicting results, but the one-minimum curve was judged [28] to be most likely by comparison with results from microwave spectroscopy, mid-range infrared spectroscopy and Raman intensities. This conclusion was later confirmed by *ab intio* computation [29]. The far-infrared data, however, are invaluable when supplemented, as in this case, by information for other sources.

2.3 Other Spectroscopies

2.3.1 Electronic Spectroscopy

A transition between different electronic energy levels in a molecule in general involves also a change in vibrational and rotational state. The spectrum for a given change in an electronic quantum number therefore consists of a series of vibrational bands, each with internal rotational structure. With the high resolution available in this spectral region, the structure can be resolved for reasonably small molecules and data obtained comparable to that coming from infrared and microwave spectroscopy. There is the added advantage that information on molecular structure in excited electronic states is available.

The high resolution available with modern laser spectroscopy carries an accompanying disadvantage in that it is sometimes difficult to identify the spectral line which is being observed, particularly in highly excited species where little information is available from other sources. Nevertheless, this technique is growing rapidly in popularity and is yielding exciting results about types of molecular configurations that can be studied by no other method.

2.3.2 Magnetic Resonance Spectroscopy

Magnetic resonance spectroscopy is one of the most important tools of the organic chemist seeking the identity of a molecule and the order of attachment of atoms within it. Information about molecular bonding is available from interpretation of the coupling constants related to the chemical shifts and splittings seen in the spectra. Of more direct interest in the present context, however, is the potential use of magnetic resonance to obtain quantitative information about the geometric structure of a molecule. As was first shown by Saupe and Englert [30] the highly resolved spectrum of a solute molecule in a liquid crystal solvent has a structure due largely to intramolecular dipole-dipole interactions that can be analyzed to give highly precise relative intermolecular distances. The distances are only relative, since they are obtained with an arbitrary, unknown scale factor.

The spin Hamiltonian which describes the spectrum of a solute molecule in a liquid crystal solvent contains three terms: 1) a term for the Zeeman interaction of the nuclei with the applied field, 2) a usually small term for the isotropic part of the indirect spin-spin coupling of the nuclei, and 3) a term describing the direct magnetic dipole interactions of the pairs of nuclei. It is the last term which contains the information from which molecular structural parameters can be derived. The theory is summarized well in several reviews [31, 32].

Determination of reliably accurate molecular geometrical parameters by liquid crystal NMR, even relative parameters, has been slowed by a number of obstacles. These include the possibility of pseudodipolar interactions, the deformation that the solute may undergo because of the anisotropy of its environment, and the effect of the molecular vibrations. The question of molecular vibrations is particularly important for flexible molecules for which the structure may be an average over several conformations because of the time scale of the NMR experiment, $\sim 10^{-3}$ s, which is very slow compared with other techniques with which the results may be compared. In spite of the difficulties, impressive results are beginning to be obtained. Since the method is particularly suited for determining the positions of hydrogen atoms which are difficult to locate accurately by most other methods, it provides a valuable tool to use in combination with other techniques [33].

2.3.3 Photoelectron Spectroscopy

Electrons are emitted when molecules are photoionized by the absorption of ultraviolet or x-ray photons. Photoelectron spectroscopy measures the kinetic energy carried off by the electrons in the process $M + h\nu \rightarrow M^+ = e^-$. This energy can be expressed as

$$E_{e^-} = h\nu + E_M - E_{M^+} \,. \tag{6}$$

ince the energy of the incident photon is known, the measurement gives the energy of ionization of the molecule. With photons of sufficient energy, a number of photoelectron energy peaks are observed, corresponding to the energy required to remove electrons from various molecular orbitals.

Although photoelectron spectroscopy does not provide information about molecular geometric parameters, it does give very valuable insight into the character and energetics of the individual occupied molecular orbitals.

3 Diffraction Methods

The second broad class of experimental methods that can provide information on molecular geometries involves diffraction of a beam of light or particles passing through the sample. Since a simple diffraction experiment requires an incident wavelength comparable to the repeated dimensions of the scattering object, determination of interatomic distances requires light in the x-ray region or electron beams with energies of the order of 40 keV. The negative charge on an electron causes it to be easily scattered by the electron cloud of a molecule, so an electron beam has little penetrating power and scattering occurs readily from gaseous samples. Although x-rays are also scattered by atomic electrons, there is no charge-charge repulsion and they are scattered less strongly than are electrons. Except for quite difficult experiments, the sample in an x-ray diffraction experiment must be in the solid phase. A beam of neutrons is scattered from the nuclei in a molecule and this diffraction experiment requires an even larger crystalline sample than is required for x-ray studies.

3.1 X-Ray Crystallography

Far more molecular structures have been determined by x-ray diffraction than by any other technique. Perhaps the greatest advantage of the method is its applicability to very large molecules, even proteins. There are, of course, accompanying disadvantages, including the requirement for a good crystal of the sample and the ill-defined character of the internuclear distance measured unless very special precautions are observed.

3.1.1 Molecular Structures from Crystal Data

The basics of x-ray diffraction are described in every physical chemistry or physics textbook. Briefly, a beam of x-ray is diffracted by a grating consisting of the repeated geometrical pattern of atoms in a crystal, the angular distribution of the intensity of the diffracted x-rays is measured, and by appropriate analysis the pattern of the 3-dimensional grating responsible for the diffraction is reconstructed. The scattering pattern consists of a summation of waves originating at each of the diffraction points. Measurement of the total scattered intensity as a function of angle provides information on the amplitude of the individual scattered waves. Unfortunately, all knowledge of their phases is lost in the experiment, and this information must be regained in some manner before the geometrical pattern of the original scatterer can be deduced.

Since the molecular electron density in a periodic crystal lattice, $\varrho(x, y, z)$, must be a periodic function, it is convenient to represent it by a Fourier series

$$\varrho(xyz) = \frac{1}{V} \sum_{h=-\infty}^{\infty} \sum_{k=-\infty}^{\infty} \sum_{l=-\infty}^{\infty} F(hkl) \exp\left[-2\pi i(hx + ky + lz)\right] \tag{7}$$

where V is the volume of the unit cell and x, y, and z are fractional coordinates of a point within the cell. Each Fourier coefficient, F_{hkl}, called a structure factor, pertains to a specific reflection, the one from the plane labeled with the Miller indices hkl. A given structure factor may be expressed as a sum over the contributions from the various atoms in the plane and their coordinates

$$F(hkl) = \sum_j f_j \, e^{2\pi i(hx_j + ky_j + lz_j)} \tag{8}$$

where the atomic scattering factors, f_j, depend on the atom involved and are functions of $(\sin\theta)/\lambda$, involving the scattering angle and the x-ray wavelength. Experimentally, the measurement gives the scattered intensity for a given plane, I_{hkl}, which is related to the structure factor by

$$I_{hkl} = |F_{hkl}|^2 . \tag{9}$$

The measurement thus gives only the magnitude of the desired structure factor while the structure factor itself is required to solve the structure problem. This is the phase problem alluded to above. A variety of ingenious techniques is available in addition to "direct methods", which make use of the fact that the electron density is nowhere negative and therefore negative contributions in the Fourier expansion must be eliminated by appropriate adjustment of the phases.

Because of the great demand for the results of x-ray diffraction analysis of structure, highly sophisticated instrumentation is now available commercially. Once the chemist has prepared a crystal of adequate quality, the remainder of both the experiment and the data analysis is handled nearly automatically by computer-controlled diffractometers. The results obtained are good enough for many purposes where accuracy at a level better than a few picometers is not required.

3.1.2 Comparison with Structures from Other Methods

Since x-rays are scattered by the electrons in matter, the result of the analysis of a diffraction experiment gives the distribution of the electrons, not of the nuclei. For this reason, a "bond length" determined by x-ray diffraction can never be exactly comparable to one obtained from an experiment which depends on any kind of internuclear distance parameter.

In the crudest analysis, it can be assumed that the crystal is made up of spherical atoms, i.e. the electrons are assumed to be arranged spherically around the separate nuclei. In more sophisticated models, allowance is made for the anisotropy of the electron distribution. Still, however, the commonly used definition of a "bond length" obtained from an x-ray experiment is the distance between peaks in the electron density distribution. Especially in highly polar bonds, these peaks may be significantly offset from the nuclear positions. Some compensation for this effect can be obtained by basing the experimental refinement only on high-angle scattering data which are most sensitive to electrons near the nuclei [34].

To obtain structural parameters accurate to a few tenths of a picometer by x-ray diffraction requires more careful attention than is normally given in routine studies to a variety of experimental conditions and details in the data refinement. The quality of the crystal is fundamental. Factors such as multiple reflection, extinction, scan truncation errors, thermal diffuse scattering, radiation damage, selective exclusion of certain intensities, and many others must be thoughtfully considered [34].

A major problem in achieving higher accuracy in structural parameters obtained from x-ray diffraction experiments is that of thermal motion in the sample. There is no completely unambiguous method available for correcting for vibrational motions in the crystal (as there is, for example, for electron diffraction results in the gas phase), but many procedures are available that can at least ease the difficulty, and there are numerous sources describing these more fully than is possible here [35–39]. A first, and fairly effective, approach to reducing the difficulties associated with thermal motion is to conduct the experiment at as low a temperature as possible. If the temperature is low enough, only zero-point vibrations remain. Temperatures obtainable with liquid nitrogen are generally adequate, although there are investigations, such as those involving examination of crystalline forms which are only stable at higher temperatures, for which such an approach obviously cannot be used.

3.2 Neutron Diffraction

Diffraction of a neutron beam is fundamentally different from diffraction experiments using either x-rays or electrons since the neutrons are diffracted by the atomic nuclei. The technique, therefore, has the great advantage that it gives a direct measure of the average nuclear positions rather than of the electron distribution and is, in this sense, complementary to the other kinds of diffraction in providing a view of the complete structure of the target material. A disadvantage is that the scattering power of matter for neutrons is relatively weak so that only solid samples can be used and, in fact, considerably larger crystals are required than in an x-ray diffraction study of the same substance. Of course, there is the further consideration that sources of the high-flux neutron beams made necessary by the low scattering efficiency are relatively scarce and time for experimentation with them is limited.

3.2.1 Location of Hydrogen Atoms

Since the scattering of x-rays or electrons from an atom is from the electrons it contains, the scattering power is, to a first approximation, proportional to the atomic number of the atom. By contrast, the scattering of neutrons from nuclei has nothing to do with the atomic number but is dependent entirely on details of the nuclear structure. Hydrogen, the lightest element with the lowest scattering power for x-rays or electrons, is a relatively efficient scatterer of neutrons, and hydrogen atoms can be located with high accuracy by neutron diffraction experiments.

Neutron diffraction is nearly unique in its ability to locate hydrogen atoms accurately. In many x-ray or electron diffraction experiments they are not seen at all, and in others their positions can be determined only with very high uncertainties. Microwave spectroscopy determines molecular moments of inertia, which depend on hydrogen atom locations, but only weakly so since the hydrogen atoms are so light. Furthermore, the vibrational motions of hydrogen atoms are of considerably larger amplitude than those of other atoms and even the microwave substitution method has a lower accuracy in determing hydrogen positions. It is probably fair to say that, in most cases, the considerable difficulty involved in neutron diffraction experiments is undertaken with the primary objective of answering structural questions that hinge on the positions of the hydrogens in a molecule. Examples of such questions might be those related to the existence and character of either intra- or inter-molecular hydrogen bonds.

3.3 Electron Diffraction

An electron beam, like a neutron beam, can be scattered from a material sample in a manner to provide structural information about the scatterer if it has a DeBroglie wavelength comparable to the structural distances to be investigated. The wavelength, λ, of an electron beam, ignoring relativistic effects, is given by

$$\lambda = h/mv \tag{10}$$

where h is Planck's constant and m and v are the mass and velocity of the electrons. In terms of the potential, E, by which the beam is accelerated, this becomes

$$\lambda = h/(2mqE)^{1/2} \tag{11}$$

where q is the electronic charge. For a wavelength comparable to typical internuclear distances, an accelerating voltage on the order of 40 kV is required.

Because of the Coulomb interaction, the scattering of an electron beam by the electrons in a sample is so efficient that the beam has very little penetrating power into a solid or liquid. Consequently, diffraction from a solid surface provides a useful tool for studying surface structure and the nature of adsorbed materials. The high scattering efficiency also makes possible the use of gas samples, and the first gas-phase electron diffraction experiments were carried out, giving molecular structural data [40], only three years after the first example of electron beam scattering from a solid surface was observed.

At the present time, the most complete and accurate experimental structural information on isolated molecules, as required for investigations of intramolecular chemical binding, is obtained from gas-phase electron diffraction experiments and from microwave spectroscopy. Powerful as these techniques are, they are even more valuable when used together or in combination with still other techniques, as described in Sect. 4.

3.3.1 Nature of the Gas-Phase Electron Diffraction Experiment

It might be naively expected that interference effects due to scattering from an assembly of gas-phase molecules would cancel because of the random orientations of the individual scattering centers. In fact, it was shown by Debye [41], even before the wave nature of electrons was first proposed, that the interference effects resulting from the scattering of x-rays for a randomly oriented rigid system of electrons do not cancel. The theory, as applied to electron scattering, is presented at various levels of sophistication in various references [42–44] in which the interested reader can follow the details of the development.

Briefly, the intensity of the electrons scattered from a gaseous sample is the sum of molecular scattering, atomic scattering, and incoherent scattering. The latter two terms do not contain the interference effects and are generally lumped together as background scattering. The molecular scattering intensity from an assembly of vibrating molecules with random orientation is given by

$$I_m(s) = \frac{K^2 I_0}{R^2} \sum_{i=1}^{N} \sum_{\substack{j=1 \\ i \neq j}}^{N} g_{ij}(s) \exp\left(-\frac{1}{2} l_m^2 s^2\right) \frac{\sin\left[s(r_a - \varkappa s^2)\right]}{s r_e}. \quad (12)$$

In this expression, the variable s is related to the experimental scattering angle θ by

$$s = (4\pi/\lambda) \sin(\theta/2) \quad (13)$$

where λ is the electron wavelength. I_0 is the intensity of the incident beam, $K = 2\pi m e^2/h^2 \varepsilon_0$ is a collection of constants including the mass m and charge e of an electron and the vacuum permittivity ε_0, R is the distance between the source and the detector, l_{m^2} is an effective mean square vibrational amplitude which is nearly identical with the mean vibration square harmonic vibrational amplitude,

$$l_{h^2} = l_{a^2} \operatorname{th} |hc\nu/2kT| \quad (14)$$

at temperature T, $\varkappa$ is an asymmetry constant related to the constant a of the Morse potential by

$$\varkappa = a l_{h^{4}/6}, \quad (15)$$

and

$$g_{ij}(s) = |f_i(s)|\,|f_j(s)| \cos[\eta_i(s) - \eta_j(s)] \quad (16)$$

where $|f_i(s)|$ and $\eta_i(s)$ are the absolute values of the amplitude and the phase of the atomic electron scattering amplitude from nucleus i. The particular type of internuclear distance, r_a, measure directly by electron diffraction is a somewhat complex average and is described more fully in Sect. 3.3.2.

Measurement of the scattered electron intensity can, in principle, lead to a determination of the average internuclear distances, r_a, the mean vibrational amplitudes, l_m, and the asymmetry constants, $\varkappa$. In practice, for larger molecules, the asymmetry constants and many or all of the vibrational amplitudes are taken from other sources. Data from vibrational spectroscopy or from quantum chemical computations can often provide estimates of the vibrational amplitudes, leaving the information content of the scattering experiment for use in evaluating the internuclear distances.

A highly schematic diagram of the electron diffraction experiment is shown in Fig. 2. The electron beam must be monochromatic, tightly focused, and arranged so that the scattering occurs in the smallest possible volume as the vapor emerges from the nozzle. The rotating sector provides a carefully calibrated filter to compensate for the very high atomic scattering at low angles, which carries no information about the structure of the molecule. Special instruments have been constructed [45, 46] which record the scattered electron intensity by electronic means, but most instruments use photographic recording, as suggested in the figure. The exposed plate is read in a microphotometer and the digitized data is analyzed.

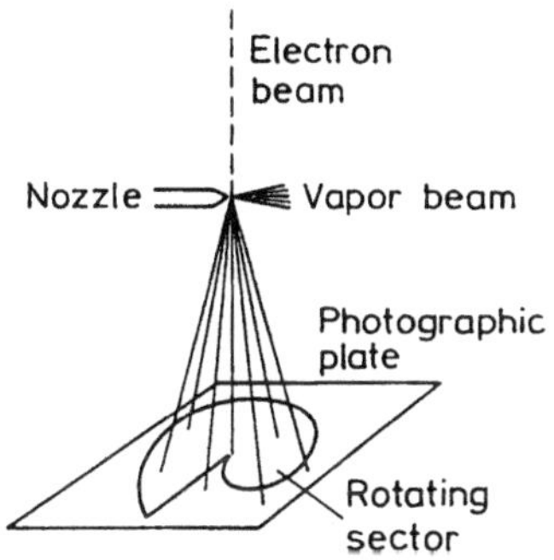

Fig. 2. Schematic diagram of an electron diffraction apparatus. A highly monochromatic electron beam is scattered in a small scattering region from the gaseous sample. The rapidly rotating sector eliminates the very strongly angle-dependent atomic scattering that would otherwise obscure the desired molecular information.

The Fourier transform of the molecular scattering intensity M(s), after the background has been eliminated, gives the radial distribution function, f(r):

$$f(r) = \int_0^{s_{max}} sM(s) \exp(-as^2) \sin(sr)\, ds\,. \tag{17}$$

The exponential term is introduced artificially to compensate for the truncation of the experimentally observed scattering angle at some upper limit s_{max}. The Fourier transform in Eq. (17) is a close approximation to a distribution function of internuclear distances. A plot of it, as shown in Fig. 3 [47] for SO_2Cl_2, shows peaks at distances corresponding to all internuclear distances, bonded or nonbonded, in the molecule. The width of the peaks gives an indication of the amplitudes of vibration of the nuclei involved. For larger molecules with less symmetry, a great deal of overlapping occurs and only a limited number of such peaks can be resolved. In any case, the final determination of the internuclear distances is normally made by a least-squares fitting of a model molecular scattering curve to the observed scattering function. In cases where the fitting procedure leads to multiple minima with fits that are of nearly equal quality, it is often desirable to resort to supplementary information from other sources to augment the information from the diffraction experiment.

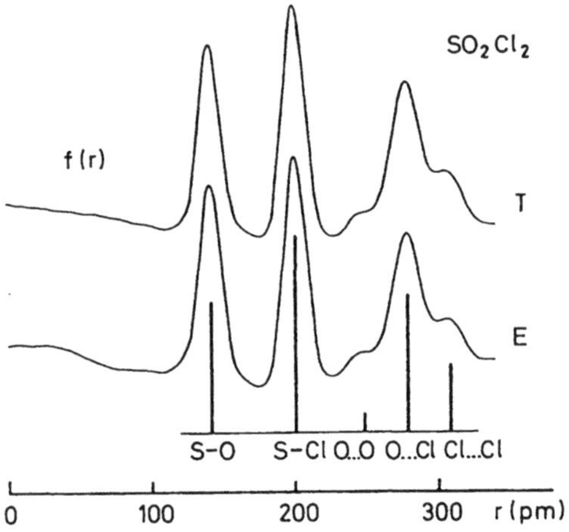

Fig. 3. Experimental (E) and theoretical (T) radial distribution curves for SO_2Cl_2 showing peaks obtained for all internuclear distances. The theoretical curve was calculated by fitting structural parameters and vibrational amplitudes to give the best fit to observed electron scattering. Taken from Ref. 47

The actual electron diffraction experiment is quite rapid, although the analysis requires considerably more time. For this reason, one apparatus is capable of providing data for a fairly large number of investigators, and the total number of instruments in the world is rather small, probably no greater than 20. Most experimental work is done at room temperature, although the groups in Moscow [48] and in Budapest [49] have done many valuable studies of species that are volatile only at high temperature and a few such studies have been done elsewhere [50].

Electron diffraction and microwave spectroscopy are currently the two most widely used experimental techniques for obtaining highly accurate and detailed information on the structures of relatively small molecules in a state where they are unperturbed by intermolecular interactions. As such, they are powerful tools to obtain the basic data needed to formulate and test concepts regarding the nature of chemical bonding.

3.3.2 Relation to Structures Obtained by Other Methods

As mentioned above, every experimental method measures some sort of average internuclear distance, but there are many sorts of averages. The most readily defined internuclear distance, the equilibrium distance r_e corresponding to the minimum on the vibrational potential surface, does not come directly from any experiment, although it can be obtained by computational methods as described in the following chapter. If sufficient information is known about the vibrational potential surface, the various vibrational averages obtained from different experiments can be corrected back to r_e structures, or they can be compared directly with each other. It cannot be overemphasized, however, that the different experiments measure different quantities, and their results should not be identical unless they are corrected to some common ground.

The average of an internuclear distance over some one vibrational state, r_0, r_1, ... , was mentioned in Sect. 2.1 in the discussion of microwave spectroscopy, as was the substitution r_s structure determined uniquely by microwave spectroscopy but not precisely convertible into any other type of distance. There are a number of other well-defined internuclear distances commonly associated with electron diffraction work and used in making comparisons between techniques.

First, the internuclear distance averaged over thermal vibrations at a given temperature is usually denoted as

$$r_g = r_e + \langle \Delta r \rangle_T , \tag{18}$$

where r_e is the equilibrium internuclear distance. The average of the displacements can be expanded as

$$\langle \Delta r \rangle_T = \langle \Delta z \rangle_T + (\langle \Delta x^2 \rangle_T + \langle \Delta y^2 \rangle_T)/2r_e + \ldots \tag{19}$$

where the z direction is between the nuclei concerned. The quadratic averages are simply related to the harmonic force constants f_{ij}, which can be obtained from either computation or from an empirical force field constructed from data from vibrational spectroscopy. The linear average, however, requires at least approximate knowledge of the cubic vibrational constants. These are obtainable by computation but are seldom known from experiment. More commonly the linear average is avoided by assuming that all of the bonds in the molecule are equivalent to simple diatomic Morse oscillators with

$$\langle r \rangle = r_e + (3/2)\, a \langle x^2 \rangle + \ldots . \tag{20}$$

The Morse constant a is on the order of 2 Å^{-1} for many diatomic molecules, and this value is often assumed to be approximately valid for all bonds in polyatomic molecules.

Equation (12) above showed that a different average, r_a, is obtained directly from an electron diffraction experiment. This average is related to r_g by

$$r_g = r_a + \langle \Delta r^2 \rangle_T / r_e + \ldots . \tag{21}$$

Papers reporting electron diffraction studies frequently report r_g distances as well as r_a. Both r_a and r_g are averages over all occupied vibrational states and are consequently temperature dependent, although the temperature dependence is small in most cases. The effect is significant, however, for data at high temperatures.

Still other types of averages are in common use. Microwave spectroscopic measurements provide the temperature-independent internuclear average distance, r_z, which is defined as the average over the motion in the ground vibrational state.

$$\begin{aligned} r_z &= [(r_e + \langle \Delta z \rangle_0)^2 + \langle \Delta x \rangle_0^2 + \langle \Delta y \rangle_0^2]^{1/2} \\ &= r_e + \langle \Delta z \rangle_0 + (\langle \Delta x \rangle_0^2 + \langle \Delta y \rangle_0^2)/2r_e + \ldots . \end{aligned} \tag{22}$$

In practice, it is common to assume that the perpendicular averages are negligibly small and approximate the relationship as

$$r \;= r_e + \langle \Delta z \rangle_0 . \tag{23}$$

A related average obtainable from electron diffraction is r_α, obtained by averaging the nuclear positions over the thermal vibrations at some particular temperature. After numerous small approximations, it can be shown that

$$r_\alpha = r_g - (\langle \Delta x^2 \rangle_T + \langle \Delta y^2 \rangle_T / 2r_e \,. \tag{24}$$

By extrapolation to a temperature of 0 K, one obtains

$$r_\alpha^0 = r_z = r_g - (\langle \Delta x^2 \rangle_0 + \langle \Delta y^2 \rangle_0)/2r_e - (\langle \Delta r \rangle_T - \langle \Delta r \rangle_0) \,. \tag{25}$$

The last term requires a knowledge of the cubic anharmonicity constant or else the approximation that the vibration is similar to that of a Morse oscillator. Thus a rigorous comparison of electron diffraction bond distances with those obtained from microwave spectroscopy requires conversion to the common r_α^0 or r_z definitions.

The differences in magnitude of the variously determined experimental bond distances are far from trivial. Table 4 shows the BF bond length in BF_3 in terms of the different definitions of an internuclear distance. Similar information is available in a number of other cases, all of which show variations up to a picometer in reliable measurements. These differences are significant if it is desired to combine partial information from several techniques, as discussed in Sect. 4, or if the accuracy of a result from one method is to be compared with a value obtained by a different experimental technique.

Table 4. Structure of BF_3 according to various definitions

Type of Structure	r
r_g	132.33 (10)[a]
r_α	131.09 (10)
r_z	131.103 (5) for $^{11}BF_3$[b]
	131.110 (6) for $^{10}BF_3$[b]
r_e	130.71 (1)[b]

[a] Ref. 51. [b] Ref. 52.

4 Combination of Methods

Increasingly, modern investigations of molecular structure tend to focus on a system of interest and bring to bear whatever experimental or theoretical methods can combine to clarify the problem. This development is partly due to the increased maturity of the techniques so that, at least in the case of many of them, an investigator does not have to devote his entire career to understanding and improving the methodology but can direct his attention to the chemistry of the system being investigated. Moreover, the nature and accuracy of the structural parameters measured or computed by various methods is now better understood, making it possible to bring results from measurements that give data on differently defined molecular properties to a common basis and utilize all of them to give the best available estimate of the actual nature of the molecule being studied.

Whatever technique is being considered, there seems always to be an interest in extending its range of applicability to larger molecular systems. New applications, then, tend to hover on the brink of having a larger number of unknowns than can be determined by available experimental data. When several independent methods can be brought to bear on the same problem, and full account is taken of their differences, more reliable information can be obtained on larger systems.

4.1 Diffraction and Spectroscopy

The major hurdle in the direct comparison of final results obtained from spectroscopy and from diffraction arises from the need to convert the quantities measured to a common ground. It might seem to be simplest to convert whatever vibrational average is measured by a given technique to the equilibrium structure, which at least is the most simply defined meaning of structure. Some of the difficulties in reaching that goal are evident from the discussion above of the different experimental methods, where it is also shown that the more usual approach is only to convert the experimental results to a common type of vibrational average.

Beyond the simple comparison of molecular structures derived from different methods, there are important ways in which information from different sources can be incorporated into a single analysis. Some of these are described in this section.

4.1.1 Electron Diffraction and Vibrational Spectroscopy

The scattering data available from an electron diffraction experiment on anything but a very small molecule is inadequate for a complete determination of all of the internuclear distances and internuclear vibrational amplitudes of the target molecule. It is common practice to reserve the experimental information for determination of the distance parameters and obtain all or some of the vibrational amplitudes from infrared and Raman spectroscopy. If the vibrational spectrum of the substance being studied has been fully analyzed (a relatively uncommon situation), the harmonic force field obtained can be used directly to calculate the mean amplitudes. It must be kept in mind, of course, that for all but the smallest molecules, the experimental force field is itself underdetermined, especially in terms of its off-diagonal elements, and so is not a unique representation of the vibrational spectrum.

If the molecule for which the electron diffraction study is being carried out has not had its vibrational spectrum observed, it is still possible to carry over the results of vibrational analyses of related molecules to construct a probable vibrational force field which may have sufficient accuracy to give useful predictions of mean vibrational amplitudes. Fortunately, high accuracy in the force field is not required. Methods devised and developed by Cyvin [53] have become a standard procedure in many electron diffraction experiments.

A further way in which information from vibrational spectroscopy can be incorporated into the analysis of an electron scattering experiment is in the use of conformational information often obtainable from the spectrum even without complete evaluation of the vibrational force field. Also, if the vibrational selection rules can establish the symmetry of the molecule, this can be used as a constraint. Because of the vibrational "shrinkage effect" [54], electron diffraction has much difficulty in deciding whether of not a molecule is exactly linear or planar. If spectroscopy can furnish symmetry arguments that establish linearity or planarity, the electron diffraction analysis is greatly simplified.

Vibrational spectroscopy, particularly in the far infrared region, can also simplify the diffraction analysis by furnishing evidence as to whether the substance under investigation consists of only one conformer or is a mixture of two or more conformers under the conditions of the investigation. Gas-phase spectroscopic evidence is much more compelling than that from condensed phase studies in such cases because of the often rather low barriers between conformations and the ease with which conformational populations are shifted by interaction with neighboring molecules.

4.1.2 Electron Diffraction and Microwave Spectroscopy

Particularly impressive examples have been seen in recent years of the combined use of electron diffraction and microwave spectroscopy. At first, the interaction consisted mainly of the use of one technique to confront and check the structural results obtained by the other. As a general rule, the cases in which large disagreements were found did not so often reflect an improper application of one of the methods as they did an over-optimistic assignment of uncertainties to the resulting structures. This should not be considered a trivial matter, however, because a chemist using the results of a structural analysis may draw quite unwarranted conclusions about bonding and properties if the data are represented as having a greater reliability than they really do. It is not always an easy matter to assign valid uncertainty limits to structural results, primarily because the often unknown effects of molecular vibrations are generally larger than the better known errors resulting from the instrumentation and mathematical procedures used in treating the directly measured numbers. A direct comparison of presumably comparable quantities measured by two separate techniques with susceptibilities to different sources of error can be of great help in a realistic evaluation of the accuracy.

The direct measurement in a microwave spectroscopic experiment is a set of frequencies, and from these not more than three rational constants can be derived for every isotopic species assigned. Provided that the rotational transitions used were ones of reasonably low quantum number or that small corrections have been made for centrifugal distorion, the resulting rotational constants are of extremely high accuracy and have a thoroughly-understood definition in terms of averages over the ground, or other, vibrational state. The uncertainties in the analysis of a microwave experiment arise in the next step, when complex methods are used to obtain structural parameters from the moments of inertia. An extremely useful method for combining diffraction and spectroscopic measurements is to include the microwave moments of inertia, properly treated to adjust for the difference in meaning of "internuclear distance" as additional observations to be fitted along with the scattering data from electron diffraction.

An elegant, but typical, example of this approach is shown in a recent paper by Kuchitsu and coworkers [55] on the structure of allene. The molecular intensities obtained from the electron diffraction experiment were converted to an r_α^0 structure while the three rotational constants obtained from the microwave spectrum of each of eight vibrational species of C_3H_4 and seven vibrational species of C_3D_4 were converted to A_z, B_z, and C_z values (see above). Since these two structural definitions are nearly equivalent, all of the information from both experimental sources could be combined in a single fitting procedure and a highly accurate and reliable structure obtained. Some of the most trustworthy molecular structures currently obtainable are derived from similar experiments and data analyses.

4.1.3 Electron Diffraction and Nuclear Magnetic Resonance

The difficulties that electron diffraction has in locating light atoms in the presence of heavier ones and in distinguishing between similar internuclear distances are partially countered by inclusion of microwave spectroscopic data as described in Sect. 4.1.2. Still an additional source of input information, and one that seems peculiarly suited to overcoming the limitations of electron diffraction is the use of structural data from liquid crystal nuclear magnetic resonance experiments. While the possibility of even considering the combination of information from such disparate sources is based on the pioneering work of a number of NMR practitioneers in developing understanding of the refinements of the experiment that must be controlled, the recent development of the combined method owes a great deal to the work of Rankin and his collaborators [56].

Dipolar couplings obtained from liquid crystal NMR give information most clearly about the positions of hydrogen atoms, which are difficult to locate accurately by either electron diffraction or microwave spectroscopy, although information on the positions of nuclei such as carbon and nitrogen is also readily obtained. Unlike electron diffraction, NMR data are obtained independently for each separate distance in the molecule, even though these distance may be nearly equal. The disadvantage of NMR in giving internuclear distances only relative to an orbitrary scale is readily overcome by the clear indication of total molecular size readily available from spectroscopy or diffraction. The highly useful complementarity of the information inherent in the different techniques is countered only by convern as to whether the condensed phase data from NMR couplings, with possible questions about degree of alignment and interaction with neighboring molecules, can be made truly comparable with gas phase data. Rankin believes that it can, and has presented [56] structural analyses of a number of molecules including pyrazine, pyrimidine, thiophene, furan and several polychlorobenzenes. For chlorobenzene, all seven bond lengths and four angles (after assuming planarity) were determined independently with standard deviations being no greater than 0.08 pm and 0.08°.

4.2 Experimental and Computational Results

Clearly, theory is very closely interwoven with any experimental method. A great deal of theoretical understanding and computation is involved in the transition between the swing of a meter needle and a final statement about an internuclear distance in a molecule. This section is not concerned, however, with such use of theory and computation, but rather with the possibilities that have become available in recent years to compute structural and vibrational behavior directly at a level that is useful as input into the analysis of an experiment or even to replace certain aspects of experimentation. Full discussion of the possibilities is left to the following chapter after the methods of computation have been described and their power and limitations outlined. At this point, however, it seems worthwhile to present several examples of simple cases where computational results have been useful in interpretation of an experiment.

4.2.1 Electron Diffraction and Computation

It sometimes happens that it is difficult to choose between two or more alternative minima which result from the least squares fitting procedure in an electron diffraction analysis. It can be known with good certainty that the structure is one of several possibilities, not somewhere in between, but there is no physical evidence from the experiment to guide the investigator's intuition in making the right choice among them. Even a computed structure with limited accuracy can then be of great help in pointing to the correct one of the alternative structures. An example of such an application appeared in the resolution of a question concerning alternative minima in diffraction studies of F_3COOCF_3 [57].

Lothar Schäfer has published a lengthy series of papers [58] describing what he has called the MOCED (molecular orbital constrained electron diffraction) method. In this approach, relative bond lengths computed at a low quantum chemical level are included, sometimes with data from other sources as well, in the electron diffraction analysis. The results appear to be highly successful. Numerous other workers now use similar methods, although the MOCED name does not appear to have as yet come into common use.

The computationally derived information carried over as input for an electron diffraction experiment may be of several types. Perhaps the most important provides a remedy to the fundamental difficulty that electron diffraction experiences in distinguishing between bonds of nearly but not exactly equal length. If a molecule contains, for example, two slightly different C—C bond lengths, analysis of electron diffraction data alone can normally provide only an average of the two distances. The next level of sophistication in refinement would be to incorporate two parameters, the average of the two distances and their difference, into the least squares analysis of the scattering data. However, the difference between the bond lengths normally comes out of the refinement only with a very large uncertainty, often larger than the quantity itself. As discussed in detail in the following chapter, the difference between two similar bond lengths in a molecule is usually obtained by quantum chemical calculation with much greater accuracy than is the absolute value of either one, so a very practical solution is to compute the bond length difference, use the computed value as a fixed constraint in the electron diffraction analysis, and then use the refined average bond length to obtain separate values for the two individula bonds.

It might at first be thought that it is of little practical interest to know that one bond in a molecule is very slightly longer than another, but, on the contrary, it is such fine details of structure that contain the chemically useful information. There are really no chemical conclusions that can be derived, in the absence of additional information for comparison, from the knowledge that some particular C—C bond distance is 154.0 pm. However, if it is known that one C—C bond in a molecule is 0.8 pm longer than another, one can reason about the substituent and other structural effects that have induced the electron shifts resulting in the bond length differences. An excellent illustration of this is found in a series of papers [59] on the effect of substituents on the bond lengths, and particularly on the ring bond angles, in substituted benzenes. It is also of interest to read a paper by Hargittai [60] on the importance of small structural differences.

The preceding discussion has been couched in terms of using computed differences in C—C bond lengths as fixed parameters in the analysis of electron diffraction data, but it is obviously equally possible to use any other structural feature, or relative parameter values, that are thought to be obtained more accurately from the computation than they can be derved from analysis of the experimental data. This may include such large scale structural information as molecular conformation or ring planarity as well as specific numerical values.

Computational results are often impossed as pieces of fixed information in the electron diffraction refinement, but there have also been efforts to include them, along with an assigned estimated uncertainty, in the least-squares refinement procedure. There is no truly objective way to determine the uncertainty in a calculated structural parameter (nor in the experimental data, either, if allowance is being made for non-instrumental, i.e. vibrational, effects), so this procedure has a large subjective element in spite of its objective appearance.

A type of input from quantum chemical computation that has not yet been widely utilized in electron diffraction studies is the computed vibrational force fields, either at the harmonic oscillator level or including vibrational anharmonicity. Such information is readily computed, and a chapter in a later volume of this series will deal with the accuracy that can be obtained and the precautions that are necessary. The mean amplitudes of vibration that are needed in all electron diffraction refinements could be obtained from this source as an alternative to the present custom of calculating them from empirical force fields derived from whatever experimental vibrational frequencies may be available or transferred from other molecules. The cubic and higher force constants needed for conversion between differently defined "bond lengths" could also be obtained by ab initio computation.

4.2.2 Microwave Spectroscopy and Computation

Unfortunately, there is not as extensive a record of constructive contributions of results from computational quantum chemistry to the analysis of microwave spectroscopic experiments as is the case with electron diffraction. There have been quite a number of instances in which computational studies have indicated that the results reported from microwave experiments were not as precise as hoped, but very few in which computational results were used as input in the spectroscopic analysis.

One major area in which computational studies have assisted in appreciating the limitations of conventional analysis of the results of microwave spectroscopy is in understanding the accuracy of the substitution method for determining atomic coordinates (see above). The substitution method would be exact if ground-state nuclear positions were invariant with respect to isotopic substitution. Unfortunately, very minor deviations of the ground state coordinates can accumulate in unexpected and unpredictable ways in the substitution procedure. It should be made clear that this problem is entirely separate from the predictable way in which the substitution method has difficulty with location of atoms that are near the inertial axes. The present discussion deals strictly with the effect of ground-state vibrational behavior on the derived substitution coordinates.

The effect is most pronounced in the location of hydrogen atoms, although it is not restricted to hydrogen. A large number of instances have been reported, but the most extreme example has appeared in the location of the hydrogen atoms in a series of substituted dimethyl ethers [61]. The microwave spectra of a large number of isotopic species were investigated with great care, and full substitution structures were obtained by standard procedures. Quite extreme, and completely unreasonable, variations were found in the C—H bond lengths, and it was clearly recognized that the full substitution structures obtained could not correspond to reality. The most extreme variation was in the case of the —OCH2— group of chloromethyl ethyl ether for which one C—H bond came out to be an impossible 13.3 pm longer than the other [62]. It was subsequently shown by ab initio computation [15] that variations in the C—H bond lengths fell in the normally expected range of about 1 pm. Variations in C—H bond lengths of this magnitude can be interpreted in terms of electronic interactions and are fully compatible with experimental results from other techniques [63].

One area in which computational results have been helpful as input for the analysis of results from microwave spectroscopy is in the study of the variation of rotational constants with torsional state. The conventional treatment of internal rotation in microwave spectroscopy is as a one-dimensional problem with the assumption that all geometrical parameters other than the torsional angle remain fixed during the rotation. This, of course, is not true, and the variation of other bond distances and angles can sometimes be of appreciable magnitude. It has recently been shown [64] that incorporation of computed information [65] on such variations for various ethene thiols can lead to a more consistent interpretation of the microwave spectra.

4.2.3 Other Combinations

Any attempt to classify the different combinations of experimental and theoretical methods that have been, or can be, used to provide the best chemical information from studies of molecular structure and bonding is inherently artificial. The important point is that improved understanding of the nature of the vibrational averaging effects incorporated in the various techniques has made it possible for the researcher to concentrate more fully on the chemical problem at hand and bring the full arsenal of experimental methods to bear on its solution. Since it is not practical to develop expertise in the use of all of the available methods in a given laboratory, increased cooperative effort between various groups has become the rule. It is likely that this trend will continue.

5 Prospects for the Future

The basic techniques of spectroscopy and diffraction are now reasonably mature with a well-developed theoretical background of undersranding of the experiments. Instrumental developments will certainly continue, and especially the major modifications that are needed to apply the standard methods to unusual species. This development reflects the growing shift in interest of chemists from the study of stable, ground-state molecules to the investigation of reactive substances under extreme conditions.

Both spectroscopy and electron diffraction from species in supersonic jets have begun to widen the opportunities for the study of unstable transients. The structure of ions is a major challenge which a number of groups are accepting. Some successful studies have been completed in this field, but there is not yet the base of structural information needed to build a solid theoretical framework of bonding theory in ions. A great deal of basic work needs to be done in the development of methods to obtain detailed and accurate structural information on the electronically excited species which are so important in chemical reactions.

A special challenge comes from the explosively rapid development of highly accurate computational methods for determining molecular structural features. Only a few years ago, computational accuracy at the level of the best experiments was impossible for anything but extremely small molecules. Improvements in computers and, especially, in quantum chemical methods, has altered the situation, and the trend will certainly continue. Chemistry is still an experimental science, but it will become increasingly important to incorporate information from computation along with data from high-quality experiments in order to penetrate as deeply as possible into the fundamental nature of matter.

Acknowledgement. The portion of the work described here that was done at The University of Texas was supported by a grant from The Robert A. Welch Foundation.

6 References

1. Barrow GM (1962) Molecular Spectroscopy. McGraw-Hill, New York
2. King GW (1964) Spectroscopy and Molecular Structure. Holt, Rinehart and Winston, New York
3. Walker S, Straw H (1966) Spectroscopy. Chapman and Hall, London, 2 volumes
4. Levine IN (1975) Molecular Spectroscopy. John Wiley, New York
5. Graybeal JD (1988) Molecular Spectroscopy. McGraw-Hill, New York
6. Schwendeman RH, Nygaard L (1988) Microwave Spectroscopy Information Letter No. XXXI. Department of Chemistry, Michigan State University, East Lansing, Michigan 48824, U.S.A.
7. Townes CH, Schawlow AL (1955) Microwave Spectroscopy. McGraw-Hill, New York
8. Wollrab JE (1967) Rotational Spectra and Molecular Structure. Academic
9. Gordy W, Cook RL (1970) Microwave Molecular Spectra. Wiley-Interscience, New York
10. Kroto HW (1975) Molecular Rotation Spectra. John Wiley, New York
11. Stiefvater OL, Wilson, Jr. EB (1969) J. Chem. Phys. 50: 5385

12. Bastiansen O, Seip HM, Boggs JE (1971) in: Dunitz JD, Ibers JA (eds) Perspectives in Structural Chemistry, vol IV. John Wiley, New York, p 60
13. Kraitchman J (1953) Am. J. Phys. 21 : 17
14. Costain CC (1958) J. Chem. Phys. 29: 864
15. Boggs JE, Altman M, Cordell FR, Dai Y (1983) J. Mol. Struct. (Theochem) 94: 373
16. Rosenblum B, Nethercot AH, Townes CH (1957) Phys. Rev. 109: 400
17. Pearson EF, Gordy W (1969) Phys. Rev. 177: 52, 59
18. Herschbach DR, Laurie VW (1962) J. Chem. Phys. 37: 1668
19. Morino Y, Kikuchi Y, Saito S, Hirota E (1964) J. Mol. Spectry. 13: 95
20. Morino Y, Saito S (1966) J. Mol. Spectry. 19: 435
21. See, for example, Stitch ML, Honig A, Townes CH (1954) Rev. Sci. Instr. 25: 759
22. Brown RD, Godfrey PD (1988) Proceedings of the Twelfth Austin Symposium on Molecular Structure, p 118, 28 February–3 March, Austin, Texas
23. Herzberg G (1945) Molecular spectra and molecular structure. II. Infrared and Raman spectra of polyatomic molecules. Van Nostrand, New York
24. Califano S (1976) Vibrational states. John Wiley, New York
25. Annual Review, Institute for Molecular Science (1988), Myodaiji, Okazaki 444, Japan
26. Durig JR, Will, Jr. JN, Green WH (1971) J. Chem. Phys. 54: 1547
27. Blackwell CS, Carreira LA, Durig JR, Karriker JM, Lord RC (1972) 56: 1706
28. Durig JR, Carreira A, Willis, Jr. JN (1972) J. Chem. Phys. 57: 2755
29. Jonvik T, Boggs JE (1981) J. Mol. Struct. 85: 293
30. Saupe A, Englert G (1963) Phys. Rev. Lett. 11: 462
31. Diehl P, Khetrapal CL (1969) in: Diehl P, Fluck E, Kosfeld R (eds) NMR, Basic principles and progress, Springer, Berlin Heidelberg New York
32. Snyder LC, Meiboom S (1974) in: Lide DR Jr, Paul MA (eds) Critical evaluation of chemical and physical structural information, National Academy of Sciences, Washington, D.C.
33. Rankin DWH (1988) Proceedings of the Twelfth Austin Symposium on Molecular Structure, p 40, 28 February–3 March, Austin, Texas
34. Seiler P (1988) in: Domenicano A, Hargittai I, Murray-Rust P (eds) Accurate molecular structures, Oxford University Press, Oxford
35. Dunitz JD (1979) X-Ray Analysis and the Structure of Organic Molecules, Cornell University Press, Ithace, New York, p 244
36. Johnson CK (1980) in: Diamond R, Ramaseshan S, Venkatesan K (eds) Computing in crystallography, Indian Academy of Sciences, Bangalore, p 14.01
37. Willis BTM, Pryor AW (1975) Thermal vibrations in crystallography, Cambridge University Press, London
38. Johnson CK, Levy HA (1974) in: Ibers JA, Hamilton WC (eds) International Tables for X-Ray Crystallography, vol IV, Kynoch, Birmingham, New York
39. Trueblood KN (1988) in: Domenicano A, Hargittai I, Murray-Rust P (eds) Accurate molecular structures, Oxford University Press, Oxford
40. Mark H, Wierl R (1930) Z. Physik 60: 741
41. Debye P (1915) Ann Physik 46: 809
42. Seip HM (1967) in: Andersen P, Bastiansen O, Furberg S (eds) Selected topics in structure chemistry, Universitetsforlaget, Oslo
43. Hargittai I (1980) Z. Chem. (Leipzig) 20: 248, 406
44. Vilkov LV, Mastryukov VS, Sadova NI (1983) Determination of the geometrical structure of free molecules, MIR Publishers, Moscow
45. Fink M, Moore PG, Gregory D (1979) J. Chem. Phys. 71: 5227
46. Ewbank JD, Schäfer L, Paul DW, Benston OJ, Lennox JC (1984) Rev. Sci. Instr. 55: 1598
47. Hargittai M, Hargittai I (1981) J. Mol. Struct. 73: 253
48. Spiridonov VP et al., Laboratory of Electron Diffraction, Department of Chemistry, Moscow State University, 119899 GSP Moscow V-234, U.S.S.R.
49. Hargittai I et al., Structural Chemistry Research Group, Hungarian Academy of Scineces, Eötvös University, Budapest, Pf. 117, H-1431, Hungary
50. Fink M et al., Department of Physics, The University of Texas, Austin, Texas 78712, U.S.A.
51. Kuchitsu K, Konaka S (1966) J. Chem. Phys. 45: 4342

52. Yamamoto S, Kuwabara R, Kuchitsu K, Takami M, to be published
53. Cyvin SJ (ed) (1972) Molecular structures and vibrations. Elsevier, Amsterdam
54. Bartell LS (1972) in: Weissberger A, Rossiter BW (eds) Techniques of chemistry, IIID, 125. Wiley-Interscience, New York
55. Ohshima Y, Yamamoto S, Nakata M, Kuchitsu K (1987) J. Phys. Chem. 91: 4696
56. Cradock S, Liescheski PB, Rankin DWH, Robertson HE (1988) Proceedings of the Twelfth Austin Symposium on Molecular Structure, p 40, 28 February–3 March, Austin, Texas
57. Gase W, Boggs JE (1984) J. Mol. Struct. 116: 207
58. See, for example, Van Hemelrijk D, Van den Enden L, Geise HJ, Sellers HL, and Schäfer L (1980) J. Amer. Chem. Soc. 102: 2189
59. Summarized by Domenicano A (1988) in: Domenicano A, Hargittai I, Murray-Rust P (eds) Accurate molecular structures, Oxford University Press, Oxford
60. Hargittai I (1986) in: Liebman JF, Greenberg A ϱeds) Molecular structure and energetics, Vol 2, VCH Publishers, Deerfield Beach, FL
61. Summarized in Hayashi M, Kato H (1980) Bull. Chem. Soc. Jpn. 53: 2701
62. Hayashi M, Kato H (1979) J. Mol. Struct. 53: 179
63. McKean DC, Boggs JE, Schäfer (1984) J. Mol. Struct. 116: 313
64. Almond V, Permanand RR, Macdonald JN (1985) J. Mol. Struct. 128: 337
65. Plant C, Macdonald JN, Boggs JE (1985) J. Mol. Struct. 128: 353

Interplay of Experiment and Theory in Determining Molecular Geometries
B. Theoretical Methods

James E. Boggs

Department of Chemistry, The University of Texas, Austin, Texas 78712, USA

1 Introduction

The previous chapter (Chap. 6) was primarily concerned with experimental methods which can provide the highest presently possible absolute accuracy in the determination of molecular geometry parameters. In favorable cases, this may involve uncertainties of only a very few tenths of a picometer or tenths of a degree. As a parallel presentation, this chapter concerns theoretical methods that can hope to approach comparable absolute accuracy. The importance of accuracy at this level was stressed in Chap. 6. Structural information is used in chemistry as a probe of the small shifts of electron density that distinguish one molecule from closely related ones, and also to suggest interpretations for molecular individuality in properties and reactivity. Substituent effects and interactions between component parts of a molecule typically cause geometry alterations on the order of no more than a few picometers or degrees, so very high accuray is required if structural data, experimental or theoretical, are to provide the maximum chemically useful information.

While the present work deals primarily with attempts to reach the highest levels of accuracy and reliability, it must be emphasized that there are certain chemical purposes for which this degree of accuracy is not needed. (There are also numerous problems for which it is needed but not possible, for various reasons discussed in detail below.) Occasionally, chemical interest centers around such questions as the possible stability of a proposed new compound involving a novel chemical bond or the nature of the intermediate in the course of a reaction. In these cases, accuracy to one or two picometers may often be adequate.

Some people believe that one should speak of the "determination" of a structure only when referring to the result of an experimental evaluation of a molecular geometry and prefer to use a word like "prediction" when referring to a computational result. It does not seem to this author that the two approaches are so logically different that it is worthwhile to quibble over the terminology. An experiment uses a great deal of theory to bridge the gap between the swing of a meter needle and an average distance between two nuclei in a molecule of the substance placed in a sample holder. Many computational steps whose accuracy is essential to the reliability of the result are needed to process the direct experimental information. Any good scientist will rely on the experimental result only in a tentative fashion, subject to refutation or modification by future information. The same attitude should apply to a computational evaluation of molecular geometry. Either the experimental or the theoretical input to the chemist's knowledge is a bit of inexact information, the reliability of which must be accepted only within certain limits and with the understanding that the initial judgment of reliability may be subject to revision.

A number of factors have combined during very recent years to create a tremendous surge in interest in the use of computational approaches to molecular geometry evaluation. All of these have worked in the direction of making possible determinations of greater reliability for progressively larger molecular systems. First, and most obvious, is the explosively rapid growth in both the power and availability of computers. The great increase in both speed and, at least as important, storage capacity has made possible the application of more rigorous ab initio methods to molecules of a size that could only be treated by semi-empirical techniques a few years ago. However, it must be realized that the computer time requirement of various ab initio procedures varies as the fourth to seventh power of the number of electrons in a molecule, so that increasing the raw speed of computers by many orders of magnitude can lead to only relatively minor increases in the size of molecules that can be studied.

The increasing availability of computers is perhaps even more important than the continual change in the capability of the currently most powerful computer. The introduction of so-called superminicomputers (a term which arouses curiosity about the future of names as well as computer design) at prices comparable to those of many common laboratory instruments promises to bring the computational capability needed for the most sophisticated quantum chemical work into every interested laboratory.

Of even greater importance than developments in computer technology have been the advances made in recent years in quantum chemical methodology, and this area presents at least as great hope for order of magnitude forward leaps in the near future. Perhaps the single most important contribution has been the introduction and elaboration of energy derivative techniques, as discussed below. Furthermore, there have been dramatic new developments in methods for treating dynamic electron correlation. Improvements in computer algorithms, the methods by which theory is implemented to carry out the computation, have been highly significant, and much more can be done in this area to take full advantage of the vectorization and parallel processing available in modern computers. There have also been significant improvements in semi-empirical methods and in techniques for compensating for the residual errors of ab initio calculations.

The popularization of the applications of quantum mechanics to chemical problems also owes a tremendous debt of gratitude to certain workers who have created standard programs and, over a period of many years, maintained them, introduced steady improvements, and made them freely available to workers everywhere. The GAUSSIAN family of ab initio programs arising from the group of John A. Pople and the family of semi-empirical programs including MINDO, MNDO, IAM, etc., coming from Michael J. S. Dewar and his coworkers must especially be mentioned.

Before delving into the details of quantum chemical methodology and the current successes and problems of the procedures, it is worthwhile to obtain an overview of how the geometry of a molecule is determined by computation and the nature of the energy derivative methods that now play such a dominant role in the field.

1.1 Theoretical Prediction of Molecular Geometry

Quantum chemical determination of molecular geometry normally begins with adoption of the Born-Oppenheimer approximation, in which nuclear and electronic motion can be treated as separable problems. There are rare occasions where this approximation may be questionable, but such cases seldom arise. An initial nuclear configuration of the molecule is assumed, and the electronic Schrödinger equation is written and solved to some level of accuracy. It is in this step that all of the approximations and neglect of various effects are made that may introduce significant uncertainty into the final answer. Next, in the conceptually simplest approach, the total energy arising as the lowest eigenvalue of the Sdhrödinger equation is noted, the nuclear geometry is altered, and the process is repeated. Finally, a nuclear configuration is obtained which corresponds to the lowest total energy. By definition, this is the equilibrium geometry of the molecule.

It is important to recognize that, unlike the various experimental methods described in the previous chapter, the computational approach always predicts an equilibrium, r_e geometry uncontaminated by any sort of averaging over vibrational states. This is an important advantage in that it simplifies comparison of structural trends in different molecules, unaffected by the vibrational behavior of the molecule as a whole. It may be objected that the resulting bond distances and angles are not those of the real, vibrating molecule, but at least the equilibrium structure is clearly and unequivocally defined and easier to work with than such concepts as the inverse root mean square average over the oscillating internuclear distance introduced by certain experimental methods (see Chap. 6). The differences are largest for bonds to hydrogen; the r_e C—H bond distance in a typical molecule differs by as much as one or two picometers from the various kinds of averages observed experimentally. It is important to remember this consideration when reading papers in which exact agreement is found between experimental and theoretical determinations — this should not be so!

1.2 Energy Derivatives

The technique outlined above is effective for diatomics and other quite small molecules in which there are very few geometrical parameters that must be varied independently. For larger molecules, however, the problem becomes hopelessly complicated. For a 15-atom molecule with no symmetry, for example, there are 39 independent parameters for which optimum magnitudes must be found simultaneously to produce the minimum energy and hence the equilibrium geometry. By variation of one of the parameters, its optimum value with respect to the pre-selected values of all the others could be determined. This would then need to be repeated for all the others, in turn. Now the complete cycle should be repeated until an overall minimum is reached. The computation-intensive solution of the Schrödinger equation, repeated so many times, becomes impossibly expensive.

The solution to the difficulty came with the introduction of gradient techniques, first formulated in detail by Péter Pulay [1] in 1969. After the first solution of the electronic Schrödinger equation, the resulting wavefunction is used to calculate the expectation value of the derivative of the total energy with respect to displacement of each of the molecular parameters, a quantity which can be calculated analytically within the limitation imposed by the accuracy of the available wavefunction. This single calculation yields both the magnitude and direction of the force acting to direct each atom toward its equilibrium position. A set of force constants is now assumed, possibly by reference to similar motions in related molecules. A second trial geometry is next determined from the computed forces and assumed force constants. Since neither the computation nor the assumption are, in general, exactly correct, the process must be recycled several times. Nevertheless, the number of times the

Schrödinger equation must be solved is greatly reduced and, most importantly, it does not increase drastically with increase in size of the molecule.

A further alternative is possible. From the wavefunction obtained after the first solution of the Schrödinger equation, the matrix of second derivatives of energy with respect to nuclear displacement can be calculated in addition to the first derivatives. Since the second derivatives are the force constants, they can be combined with the computed forces to give a more accurate prediction of the displacements from the original assumed geometry to obtain the final equilibrium geometry. Again, with the necessarily approximate solution of the Schrödinger equation, a few cycles may be needed to obtain convergence. This is especially true if some of the molecular vibrational motions are highly anharmonic, since higher derivatives of the energy would be needed to compensate for the effect of anharmonicity.

2 Theoretical Models

No full exposition of the mathematical basis of quantum chemistry is possible here. General background can be found in the intermediate level text by Szabo and Ostlund [2]. More specialized discussions of direct relevance to the present topic are found in Vols 3 and 4 of Schaefer's series on Modern Theoretical Chemistry [3, 4] and vols 67 and 69 of Advances in Chemical Physics [5, 6].

2.1 Ab Initio Methods

It is impossible to solve the molecular electronic Schrödinger equation without making approximations of various sorts. Those procedures which utilize no experimental data beyond the numerical values of physical constants (Planck's constant, the charge on the electron, etc.) and the comparisons which may have been involved in optimization of basis sets are known as ab initio methods. Still, there is a broad range of approximations possible within the ab initio family of procedures, and the effect of these on the accuracy of computed molecular geometries can range from negligible to unacceptably large. To characterize a computed results as an ab initio result without further clarification is nearly meaningless.

2.1.1 The Hartree-Fock Approximation

In the usual type of Hartree-Fock self-consistent field (HF—SCF) approach to the solution of the Schrödinger equation, the Born-Oppenheimer separation of nuclear and electronic wavefunctions is assumed, all relativistic and spin coupling terms are omitted from the Hamiltonian, and the total electronic wavefunction is written as a single determinant in a set of occupied spin orbitals subject to the usual condition of orthonormality:

$$\Psi = A \det \{\chi_1(1)\, \chi_2(2) \ldots \chi_n(n)\} . \qquad (1)$$

The spin orbitals, χ_i, are one-electron functions which may be written as products of spatial and spin functions $\psi_i^\alpha\alpha$, $\psi_i^\beta\beta$, etc. If the molecular system does not contain unpaired electrons, it is often considered adequate to use the restricted Hartree-Fock (RHF) procedure in which a pair of electrons with opposite spin is associated with a single spatial orbital, i.e., $\psi_i^\alpha\alpha = \psi_i^\beta\beta$. Although difficulties in convergence are sometimes encountered, it is safer to avoid this further approximation and use the spin-unrestricted Hartree-Fock (UHF) method when spin multiplet states are sought.

Next the molecular orbitals are expanded as a linear combination of some finite set of basis functions:

$$\psi_i = \sum_\mu c_{\mu i}\varphi_\mu, \tag{2}$$

where the expansion coefficients are evaluated by the variation principle to produce the lowest molecular energy. Equation (2) would, of course, be exact if an infinite basis set were used, but the necessary requirement for a finite basis set introduces one of the major errors of this, and all other, ab initio methods.

The basis functions, φ_μ, may be Slater-type functions [7], constructed to mimic the expected atomic orbitals:

$$\begin{aligned} \varphi_{1s} &= (\zeta_1^3/\pi)^{1/2} \exp(-\zeta_1\mathbf{r}) \\ \varphi_{2p} &= (\zeta_2^5/\pi)\,(x, y, z) \exp(-\zeta_2\mathbf{r}) \\ &\text{etc.} \end{aligned} \tag{3}$$

A minimal basis set is one using only the same number of such functions as is required to correspond to the filled or partially filled shells of the atoms. If two sets of Slater-type functions are used for each populated or partially populated shell, the basis set would be called a double-zeta basis, etc. Only a relatively small number of Slater-type orbitals are required to develop a reasonably good approximation to the true molecular wavefunction. They present a computational difficulty, however, in that they lead to two-electron repulsion integrals which are difficult to evaluate.

As described in more detail by Pople [8], it has been found to be more efficient to approximate the Slater functions by the use of linear combinations of Gaussian-type functions:

$$\varphi'_\mu(\xi, \mathbf{r}) = \xi^{3/2}\varphi'_\mu(1, \xi\mathbf{r}) \tag{3}$$

where

$$\begin{aligned} \varphi'_{ns}(1, \mathbf{r}) &= \Sigma\, d_{ns,k}\, g_{1s}(\alpha_{nk}, \mathbf{r}) \\ \varphi'_{np}(1, \mathbf{r}) &= \Sigma\, d_{np,k}\, g_{2p}(\alpha_{nk}, \mathbf{r}) \\ &\text{etc.} \end{aligned} \tag{4}$$

The normalized Gaussian functions g_{1s} and g_{2p} are:

$$\begin{aligned} g_{1s}(\alpha, \mathbf{r}) &= (2\alpha/\pi)^{3/4} \exp(-\alpha\mathbf{r}^2) \\ g_{2p}(\alpha, \mathbf{r}) &= (128\alpha^5/\pi^3)^{1/4}\,(x, y, z) \exp(-\alpha\mathbf{r}^2) \\ &\text{etc.} \end{aligned} \tag{5}$$

Although more Gaussian than Slater-type basis functions are required to obtain the same accuracy in reproduction of the molecular wavefunction, the greater ease in evaluation leads to an overall gain.

The subject of basis set selection has been reviewed many times. Two recent surveys are those by Dunning and Hay [9] and by Wilson [10]. A commonly used type of minimal basis set is that developed by the Pople group [11] and designated as STO-KG, where K is a number commonly between 2 and 6 that shows the number of Gaussian-type functions which are combined to mimic each Slater-type orbital. The constants d and α in Eq. (4) are chosen to give the best fit to the Slater orbitals with $\zeta = 1$. The exponents ζ_1 and ζ_2 in Eq. (3) should ideally be optimized for every molecule studied, but they are normally set at pre-established values to save computer time.

Minimal basis sets provide only rough values for computed bond distances, conformations, and relative energies. Extra flexibility for the description of the true molecular wavefunction can be provided by the use of double-zeta bases in which two sets of functions with different orbital exponents are used in place of each minimal basis orbital. Triple-zeta bases constitute a similar further extension. To describe valence properties including bond lengths and angles, it is more important to have a good description of the outer region of the electron density distribution (although this is not the part which contributes most strongly to the absolute molecular energy). In recognition of this fact, it is common to use basis sets which are minimal in the inner shells but double- or triple-zeta in the valence shell. In the Pople-type bases, these may be described as, for example, a 6-311G basis for the carbon atom. In this nomenclature, the 6 shows that the Slater-type orbital describing the inner shell is approximated by a combination of 6 Gaussian functions. The outer shell is triple-zeta, one orbital being described by the sum of 3 Gaussians and each of the other two being represented by a single Gaussian function.

While splitting a Slater-type orbital into a combination of several similar orbitals increases the radial flexibility of the basis in describing the true molecular orbital, it is not an efficient way to improve the angular description. This is done by adding polarization functions to the basis, i.e., functions corresponding to a higher orbital angular momentum quantum number than actually present in a given shell, for example, adding *d* functions to the basis for an atom such as carbon. Within the Pople family of basis sets, this is denoted by an asterisk, as a 6-31G* basis for carbon would contain a set of Gaussian functions reproducing an atomic *d* orbital. Several *d* orbitals, or even *f* orbitals or ones of higher quantum number may be used in very exact calculations on small molecules. At least *d* orbitals are needed in the valence shell for highly accurate reproduction of molecular geometry, especially the angular geometry, of most molecular systems.

Aside from the necessary use of a finite basis set, the most important error in the Hartree-Fock approach is neglect of dynamic electron correlation, which is inherent in the fundamental expansion of the molecular wavefunction in terms of one-electron functions. The method describes the motion of each electron in the average field of all the others, but it does not reproduce the instantaneous avoidance of close proximity by two electrons. As a result, electron distributions calculated by the Hartree-Fock method give somewhat too great a concentration of electron density in crowded regions such as multiple bonds. Bond lengths computed at the Hartree-Fock limit (an infinitely large basis set) are therefore typically shorter than the true value, with the effect being most marked for multiple bonds such as C≡C or C≡N.

2.1.2 Post-Hartree-Fock Methods

The Hartree-Fock approximation is reasonably adequate for evaluating the geometries of most molecules at or near a state of equilibrium, although lengths of multiple bonds can be several picometers in error and even larger errors are encountered in calculations on molecules involving heavier atoms. Its flaws become most serious for molecular systems far from stable equilibrium conditions, exactly the conditions which are of most interest in following the progress of chemical reactions involving bond breaking and formation.

The development of cost-effective methods for overcoming the limitations of the Hartree-Fock procedure has become one of the most active fields of research in quantum chemistry. No consensus of opinion has yet been reached as to the best method, and indeed it is unlikely that any single method can be acclaimed as best for all applications.

Perhaps the most direct approach is use of the configuration interaction (Cl) method, described in a fundamental way by Szabo and Ostlund [2], and for which Shavitt [12] has presented an excellent and extensive recent review. The method is attractive in that it is, in principle, capable of providing an exact solution of the non-relativistic Schrödinger equation subject only to the limitation imposed by the Born-Oppenheimer approximation. Achievement of its potential is, however, extremely difficult computationally.

The basic idea of the CI method is very simple. A trial wavefunction is written as a linear combination of "configurations", i.e., products of one-electron orbitals. The variational principle is then applied to obtain the coefficients resulting in the lowest system energy. The number of configurations is finite for a set of one-electron orbitals coming from a finite basis set, but unfortunately, the number is extremely large, even when only a minimal basis set is used for a small molecule. Such a full CI result has been obtained only in a very few cases, but it is exact for the specified one-electron basis. For any but the smallest molecules and most elementary basis sets, a selection of configurations must be made. Needless to say, the difficulties with the method arise in the selection of those terms which are most significant and in concern over the cumulative effect of small contributions from all the eliminated terms.

Most commonly, a restricted Hartree-Fock wavefunction with an appropriate basis set is chosen as the zero-order wavefunction. The other configurations used in the CI expansion can then be classified in terms of the number of electrons promoted from the occupied SCF orbitals to virtual orbitals. It can be shown [13] that the first-order correction to the wavefunction contains only double excitations and these determine the energy through third order. The second-order wave function correction contains single, triple, and quadrupole excitations. Thus, a CI(SDTQ) calculation means a configuration interaction calculation in which all corrections to the wavefunction through second-order are retained. Such a calculation, based on an adequate basis set, would give a very good evaluation of most molecular properties, but it would also require so much computer time it is not currently practical for molecules containing more than at most half a dozen non-hydrogen atoms.,

A different approach that is of considerable importance is the multiconfiguration self-consistent field (MCSCF) method. A good summary has been given by Wahl and Das [14]. In this technique several configurations as well as the orbitals of which they are composed are mixed, subject to optimization using the variational principle. Again, an exact solution of the Schrödinger equation could be obtained by use of an infinite set of configurations, but this is impossible. A selection of configurations must be chosen.

In the simplest formulation of MCSCF theory, only two configurations may be used; for example, corresponding to the initial and final states when calculations are to be made on the transition state of an isomerization reaction. In other cases, the configurations to be mixed are not so obvious and their selection is a cause for concern.

Coupled-pair and coupled-cluster methods are size-consistent generalization of the configuration interaction method. The method was first introduced in 1966 by Cizek [15] and the first implementation of a simplified version (the coupled electron-pair approximation, CEPA, was made by Meyer [16]. An introductory summary of such methods is given by Szabo and Ostlund [2] and a further review with a large number of references has been provided by Kutzelnigg [17]. These techniques are the subject of much current interest, and they appear to be achieving ever wider utility.

A completely different procedure that is coming into increasingly more common use is based on Rayleigh-Schrödinger perturbation theory [2]. Many-body perturbation theory (MBPT) uses an average-field, one-electron Hamiltonian and calculates the correlation energy by standard perturbation theory techniques. The approach is computationally efficient since it is not an iterative procedure. The formulation by Møller and Plesset [18] has been included in the popular GAUSSIAN family of computer codes as well as a number of others which are in general use. A review covering both MBPT and coupled-luster methods has been presented by Bartlett [19].

A reasonably large number of geometry predictions using second-order Møller-Plesset perturbation theory (MP2) has now appeared in the literature, and the results show a marked improvement over the SCF results. For accuracy approaching a few tenths of a picometer in bond length or a few tenths of a degree in bond angle, it has become apparent [20] that MP4 is required, i.e., perturbation theory through fourth order, as well as a large SCF basis set, at least triple zeta with multiple sets of polarization functions. Such calculations for medium-sized molecules are expensive, but they probably represent the cheapest way that such absolute accuracy can be currently obtained.

One recent innovation that promises to create a breakthrough in electron correlation treatments for larger molecules is the use of a basis set of localized orbitals for the correlation calculation [21]. The fundamental idea is very simple. There is little interaction between electrons which spend most of their time in parts of the molecule remote from each other. While the canonical one-electron orbitals obtained directly from solution of the Hartree-Fock equations are highly delocalized over the entire molecule, any combination of the complete set of such orbitals provides an equally valid solution to the Schrödinger equation. Thus one can devise criteria [22] for finding equivalent orbitals that have minimum overlap with each other and give the closest approximation to the usual chemist's concepts of lone pairs, bonding pairs, inner shells, etc. Such a set of localized orbitals can then be used as the basis set for any of the common treatments of electron correlation. A preliminary test is used to indicate whether there is a significant contribution to the total correlation energy from interaction between any given pairs of electrons. If the contribution is extremely small, it is neglected. If it is marginal, it is treated by a less accurate method. The great advantage of this approach is that the computer time required for the electron correlation treatment is essentially independent of the size of the molecule, above a certain minimum. The technique has been applied for molecules as large as benzene [23] using a large basis set and MP4 correlation treatment.

An entirely different sort of error becomes important for geometrical predictions in molecules containing heavier elements — the common neglect of relativistic effects in the molecular Hamiltonian. The direct solution of the relativistic Schrödinger equation is impossible for any atomic or molecular system, but corrections can be made to calculations based on a non-relativistic Hamiltonian. There are a number of types of relativistic effects that may be of importance. First, the mass-velocity correction alters the kinetic energy due to the variation of electron mass with its velocity. Of special importance for heavier atoms with open shell configurations is the spin-orbit correction arising from the strong coupling of the spin of the electron with the orbital angular momentum. An analogous two-electron effect is called the Breit interaction. There is also a Darwin correction which comes from the Dirac relativistic equation but has no obvious physical interpretation.

Since only the inner-shell electrons in atoms with large nuclear charge are moving at relativistic velocities, it might at first be thought that chemical properties, dependent as they are on the electron density distribution in the outer regions of an atom, would be immune to relativistic effects. Regrettably, this is not the case. The inner-shell orbitals shrink due to the mass-velocity correction, thus shrinking in turn the *s* orbitals of the outer shells. This shrinkage produces variations in predicted bond distances as well as in properties such as ionization potential. The *p* orbitals are less affected by the mass-velocity correction, but the spin-orbit interaction splits them into $p_{1/2}$ and $p_{3/2}$ subshells and makes possible the mixing of states that would otherwise be forbidden. The chemical consequences of relativistic effects are of sufficient magnitude that their consideration is essential in any accurate calculation on heavier atoms. For details, the reader is referred to the excellent review by Balasubramanian and Pitzer [24].

2.1.3 Derivative Procedures

Solution of the electronic Schrödinger equation at some chosen level of approximation leads to an approximate energy and an approximate wavefunction for the molecular system. From the wavefunction, any observable property of the molecule can, in principle, be calculated, including the energy gradients, i.e., the forces acting on the nuclei tending to move them toward their equilibrium positions. If the wavefunction were exact, these forces could be evaluated in a trivial manner from the Hellmann-Feynman theorem [25, 26] but the treatment is more complicated when an approximate wavefunction is used [1], as it must be in any real situation.

The mathematical formulation of the procedure required for evaluation of the first derivative of energy with respect to nuclear displacements is clearly presented in two excellent reviews by Pulay [27, 28]. The first of these considers only SCF wavefunctions, but Ref. 28 also discusses the rapidly developing field of evaluation of energy derivatives using wavefunctions that include electron correlation. The first efficient method for evaluating derivatives from correlated wavefunction was presented in a classic paper by Pople et al. [29], and a number of significant improvements in the implementation have subsequently appeared [30, 31]. A recent paper by Rice and Amos [32] presents a simplification in the general formalism for correlated wavefunctions. The field is a very active one.

Essentially all computational evaluations of molecular geometry now use the gradient method because of the tremendous saving in computer time which it provides. Up until very recently, such calculations were largely restricted to SCF wavefunctions, but the recent implementation of gradients at least through the MP2 level into a number of commonly used computer codes, including the recent versions of the GAUSSIAN series, has removed this limitation.

2.2 Semi-empirical and Empirical Methods

Historically, semi-empirical methods have played an important role in the computational prediction of molecular geometries. While such techniques have now largely been supplanted by ab initio calculations for small and medium-sized molecules, they remain the only practical alternative for very large systems. As such, they continue to play a vital part in studies of many important biological and pharmaceutical problems. While their reliability cannot approach that of the higher level ab initio methods, they can give satisfactory answers in appropriate cases.

The time-saving features of semi-empirical methods center on their treatment of the overlap integrals

$$S_{12} = \int \varphi_1^*(\mathbf{r})\, \varphi_2(\mathbf{r})\, d\mathbf{r} \tag{6}$$

that arise in the Hartree-Fock method. The various methods may be thought of as derived from Hartree-Fock type calculations by neglecting certain of the matrix elements and estimating the majority of the remaining integrals empirically, the manner in which they do this serving to distinguish one method from another. Procedures are variously designated as CNDO (complete neglect of diatomic overlap), INDO (intermediate neglect of diatomic overlap), MINDO (modified intermediate neglect of diatomic overlap), or NDDO (neglect of diatomic differential overlap). The nature of the parameterization also varies significantly. Early development of the methodology is largely attributable to Pople and his coworkers while more recent developments, particularly improvements in parameterization and extensive application to problems of organic chemistry, are associated with the fine work of the group associated with Dewar.

Completely empirical predictions of molecular structure are based entirely on classical mechanics and involve the use of molecular vibrational potential functions generated from empirically determined terms for such factors as interatomic repulsion, vibrational force constants, torsional constants, etc. Within their range of applicability, such predictions are often surprisingly accurate. They cannot, of course, be used for predictions on molecules which vary seriously from the molecules which were used in deriving the input parameters of the method, either in the atoms they contain or in the nature of their bonding.

In spite of the usefulness of semi-empirical and empirical procedures in prediction of the structures of large molecules, the subject is not pursued in further detail here because of the lower certainty of reliability that may be attached to the results obtained.

3 Applications

Literally thousands of studies have been reported giving structural predictions obtained by means of computation. There is an unfortunate tendency among non-theoreticians when referring to such results to label them simply as "theoretical structures", in spite of the great difference in accuracy and reliability which can be attached to results from various methods. This is in close parallel with the similarly unfortunate habit of theoreticians to refer to "experimental results" and accept them as necessarily absolutely true (without even any citation of the author's claimed experimental uncertainty), especially if they happen to agree with the computed values. A more rigorously scientific discrimination on both sides would elevate the level of discourse in the literature.

3.1 Completely Ab Initio Calculations

Careful comparisons have been made between high-level ab initio geometry optimizations and correspondingly accurate experimental determinations for a number of very small molecules. The results of one such comparison [20] for HF and HCN are shown in Table 1. The calculations made use of a basis set which was triple-zeta in the valence shell (6-311G) with the addition of two sets of *d* functions for C and N and two sets of *p* functions for H.

The simple SCF result for HF, without any correlation correction, gave an H—F bond length which was too short by 1.8_8 picometers compared with the highly accurate experimental value [33], an error of 2%. When the computation also included a CI calculation at the SD (single and double excitation) level plus the Davidson correction [34] for unlinked cluster contributions, the theoretical value became indistinguishable from the experimental value.

Table 1. Experimental and Theoretical Structural Comparisons for HF and HCN[a]

	SCF (6-311G****)	+CI(SD)+Davidson	Experimental
HF, r_e	89.82	91.72	91.70[b]
HCN, r_e(H—C)	105.73	106.56	106.55[c]
r_e(C≡N)	112.36	115.38	115.32[c]

[a] Ref. 20. Distances in picometers. [b] Ref. 30. [c] Ref. 32.

For HCN, the SCF calculation at the same level as described for HF gave bond lengths too short by 0.82 and 2.96 picometers for H—C and C≡N, errors of 0.7% and 2.6%, respectively. The full theoretical calculation, including correlation, produced an H—C bond length indistinguishable from the experiment [35] and a C≡N bond length 0.05% too short, which may or may not be significant. The larger effect of correlation on the C≡N triple bond is expected, since ignoring dynamic electron correlation tends to predict too great a concentration of electrons in densely populated regions such as multiple bonds and therefore to make predictions of such bond lengths excessively short.

Geometry optimizations at the quantum mechanical level expected to produce results of this accuracy have been performed for a variety of other small and medium-sized molecules, containing up to about half a dozen non-hydrogen atoms. It is very difficult to compare such results with experiment, however, because of the paucity of experimental data of comparable accuracy. As discussed in detail in Chap. 6, experimental determinations of molecular structure can, in favorable cases, produce highly accurate values for the particular type of vibrational average structure measured by the experiment. The difficulty generally arises in lack of the harmonic and anharmonic vibrational force constants required to obtain comparable accuracy in removing the vibrational contributions from the experimental result to obtain the equilibrium structure, found directly by computation, that is most simply related to chemical bonding effects.

It is now generally assumed that a theoretical geometry optimization which is expected to produce bond lengths accurate to one or two tenths of a picometer must use a basis set that is at least triple-zeta in the valence shell augmented by at least one set, and preferably two sets, of polarization functions on all atoms, and include a high level correlation correction, perhaps an MP4 perturbation treatment or a comparable variational procedure. A common error has involved emphasis on use of high level treatment of electron correlation with an inadequate basis set. It has been known for a long time [20] that a basis set of the size described above is more important than extension of the correlation treatment beyond the MP2 or corresponding variational level. In fact, it has even been shown [36] that the inclusion of *f* functions in the SCF basis is necessary for some systems such as N_2 and O_2. Special cases that require additional consideration are discussed in Sect. 4 below.

Interestingly, electron correlation generally has a much smaller effect on bond angles than on bond lengths. This, again, is understandable, since the main effect of ignoring correlation is to increase electron density in bonded regions, thereby having a larger effect on the length and strength of bonds than on the angles between them. In many cases, computed bond angles are found to agree with high-level experimental values even without the inclusion of any electron correlation treatment and with an appreciably smaller basis set than is required for the accurate determination of bond lengths. The accurate prediction of dihedral angles or out-of-plane puckering angles generally requires a larger basis set than is needed for bond angles, but even here the use of electron correlation is often unnecessary.

A very useful review of geometry optimizations of small molecules has recently appeared [37], although the degree of satisfaction expressed with MP2 results may not always be warranted, especially for multiply bonded systems.

It is important to emphasize the great value of theoretical calculations of the geometries of molecular systems that cannot be measured experimentally by any present method; for example, reaction transition states. One typical example, chosen nearly at random, is the study by Schleyer and Pople [38] of the reaction between H_2 and Li_2 to form a dimer of unknown structure. Both the detailed geometry of the dimer and of the transition state (the structure at the saddle point along the minimum-energy reaction coordinate) were obtained by MP2/6-31G** calculations. Although this level of calculation is not quite that needed to give the highest possible accuracy, it provided reasonably accurate data for this reaction that could not have been obtained in any other way. Numerous other laboratories, of which the prolific group working with H. F. Schaeffer III should certainly be mentioned, are now producing detailed information on the structures, energetics, and vibrational energy surfaces involved in the reactions of many very small organic molecules.

At the present time, calculations of the type described in this section are restricted to quite small molecules by the limitations of existing computer speed and memory and by existing computational methods. Some of the largest molecules for which geometries have been determined with triple-zeta, double polarization, MP4 or equivalent, accuracy are benzene [23] and the trifluorocyclopropenyl cation [39], $C_3F_3^+$. Work is underway which promises that somewhat larger molecules can be treated in the near future, but for the present it is necessary for larger molecular systems either to be satisfied with a lower standard of accuracy or to introduce an element of empiricism as described in Sect. 3.2.

3.2 Use of Transferable Empirical Corrections

From the earliest days of determination of molecular geometries by energy optimization it has been noted that the relative accuracy of the results obtained using some chosen computational level is better than the absolute accuracy. For example, using a basis set such as the 4-21G split Gaussian basis and no treatment of electron correlation, it was found that the C—H bond lengths as calculated in a wide variety of compounds always seem to be about 0.5 picometer shorter than the best experimental estimate of the corresponding r_e values [40]. There is an obvious reason for this in that the correlation error, produced by excessive accumulation of electron density in bonding regions, tends to be nearly identical for a given type of bond with only very minor influence from the nature of the remainder of the molecular environment. If the basis set is large enough, the error due to its finiteness becomes negligible and is therefore also essentially invariant to the molecule in which the given type of bond occurs. The contributions of the two effects have been illustrated [41] by the schematic diagram shown in Fig. 1.

In Fig. 1, the vestical axis shows the difference between the true value in a given molecule of some particular parameter, say C—C bond length, and the value computed with various sizes of SCF basis but without any consideration of electron correlation. The errors in computing the chosen parameter in many different molecules are found to fall within the shaded area of the diagram. The Hartree-Fock convergence limit, approached by very large basis sets, still differs from the true value because of the neglect of electron correlation, but this residual error is very nearly independent of the molecule studied. For basis sets which are too small, there remains a random scatter in different molecules because of the more variable basis

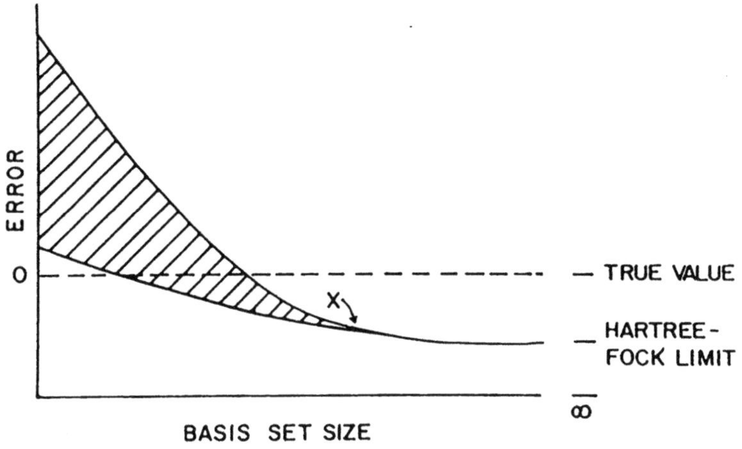

Fig. 1. Schematic representation of the error in calculating some single structural parameter (such as C—H distance of X—O—Y angle) in a variety of molecular environments as a function of basis-set size. For wide families of systems, the error is found to fall within the shaded area. Taken with permission from Ref. [41]

set error. Thus, a series of calculations with a minimal basis set, falling in the shaded part of the diagram, would show scatter in the error found for the C—C bond length in various molecules. (Some semi-empirical calculations approach the results of minimal basis set ab initio calculations very closely, and show equivalent scatter of error.) With a sufficiently large basis set, which may be as small as double-zeta plus polarization functions, the remaining basis set error is so small that its variation from molecule to molecule is negligible. It would appear most economical to make calculations at the point marked "X" on the diagram where a known error, called the offset value [41], remains and for which a simple small additive correction can be made. Some important limitations of the relationships shown in Fig. 1 are emphasized in Ref. 41.

Figure 1 shows clearly the reason for the commonly observed phenomenon that a bond length obtained from a low-level calculation may agree more precisely with the true value of the parameter than does the result from a much higher level SCF calculation. While it is not an inviolable rule, it often happens that too small a basis set causes computed bond lengths to be too long and neglect of electron correlation causes them to be too short. The resulting compensation of errors may thus, by coincidence, produce a fortuitously good agreement. One should not rely on such compensation, however.

Correction with pre-determined offset values can give the best estimate for the absolute values of bond lengths available from low-level theory, and therefore for medium-sized molecules. As mentioned above, bond angles are most commonly obtained with high accuracy even without the use of electron correlation, i.e., their offset values are zero. Most often, however, the absolute structural parameters are not the information that is required for chemical purposes since the length of a chemical bond, by itself, is seldom of interest. What is important is the difference in the lengths of similar bonds in related systems; for example how the relative C—C bond lengths in cyclopropane are affected by progressive fluorination of the ring [42]. Such structural information provides data for an understanding of the electronic effects of substitution on a given molecular framework, leading to the ability to predict the variations in properties and reactivities of individual members of various classes of compounds. Questions of this type are, happily, both the ones of most interest to the rest of chemistry and the ones most accurately answered directly by relatively modest-level ab initio calculations.

A few examples of the use of moderately low-level calculations to obtain chemical information may be of interest. In one instance, extensive microwave and infrared spectroscopic investigations [43] of ethene thiol (vinyl mercaptan) showed that it exists as a mixture of two conformers, a *syn* form and a slightly less stable *gauche* or "quasi-planar" *anti* form (the $v = 0$ torsional state lies above the small barrier at the *anti* conformation). See Fig. 2. A primary target of these studies was to determine the form of the rotational potential function controlling interchange between the two forms and to fully characterize the stable species. However a number of unsolved questions remained after exhausting the information available from the experiments. In particular, the size, and even the existence of the small barrier at the *anti* conformation could be questioned. Furthermore, there was no experimental information on the height of the barrier between the *syn* and *anti* conformations. Particularly perplexing was the inability to interpret plots of the variation of the rotational constants (inverse moments of inertia) of the stable species with increase in the torsional quantum number, a very sensitive probe of the nature of the internal rotational potential function.

In order to provide supplementary information, an *ab initio* structure determination of ethene thiol and three of its substituted derivatives was undertaken [44] with full geometry optimizations at various fixed values of the internal torsional angle. The calculations were relatively low level, using 4-21 basis sets for carbon and oxygen and a 3-3-21 set for sulfur, with complete neglect of electron correlation. Corrections for these effects were made by the use of small, predetermined offset values, as described above. The most useful contribution from the theoretical study to the experimental analysis was detailed information about the variation of other structural parameters during the internal rotation. The CCS angle was found to vary over a range of 5° and the C—S bond length by 1.6 picometers. In the absence of this sort of information, the usual procedure in the analysis of a microwave spectrum is to consider internal rotation as a one-dimensional problem, assuming that all of the rest of the molecule is strictly rigid. In general, this assumption is far from true,

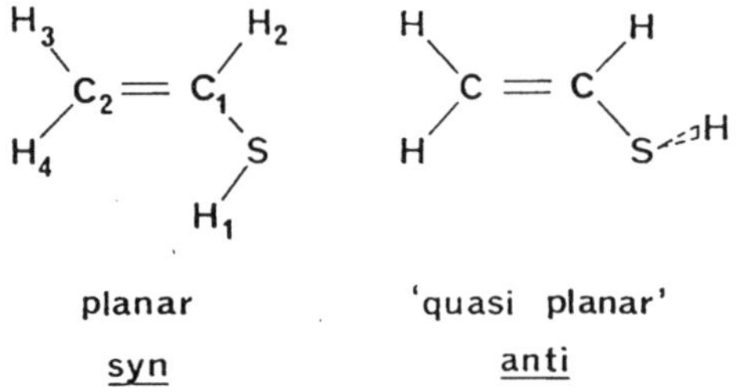

Fig. 2. The stable conformations of ethene thiol. Taken with permission from Ref. [45]

but from spectroscopic information alone there is little alternative. For ethene thiol, the framework relaxation information obtained from the theoretical study was incorporated into the analysis of the microwave data, leading to a full and presumably quite accurate potential curve for the internal rotation [45]. The sensitive plots of the expected variations of the spectroscopic constants with vibrational quantum number were now in excellent agreement with observation, and the entire understanding of the molecular motion was complete. The final internal rotation potential function is shown in Fig. 3.

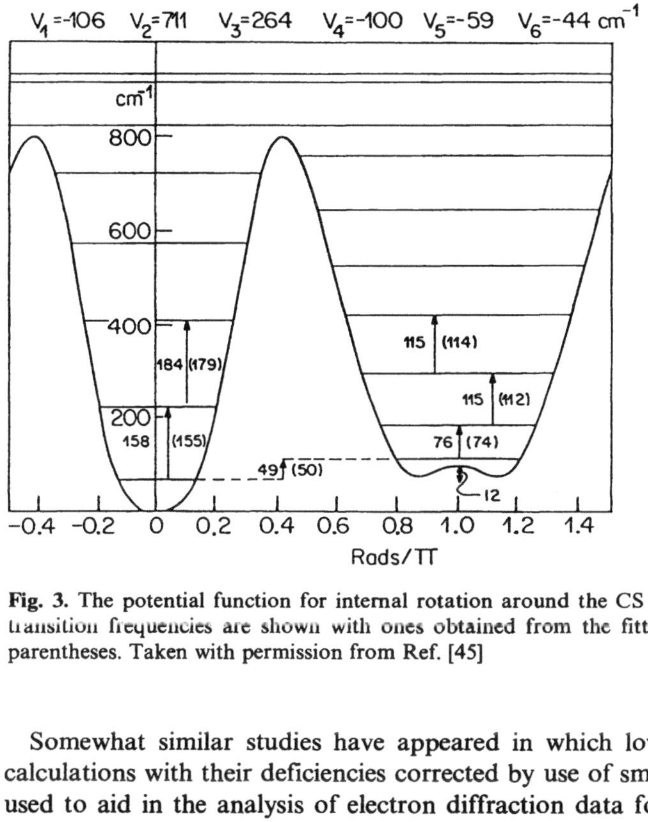

Fig. 3. The potential function for internal rotation around the CS bond in ethene thiol. Observed transition frequencies are shown with ones obtained from the fitted potential function shown in parentheses. Taken with permission from Ref. [45]

Somewhat similar studies have appeared in which low-level quantum chemical calculations with their deficiencies corrected by use of small offset values have been used to aid in the analysis of electron diffraction data for pyrrolidine [46] and for *N*-chloro- and *N*-methylpyrrolidine [47]. In these molecules, the experimental analysis was complicated by the low-barrier pseudorotational motion of the 5-membered rings. Computational determination of the potential function controlling the pseudorotation and of the variation of other geometrical parameters during the motion made possible a more complete experimental analysis.

A simpler application of such computations as an aid to experiment was mentioned briefly in Chapter 6. Analysis of an electron diffraction experiment on CF_3OOCF_3 by Marsden et al. [48], produced three minima in the standard least-squares procedure for fitting a structure to the observed scattering pattern. The authors chose one of the minima as the correct one, largely on the basis of intuitive reasoning. Their choice was challenged by analysis of a new experiment by a different investigator whose analysis of somewhat similar data favored one of the other minima. A computational study [49] quickly produced conclusive evidence that the orginal choice by Marsden et al. was correct.

The utility of low-level computations with correction by offset values is not limited to providing information to serve as a supplement to experimental data. The group headed by Lothar Schäfer has published a brilliant and extensive series of papers, many under the general heading of "*Ab Initio* Studies of Structural Features Not Easily Amenable to Experiment", in which the computed structures are taken to be quite comparable in accuracy and reliability to those which could, in principle, be obtained from much more difficult and expensive experiments. An outstanding

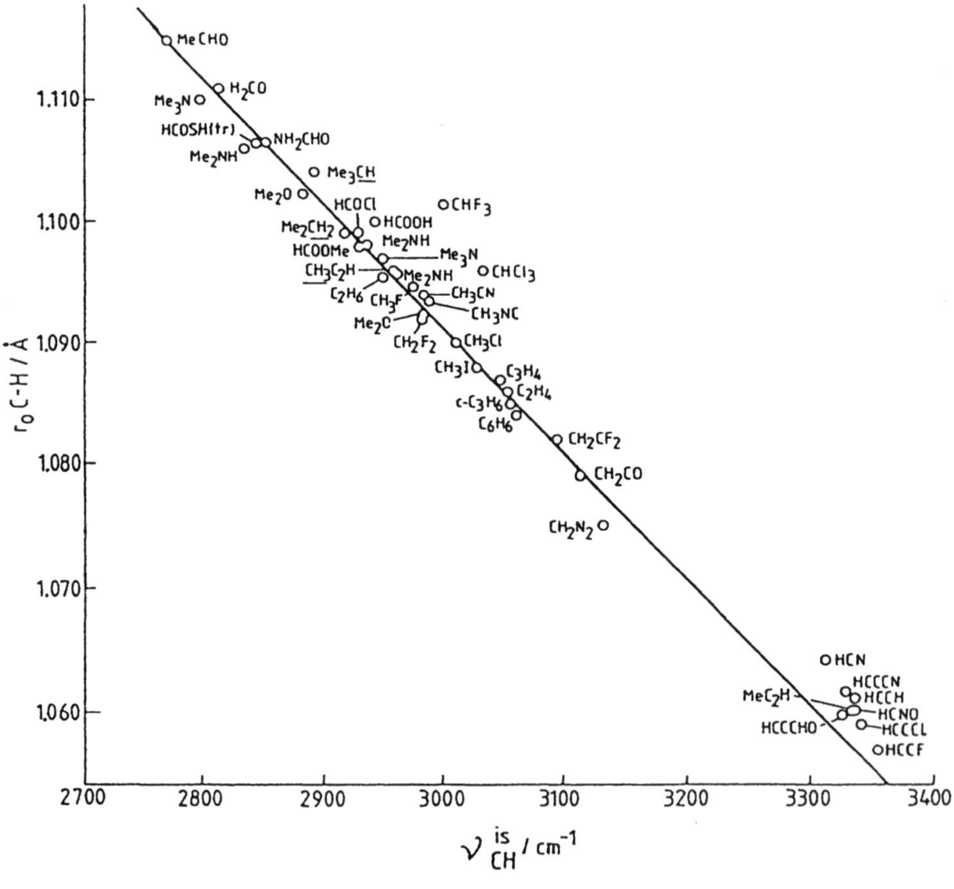

Fig. 4. Correlation of isolated CH stretching frequencies versus C—H bond lengths computed at the SCF 4-21 level. The linearity indicates that *relative* values are determined with high precision by both methods. Taken with permission from Ref. [51]

recent example is a study [50] of the dipeptide *N*-acetyl *N*′-methyl serine amide in which the geometries of five stable conformations were determined. The most stable form has a tricyclic structure held together by hydrogen bonding. The α-carbon serves as the bridgehead for a 5-, a 6-, and a 7-membered ring.

The precision that can be obtained in a comparison of *relative* bond lengths even by low-level calculations is difficult to assess directly because of the lack of experimental data of consistent accuracy on families of molecules. Perhaps the clearest indication comes from a study [51] that compares relative C—H bond lengths as computed by 4-21 basis set SCF calculations with those obtained by an experimental method that also provides relative values of extremely high accuracy, McKean's isolated CH stretching frequency method [52], which is based on infrared absorption by molecules in which all hydrogen atoms except one have been replaced by deuterium. In such a molecule, the remaining C—H bond gives a very sharp infrared stretching frequency that is exquisitely sensitive to its length. A plot, reproduced here as Fig. 4, exhibits the linearity between computed C—H bond lengths (in error by some constant amount) and the infrared C—H stretching frequency (proportional to the bond length). The linearity indicates that comparisons based on *relative* values taken from the calculation should be valid to better than 0.1 picometer.

In Sect. 3.1, a few studies of very small molecules involving extremely high-level quantum chemical calculations were reviewed, all leading to highly accurate and reliable molecular geometrical information. This section makes a considerable jump and deals with quite low-level calculations, generally double-zeta or double-zeta plus polarization SCF calculations with complete disregard of electron correlation. In this author's opinion, there is little to be gained by intermediate levels of computation. The errors in geometry remaining at the double-zeta or double-zeta plus polarization level are largely due to neglect of electron correlation. Repeated studies have shown that a good correlation treatment requires a large basis set, at least triple zeta plus polarization functions. Either a calculation should be made at such a level that the resulting geometrical parameters can be relied on as accurate for reasons other than chance internal compensation of errors, or it should be made at a convenient lower level and a correction for the residual error made by use of small offset values. The only exception to this suggestion would be in the obvious situation in which a molecular species under study involves bonding of a type which is suspected of being significantly different from that in any other known molecule for which structural data are known. In such a case, of which an example might be a compound containing a 4-membered ring containing two tin atoms and a C=P moiety [53], offset values could not be independently determined and the only recourse would be to make the study at the highest quantum chemical level conveniently available.

For a quite different reason, calculations at a lower level should be avoided if possible. Minimal basis sets are now commonly recognized as producing geometrical parameters that are unreliable in the sense that their results fall in the shaded portion of Fig. 1. Aside from a constant error, for which correction could be made by means of offset values, there is an error, random between different molecular environments, for which no compensation is possible. The same problem clearly makes their use even to compare relative bond lengths or angles somewhat unreliable. A slight extension to a 3-21 basis is not too much more expensive in computer time and it produces a considerable improvement in the reliability of the computed geometries.

4 Special Problems

It must be admitted that the present-day applicability of molecular quantum mechanics to the classical problems of organic chemistry is in a much better state of development than is its applicability to inorganic chemistry and to the novel molecular systems on which a great deal of current interest focuses. Certain types of systems present special difficulties for accurate quantum chemical geometry determinations, and these are the ones which are now attracting much of the most active research by theoreticians. Some of the problems and the degree of progress that has been made in solving them are reviewed below.

4.1 Transition States

While an understanding of the nature and consequences of chemical bonding is an essential part of the foundation of the science of chemistry, an equally vital part is concerned with details of the processes involved during chemical reactions. All current approaches to the prediction of reaction rates require an understanding of the nature of the reaction coordinate and the barrier or barriers encountered in going from reactants to products. At the moment, even though there are severe limitations in the ability of theory to provide such information, it is far ahead of experiment. The goal, of course, is first to provide a detailed map of the energetics and structural changes occurring on and near the minimal-energy path from initial to final states of the reacting system and, secondly, to be able to analyze the dynamics of the system moving along such a path. Here we are concerned with the first of the two problems.

In principle the determination of the geometry and energy of a transition-state structure is no different from that used for a stable molecule, but the details present some important special difficulties. From a theoretical viewpoint, a transition state is a configuration of the nuclei of a molecule in which one derivative of the energy with respect to a coordinate displacment is negative and all the others are positive. The negative derivative points downward along the reaction path. One might envision guessing rather blindly at possible structures for the transition state for a reaction and testing them until one is found to possess the proper derivative characteristics. The difficulty is that for anything but a trivially simple reaction more than one transition state exists, and the object is to find the lowest energy one, since it is the one which determines the reaction rate. Having identified and found the relative energy of a number of transition states does not guarantee that the lowest energy one has been discovered.

Another basic difficulty is identification of the changes in geometrical coordinates that characterize the minimal-energy reaction path between the starting materials and the transition state and between the transition state and the products. Textbook examples such as the dissociation of a diatomic molecule make the reaction coordinate obvious, but such is not generally the case for reactions of polyatomic molecules. While a reaction may start out along a normal coordinate or a given bond-stretching motion, it does not in general continue along coordinates readily identifiable with respect to the starting molecules. The complexity of the problem increases drastically as the molecular size increases.

A very extensive review of transition structure computations and their analysis has been given recently by Bernardi and Robb [54]. This article is especially helpful in providing an exhaustive list of references in the field dating up through 1986. Rather than entering here into the computational complexities involved, it may be of interest to consider a typical case of a modern investigation.

The thermal [1,5] sigmatropic hydrogen shift in 1,3-pentadiene is illustrated at the top of Fig. 5. Geometry optimizations at the minimal [55] and double-zeta [56, 57] SCF levels all led to the transition state structure shown in Fig. 5. The structure can be described as aromatic since the C—C bond lengths are very similar to those of benzene. The studies gave details of the transition state geometry, as sketched in Fig. 5, as well as the energy barrier corresponding to the activation energy of the reaction. However, the lack of complete satisfaction one may feel with the current state of the field is well illustrated by the fact that the predicted activation energies were 55 kcal mol^{-1} at the 3-21G level [57], 52.8 kcal mol^{-1} at the 3-21G level with zero-point energy corrections [56], and 59 kcal mol^{-1} at the 6-31G level [57], compared with the experimental value of 35 kcal mol^{-1}. The large discrepancy could simply be the result of the fact that the level of calculation used is too low, with a larger basis set needed to give the required flexibility for the proper description of the hydrogen bonded transition structure (see Sect. 4.3) and with treatment of electron correlation required to reproduce the bond breakage and formation. Alternatively, and less probably, the transition state structure found might not be the correct one or the experimental value derived from kinetics measurements might be incorrect.

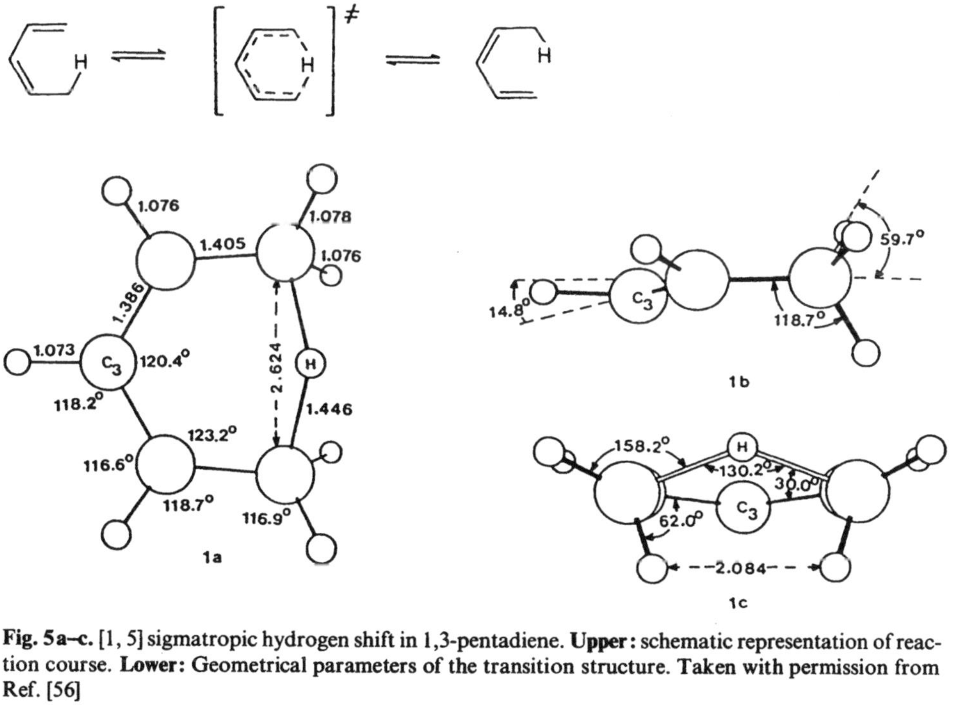

Fig. 5a–c. [1, 5] sigmatropic hydrogen shift in 1,3-pentadiene. **Upper:** schematic representation of reaction course. **Lower:** Geometrical parameters of the transition structure. Taken with permission from Ref. [56]

4.2 Anions

Calculation of the structure of closed shell positive ions such as the cyclopropenyl cation, $C_3H_3^+$, pose no particular problems beyond those found with neutral species [58]. Anions, however, have been a source of considerable difficulty. First, the extra electron causes electron correlation to be of special importance, particularly in studies directed toward calculation of electron affinities, the difference in energy between an anion and its neutral counterpart. In fact, when attempts are made to calculate electron affinities at the SCF level, the anion is frequently found to be an unbound species, i.e., to have a higher energy than the neutral form. Moreover, the extra electron causes the entire electron distribution to swell so that basis sets optimized on neutral atoms or molecules are inadequate. This is particularly true because the energy variational method used for basis set optimization lays special stress on a good reproduction of the electron distribution at small distances from the nucleus. A review of earlier work in the field has been given by Radom [59].

The obvious remedy for the need to use treatment of electron correlation is now easier than it was previously, since correlation at least through the MP2 level is available in a number of standard programs. Similarly the problem of needing a basis set that gives a better description of the electron distribution farther from the nucleus has been addressed. Basis sets including so-called diffuse functions have been devised [60] and recommended for standard use in calculations on anions as well as for the determination of properties such as dipole moments that are particularly sensitive to a good description of the outer reaches of the electronic cloud. Both of these remedies, of course, lead to an increase in the computational requirement for studies on anions.

4.3 Hydrogen Bonds and Weakly Bound Species

The applicability of quantum chemical calculations at various levels to the problem of identifying intra- and inter-molecular hydrogen bonds and determining their energies and structural details has been reviewed by Schaad [61] and by Kollman [62]. Single-determinant SCF calculations are not well suited for such studies, although they were frequently used in earlier days. Similarly, the study of weakly bound dimers requires a post-Hartree-Fock method. In fact, Margenau and Kestner [63] have proved in general that a single-determinant wavefunction cannot yield a dispersion attraction between two molecules.

In addition to the need for a treatment of electron correlation, studies of weakly bound species require large basis sets to obtain the accuracy needed to reproduce the very small interactions being sought. Perhaps the extreme situation arises in studies of rare gas dimers. For determination of the small energy variations as neutral atoms are brought together, the problem of basis set superposition error, always present but usually ignored, becomes of high importance. If an incomplete basis set is centered on atom **A** and a second atom, **B**, with a basis set centered on it is placed nearby, the basis set formally assigned to **B** will assist in the description of **A**. As **B** is moved further away from **A**, the contribution of the **B** basis set to the description of **A** decreases and an inaccurate energy variation is obtained. The effect diminishes with increasing completeness of the basis sets, but it is always present to some extent with any finite set. One approach to eliminating the effect was demonstrated in an interesting paper by Frommhold and Meyer [64] on the argon dimer. By transforming the wavefunction to an equivalent set of localized orbitals, the superposition error was effectively removed from the correlation part of the calculation where it was most significant in this study.

4.4 Electronically Excited States

The spatial extent and shape of a molecule in an excited state may be quite different than is found in the ground state. Thus, for accurate results, basis sets to be used in the description of electronically excited states of molecules should be specially optimized or at least generally require functions corresponding to higher angular momentum quantum numbers than are used for ground state calculations. Furthermore, a good treatment of electron correlation is needed since the usual purpose of computations of excited states is to make predictions about the frequencies of electronic transitions in the molecule. The difference between the two states involved is one of the electronic distribution and therefore there is generally a significant difference in the extent of electron correlation which must be accounted for in the calculation.

Recent reviews of this very active field, both with extensive bibliographies, have been given by Davidson and McMurchie [65] and by Bruna and Peyerimhoff [66].

4.5 Condensed Phases

While questions relating to chemical bonding and to structure-dependent intrinsic properties of molecules are best answered by experimental or theoretical gas-phase studies, there are numerous practical reasons for interest in condensed phase matter. After all, most natural chemical and biochemical processes on the surface of the earth occur in water solution. Important properties of solid matter are strongly dependent on the nature of the intermolecular forces binding crystals together. In the case of more complex substances, the natural minimum-energy structure of the isolated molecule may be strongly distorted in condensed phases and the substance may even exist in other conformations. Reaction mechanisms are often radically different in the presence of solvent molecules which can intervene as intermediaties in the reaction.

The additional computational problem for study of condensed-phase matter arises from the increase of the size of the system that has to be considered. Perhaps the easiest case is that of small clusters, which are also now of keen interest experimentally. A number of theoretical studies have been published, mostly on clusters containing a dozen or fewer atoms, dealing with the geometry and energetics of small clusters of the lighter elements. A valuable review of work on metal atom clusters has appeared recently [67].

The current record for size of an SCF computation is held by the carbon cluster investigations of Almlöf and coworkers. Using their direct SCF methods which minimize the computational requirements for disk storage and I/O time, they have performed double-zeta geometry optimizations on several configurations of C_{60} clusters, including the globular form known as buckminsterfullerene [68]. The largest species they studied was $C_{150}H_{30}$, using 1,560 contracted Gaussian basis functions [69].

Interest in catalysis has inspired numerous efforts to model the critical adsorption process by use of a few atoms of the substrate and a single adsorbed molecule. Success has been limited by the fact that heavier transition metals are of most interest as the substrate. Experience has also shown that the interaction force is propagated through the solid to a distance which requires inclusion of an askward number of metal atoms in the calculation.

Studies of larger molecules in the crystal phase have been rather limited. One example is an investigation of the structural changes undergone by furan when it goes from the isolated molecule to the crystalline form [70]. In this study, geometry optimizations were made for furan dimers and trimers that were fixed at the relative orientations and distances previously observed in x-ray diffraction experiments on the crystal. By comparison with similar calculations for a single furan molecule, conclusions could be drawn both about the energetics of the intermolecular interactions and about the induced changes in internal geometry.

An ingenious method for the study of crystals was introduced by Saebø and coworkers [71] in a theoretical study of the crystal structure of cyanoformamide. From a population analysis of an ab initio calculation on a single molecule, they determined the net charges on each of the atoms. They then surrounded a single cyanoformamide molecule by a set of point charges located at positions where the corresponding atoms in neighboring molecules were found experimentally. The geometry was then reoptimized in the presence of the perturbing field of point charges. A large number of neighboring molecules can be modeled in this manner with little additional computational expense since no additional electron-electron integrals are introduced. The technique has been modified slightly and further applied to H_2SO_4 crystals [72]. A variety of other approaches have been tried by many other workers [73–79].

Ab initio computational studies have also been carried out on the structure of liquids, particularly water. A full practical understanding of such systems, however, requires investigation of the dynamics as well as of the static structure of the system.

5 Computational Requirements

Advances in computer technology are coming so rapidly that it is futile to try to comment on the capability of specific current computer systems. Such remarks tend to be obsolete before the ink with which they are written dries. However, some general observations can be made.

First, the most significant advance in very recent years from the viewpoint of the computational chemist has been the introduction of powerful superminicomputers. These are systems available at a cost as much as two orders of magnitude lower than that of the most advanced supercomputers yet capable of speeds lower by less than a factor of ten. This can bring the capability of performing high-level computational quantum chemistry to many more laboratories at a cost quite in line with that of much modern experimental apparatus. The ability to run a job in ten hours on a computer in your own laboratory is much more attractive than being able to have it run in one hour after you have arranged access to some nationally supported supercomputer elsewhere. The most advanced supercomputers are still essential for the most massive calculations, and it is to be hoped that those responsible for their operations will see that they are reserved for such use as the proliferation of medium-sized computers increases still more widely.

Simple ab initio calculations are now possible on a personal computer, for which an inflation-free definition might be one costing less than a thousandth as much as a supercomputer. As the name suggests, such computers are intended for each individual researcher to have sitting on his desk and working independently, although in some laboratories these are now linked in small local networks. Especially noteworthy is the work of Susan Colwell [80] in simplifying the Cambridge ab initio package of programs to run on a personal computer. The size of an SCF calculation that can be run was limited to 63 basis functions in the 1986 version of the program, but this is subject to steady revision upward. The capacity is marginal for serious work, but it is interesting to recall that this is about the same size calculation that could be run on the world's largest computer only 15 years ago. Both program developments and improvements in computers that still fall within the "personal computer" category are currently taking large steps ahead, and it can be anticipated that in a very few years quite important ab initio calculations may be possible on such individual desk-top computers.

The above discussion has been written as though raw speed of computation were the only decisive consideration. This is far from the case. For the very largest ab initio calculations on today's supercomputers using standard programs, the limitation on what is possible is reached first by disk memory capacity. In the conventional approach, hundreds of millions of integrals must be evaluated, stored, retrieved for use, stored again, etc. Geometry optimization of $C_3F_3^+$ or of benzene, both at the 6-311G plus two sets of polarization functions plus MP4, are near the practical limit for a Cray X/MP-24 computer, the limit being set by high-speed memory capability or the impossible I/O time for slower bulk memory.

An interesting although not yet widely adopted solution to the memory problem at the SCF level is the "direct method" of Almlöf [68], mentioned above in Sect. 4.5. Saebø and Almlöf have just completed a very important extension of direct methods to electron correlation calculations. In the first report [81], they describe the procedure and give results for a 6-311G** plus MP2 computation on diphenyl acetylene with 312 contracted basis functions, an extremely large calculation.

In addition to computational speed, memory capacity and the rapidity of information transfer to memory, other aspects of computer architecture are of importance. Conventional SCF programs do not yet take as full advantage of the benefits of vectorization as do programs used in some other fields. There are inherent obstacles to getting long vector length in ab initio calculations, but more could be done. One trend in computer design is toward more parallel computational structures, but essentially nothing has been done along this line in quantum chemistry. It is likely that rethinking of the basic mathematical problem would be necessary to take advantage of massively parallel computer architecture.

6 Prospects for the Future

Forecasting the future of computational quantum chemistry and its application to studies of chemical structure and bonding is at once made both simpler and more difficult by the present explosive rate of growth of the field. The simple aspect is mere recognition that the area is so dynamic that it must expand to play a progressively greater part in the entire range of chemistry over the next few decades. It will surely be found that many practical questions can be answered by computation with adequate accuracy and as high a confidence level as could be achieved by much more expensive and time-consuming experimental investigation.

The difficult aspect of a forecast comes in trying to pick out details of what computational chemistry will look like, even in the relatively near future. The field is still so young that truly revolutionary discoveries are yet to be made, providing really major deviations from any presently attempted extrapolation of future developments.

7 Acknowledgement

The portion of the work described here that was done at The University of Texas was supported by a grant from The Robert A. Welch Foundation.

8 References

1. Pulay P (1969) Mol. Phys. 17: 197
2. Szabo A, Ostlund NS (1982) Modern quantum chemistry, Macmillan, New York
3. Schaefer HF III (1977) Methods of electronic structure theory, Plenum, New York (Modern theoretical chemistry, vol. 3)
4. Schaefer HF III (1977) Applications of electronic structure theory, Plenum, New York (Modern theoretical chemistry, vol 4)
5. Lawley KP (ed) (1987) Ab initio methods in quantum chemistry, Part I, John Schaefer, New York (Advances in chemical physics, vol LXVII)
6. Lawley KP (ed) (1987) Ab initio methods in quantum chemistry, Part II, John Wiley, New York (Advances in chemical physics, vol LXIX)
7. Slater JC (1930) Phys. Rev. 36: 57
8. Pople JA (1977) in Ref. 4, p 1
9. Dunning TH Jr, Hay PJ (1977) in Ref. 3, p 1
10. Wilson S (1987) in Ref. 5, p 439

11. Hehre WJ, Stewart RF, Pople JA (1969) J. Chem. Phys. 51: 2657
12. Shavitt I (1977) in Ref. 3, p 189
13. Brillouin L (1934) Les Champs "Self-Consistent" de Hartree et de Fock (Actualités Sci. Ind. No. 159), Hermann and Cie., Paris
14. Wahl AC, Das G (1977) in Ref. 3, p 51
15. Cizek J (1966) J. Chem. Phys. 45: 4256
16. Meyer W (1971) Int. J. Quantum Chem. Symp. 5: 341
17. Kutzelnigg W (1977) in Ref. 3, p 129
18. Møller C, Plesset MS (1934) Phys. Rev. 46: 618
19. Bartlett RJ (1981) Annu. Rev. Phys. Chem. 32: 359
20. Pulay P, Lee JG, Boggs JE (1983) J. Chem. Phys. 79: 3382
21. Saebø S, Pulay P (1985) Chem. Phys. Lett. 113: 13
22. See, for example, Boys SF (1962) In: Löwdin PO (ed) Quantum theory of atoms, molecules and the solid state, Academic, New York, p 253
23. Pulay P, Saebø S, Boggs JE (to be published)
24. Balasubramanian K, Pitzer KS (1987) in Ref. 5, p 288
25. Hellmann J (1937) Einführung in die Quantenchemie, Deuticke, Leipzig
26. Feynman RP (1939) Phys. Rev. 56: 340
27. Pulay P (1977) in Ref. 4, p 153
28. Pulay P (1987) in Ref. 6, p 241
29. Pople JA, Raghavachari K, Schlegel HB, Binkley JS (1979) Int. J. Quantum Chem. Symp. 10: 1
30. Handy NC, Amos RD, Gaw JF, Rice JE, Simandiras ED (1985) Chem. Phys. Lett. 120: 151
31. Pulay P, Saebø S (1986) Theor. Chim. Acta 69: 357
32. Rice JE, Amos RD (1985) Chem. Phys. Lett. 122: 585
33. Mills IM (1974) Theoretical Chemistry, Specialist Periodical Report (The Chemical Society, London) 1: 110
34. Langhoff SR, Davidson ER (1974) Int. J. Quantum Chem. 8: 61
35. Strey G, Mills IM (1973) Mol. Phys. 26: 129
36. Sellers HL, Almlöf JE (in press)
37. Handy NC, Gaw JF, Simandiras ED (1988) J. Chem. Soc., Faraday Trans. II, 83: 1577
38. Schleyer P von R, Pople JA (1986) Chem. Phys. Lett. 129: 475
39. Xie Y, Boggs JE (in press)
40. Pulay P, Fogarasi G, Pang F, Boggs J (1979) J. Am. Chem. Soc. 101: 2550
41. Boggs JE, Cordell FR (1981) J. Mol. Struct, Theochem 76: 329
42. Boggs JE, Fan K (1988) Acta Chem. Scand. A42: 595
43. Almond V, Charles SW, Macdonald JN, Owen NL (1983) J. Mol. Struct. 100: 223, and earlier papers cited therein
44. Plant C, Macdonald JN, Boggs JE (1985) J. Mol. Struct. 128: 353
45. Almond V, Permanand RR, Macdonald JN (1985). J. Mol. Struct. 128: 337
46. Pfafferott G, Oberhammer H, Boggs JE, Caminati W (1985). J. Am. Chem. Soc. 107: 2305
47. Pfafferott G, Oberhammer H, Boggs JE (1985) J. Am. Chem. Soc. 107: 2309
48. Marsden CJ, Bartell LS, Diodati FP (1977) J. Mol. Struct. 39: 253
49. Gase W, Boggs JE (1984) J. Mol. Struct. 116: 207
50. Siam K, Klimkowski VJ, Van Alsenoy C, Ewbank JD, Schäfer L (1987) J. Mol. Struct. 37: 261
51. McKean DC, Boggs JE, Schäfer L (1984) J. Mol. Struct. 116: 313
52. McKean DC, Duncan JL, Batt L (1973) Spectrochim. Acta Part A29: 1037
53. Dobbs KD, Boggs JE, Cowley AH (1988)
54. Bernardi F, Robb MA (1987) in Ref. 5, p 155
55. Adeny PD, Bouma WJ, Radom L, Rodwell WR (1980) J. Am. Chem. Soc. 102: 4069 and several other papers from the same group
56. Rondan NG, Houk KN (1984) Tetrahedron Lett. 25: 2519
57. Bernardi F, Robb MA, Schlegel HB, Tonachini G (1984) J. Am. Chem. Soc. 106: 1198
58. Xie Y, Boggs J (1989) J. Chem. Phys. 90: 4320
59. Radom L (1977) in Ref. 4, p 333
60. Frisch MJ, Pople JA, Binkley JS (1984) J. Chem. Phys. 80: 3265
61. Schaad LJ (1974). In: Joesten MD, Schaad LJ (eds) Hydrogen bonding, Marcel Dekker, New York

62. Kollman PA (1977) in Ref. 4, p. 109
63. Margenau H, Kestner NR (1969) The theory of intermolecular forces, Pergamon, Oxford
64. Meyer W, Frommhold (1986) Phys. Rev. A 33: 3807
65. Davidson ER, McMurchie LE (1982) In: Lim EC (ed) Excited states, Academic, New York, p 1
66. Bruna PJ, Peyerimhoff SD (1987) in Ref. 5, p 1
67. Koutecky J, Fantucci P (1986) Chem. Rev. 86: 539
68. Luthi HP, Almlöf J (1987) Chem. Phys. Lett. 135: 357
69. Almlöf J (1987) ACS Symp. Ser. 353 (Supercomputing Res. Chem. Chem. Eng.): 35
70. Cordell FR, Boggs JE (1988) J. Mol. Struct., Theochem 164: 175
71. Saebø S, Klewe B, Samdal S (1983) Chem. Phys. Lett. 97: 499
72. McCormick MA, Boggs JE (to be published)
73. Dauber P, Hagler AT (1980) Acc. Chem. Res. 13: 105
74. Pullman A, Zakrewska C, Perahia D (1979) Int. J. Quantum Chem. 16: 393
74. Smit PH, Derissen JL, van Duijneveldt FB (1977) J. Chem. Phys. 67: 274
76. Berkovitch-Yellin Z, Leiserowitz L (1980) J. Am. Chem. Soc. 102: 7677
77. Bonaccorsi R, Petrongolo C, Scrocco E, Tomasi J (1971). Theoret. Chim. Acta 20: 331
78. Caillet J, Claverie P, Pullman B (1978) Theoret. Chim. Acta 47: 17
79. Berthod H, Pullman A, Hinton JF, Harpool D (1980) Theoret. Chim. Acta 57: 63
80. Program MICROMOL available from S. M. Colwell, Department of Theoretical Chemistry, University Chemical Laboratory, University of Cambridge, Lensfield Road, Cambridge CB2 1EW, United Kingdom
81. Saebø S, Almlöf J (in press)

Molecular Mechanics Alias Mass Points and Elastic Springs Model of Molecules

A. Y. Meyer

Department of Organic Chemistry, Hebrew University, Jerusalem 91904, Israel

Molecular Mechanics is a computational tool that the organic stereochemist uses frequently. It treats a molecule as an assembly of atoms and bonds, such that the strain becomes a sum of atom-atom interaction energies, bond-stretching strains, etc. Minimization with respect to the atomic coordinates provides optimal geometries, along with a quantity that has the dimension of energy and can be used in comparing the stability of isomers. The method has found extensive fields of application, and many of its facets have been reviewed. The first four Sections of this Chapter recapitulate on the method and list leading references.

The last four Sections concentrate on novel aspects of the molecular mechanist's outlook and activities. Intramolecular electrostatic interactions are now considered more significant than they used to be, and new techniques for distributing atomic charges are described. Molecular graphics is calling attention to the envisionment of molecules as objects in solid geometry. This calls for re-consideration of notions such as atomic radii, molecular volume, packing density. Quantum Chemistry is having several types of impact on the current development of Molecular Mechanics. Relations between the two doctrines, existing and yet to be explored, are examined. We finally report that molecular mechanists are now starting to tackle questions of non-covalent bonding.

1 Introduction

> "... die von ihm [C. A. Bischoff] vorgeschlagene „Dynamische Isomerie" und „Dynamische theorie" [haben sich] in der Folge als *zu mechanistisch* erwiesen; als Arbeitshypothesen haben sie ihm selbst reichlich gedient." ("Bischoff's model of dynamic isomerism, and his dynamic theory, prove to be *too mechanistic*. As a working hypothesis, though, they served him amply").
>
> Walden, 1908 [1]

Molecular mechanics, or "the Force Field Method", comprises a group of algorithms that serves to calculate the geometry of molecules and their energy relative to a chosen state of reference. Calculation is performed by expressing the energy as a function of certain internal coordinates, and minimizing the function. The method is also made to meet heats of formation, vibrational frequencies, and saddle points on potential surfaces [2]. Molecular mechanics is the main computational tool of the organic stereochemist, and constitutes the grain of the novel art of molecular design [3]. Nowadays, it is entirely computer-dependent. The model, however, and the first applications, precede the computer age [4].

In writing this report, I have tried not to dwell upon topics that had been examined in detail by others. Aspects emphasized here are either of recent interest or such that happened to evade previous reviewers. The titles accorded this Book and this Chapter by the editor invited some stress on the conceptual model that antedates and guides the mathematical model. It is proper to admit at the start that the author is by training an organic chemist, not a theoretician, and that he is much more familiar with one school of molecular mechanics than with others. These factors unavoidably bias the exposition.

It can prove unpleasant to describe molecular mechanics to professional theoreticians. By Daudel's classification of concepts in theoretical organic chemistry [5], the concept of chemical bonds belongs in the basest class: vague and unassociable with any operator under any approximation [6]. And yet, the notion of bonds, or worse still, of bonds as mechanical springs that interconnect atoms, lies at the very heart of molecular mechanics. As to the atoms, they are regarded as indivisible points of mass with spherical domains of influence. Atomic charges or bond dipoles are recognized (in the "political" sense of the verb), but frequently as a necessary evil [7]. The existence of electrons, however, is deliberately glossed over. Developers of computational variants do resort nowadays to quantum chemistry (Sect. 7.1), but this is well disguised in the final product.

By contrast, it is a pleasure to describe molecular mechanics to organic chemists. For them, the method constitutes a numerical counterpart of the manual examination of molecular models, the ubiquitous rods and balls of structural research [8]. The method does best quantitatively where the chemist's intuition and experience work best qualitatively, e.g. in comparing the stability of stereoisomers. It ventures less where he ventures less, e.g. in comparing the height of conceivable transition states [9, 10]. Its output, in short, is the product he needs, and usually needs quickly. When quick answers are sought, one does not care too much for first principles.

Unlike quantum chemistry, which dissects a molecule into electrons and nuclei, molecular mechanics dissects molecules into atoms and bonds. Again unlike quantum chemistry, atoms are not endowed of electronic clouds that fade with distance. Rather, they are conceived as having sharp boundaries, thus demarking molecular surface areas that can be described mathematically by the formulas of solid geometry. Notions of molecular volume, shape, globularity or deviation from globularity, eccentricity, surface roughness and the like, are the unescapable outcome [11].

Recently, the author had occasion to describe molecular mechanics to an audience of chemical physicists [12]. The exposition was frequently interrupted: some listeners, presumably unacquainted with molecular modeling, seemed to consider the topic as a curious personal fad of the lecturer. One unanticipated question was "how would you calculate ammonia?" Now, molecular mechanics is constrained to have its own philosophy of targets. Amines and peptides are worthy of computation [13], while ammonia is not and cannot be. The reason is very simple: in NH_3, the number of input parameters (force field constants, Sect. 3), exceeds the number of items sought (barrier to inversion, H—N—H angle and dipole moment in ground state, N—H distance in ground and coplanar varieties). Many combinations of constants would reproduce the traits of NH_3, and all are equally uninteresting as long as ammonia alone is at stake. In quantum chemistry, simple species are paradigmatic: ammonia can serve as an archetype for the tunneling effect [14], or, alternatively, an unspecified particle in an unspecified box can serve as a prototype for ammonia [15]. In molecular mechanics, by contrast, data on simple species constitute a database, to be exploited when preparing to study more complex species. By employing the measured value of angle H—N—H in NH_3, and further background data at the same level, one may attempt, say, to predict the evolution of H—N—H, C—N—H and C—N—C along the series of amines [16]. There is, therefore, a threshold of complexity, beneath which the molecular mechanist does not act. At stake is not the unity of chemistry, but delicate variations with respect to reference molecules. In principle, quantum chemistry will deal with complex molecules, if the user is an expert and if he can spare a liberal amount of man-time and computer-time. In practice, for molecules that are not small and for users in a hurry, molecular mechanics is the method to choose.

2 Bibliographic Background

A monograph on molecular mechanics, published in 1982, is the standard source [2]. In addition, reviews are available, the bulk of them dating back to the period 1974–1983. In this period, the fundamentals of the method were scrutinized and splitting into schools took place. An historical overview has been given [17]. Since the sequence of reviews follows the evolution of the method, we now list some titles in chronological order.

a) "Calculation of the magnitude of steric effects" (1956, Westheimer's seminal work [4]);

b) "Quantitative conformational analysis; calculation methods" (1968 [18]);

c) "Empirical force field calculations. A tool in structural organic chemistry" (1974 [19]);

d) "Nonbonded interactions in organic molecules" (1975 [20]);

e) "Calculation of molecular properties using force fields. Applications in organic chemistry" (1976, stressing the mathematical background [21]);

f) "Calculation of molecular structure and energy by force field methods" (1976, wide-scope account of the state of the art in the mid-seventies [22]);

g) "Molecular mechanics: a symposium in print" (1977 [23]);

h) "Some applications of the empirical force field method to stereochemistry" (1978 [24]);

i) "Molecular mechanics calculations", and "Structures calculated by the molecular mechanics method" (1979 [25]);

j) "Molecular mechanics, the method and its underlying philosophy", (1982, aimed to increase awareness of the method [26]);

k) "Application of molecular mechanics calculations to organic chemistry" (1982, state of the art in the late seventies [13]);

l) "Examples of the usefulness of this simple non-quantum mechanical model" (1983 [27]);

m) "Molecular mechanics and conformation" (1983, stressing intramolecular electrostatics and application to haloalkanes [28]);

n) "Potential energy functions: a review" (1985, personal appreciation of schools and persons [29]);

o) "Molecular mechanics and the structure hypothesis" (1985, a quest for first principles [30]).

Activities in the mid- and late eighties concentrate more on widening the scope of application than on consolidating the fundamentals. One probable reason is the constant demand for results, by industrial units of drug design as well as by academic syntheticians and structuralists. Another may be a delayed reaction, after two decades or so of dwelling on algorithms, computer-programs and the like. In a sense, the situation parallels that of quantum chemistry, where most users nowadays are content to employ as "blackboxes" the available ab initio and other programs. The parallelism is heightened in a recent "Handbook" [31], where quantum-chemical and molecular-mechanical programs are described side by side.

Typical present-day research topics in academia are: organosilicon compounds [32], molecules with conjugation [33] or partial conjugation [34], organo-metal complexes [35], and non-covalent bonding [36–38].

3 The Force Field

The total potential energy of a molecule, relative to some reference state, is formulated as a sum of components, $E_T = \Sigma E_I$. The components are those that a classical chemist would list as contributing to what he calls "strain energy" [39]. More specifically,

$$E_T = E_S + E_B + E_{NB} + E_{ES} + E_{TOR} + \dots \qquad (1)$$

E_S (Stretch) is the energy due to the stretching or compression of bonds, relative to their lengths in the chosen state of reference. Likewise, E_B (Bend) has to do with the opening or closing of valence angles, E_{NB} (Nonbonded) with the interaction of atoms that are not bonded to each other nor to a common atom, E_{ES} (Electrostatic) with intramolecular electrostatic interactions, and E_{TOR} (Torsion) with the deviation of dihedral angles from their reference values. Some cross-terms are also included [21], the choice varying with school. The most frequently encountered are E_{SB} (Stretch-Bend) and E_{TB} (Torsion-Bend).

When the components in Eq. (1) are made explicit, it becomes the *force field* equation. The numerical parameters that intervene in the explicit version are the *force field constants.* Since many ways are conceivable to make the equation explicit, there are many *force fields.* The multiplicity of force fields is one of the factors that divide schools; the selection of force field constants is another. The best analogy is with the multitude of "techniques" in semiempirical quantum chemistry [40].

The reader may have noted that van der Waals forces are grouped under E_{NB} (dispersion, also called loosely "van der Waals") and E_{ES} (dipole-dipole interaction). Dipole-induced dipole interactions are considered to be implicitly accounted for by E_{NB}. The division is causing quite a bit of confusion, inasmuch as some fields do not include E_{ES} and consider it as also included implicitly in E_{NB} (Sect. 5).

The reference state is now easy to define. It is the geometry in which all bondlengths, valence angles and dihedral angles are equal to chosen reference values, and where all van der Waals forces have been "switched off". This definition does not coincide with any thermodynamic standard state. Likewise, the energy E_T, as calculated for a species, does not coincide with any thermodynamic measure of energy, e.g. the standard heat of formation. By contrast, the force field constants are chosen such that the energy difference between stereoisomers (ΔE_T) reproduce the standard difference in internal energy, ΔU^0. Force field calculations are primarily performed for the vapor state (but see Sect. 4.3), and molecular mechanists make the approximation $\Delta U^0 \approx \Delta H^0$ a matter of course. ΔE_T, therefore, is usually alluded to as "the enthalpy difference between stereoisomers". There are ways to convert E_T into the standard heat of formation [2, pp 169–174].

Coming back to Eq. (1), each of the intervening terms is expressed in turn as a sum of components, $E_I = \Sigma(j)\, e_{i,j}$. Each $e_{i,j}$ depends on one sole internal coordinate j. Thus, $E_S = \Sigma(j)\, e_{s,j}$, where j runs over the chemical bonds, and $e_{s,j}$ is the energy of bond j relative to its energy at the reference length. The actual formulation varies with field, but the position

$$e_{s,j} = k_{s,j}(l_j - l_{o,j})^2 + k'_{s,j}(l_j - l_{o,j})^3$$

is not uncommon [2]. Here, l stands for a bondlength. The force field constants are $k_{s,j}$, $k'_{s,j}$, and the reference bondlength $l_{o,j}$. Each type of bond (C—C, C—H, etc.), has its own set of constants. A bond is thus conceived as a mechanical spring, responding according to an unharmonic Hooke law to changes in length. The actual bondlength l_j counts among the targets of the *force field calculation.*

Likewise, $e_{b,j}$ is conceived as mimicking some sort of a pruner, with

$$e_{b,j} = k_{b,j}(\vartheta_j - \vartheta_{o,j})^2 + k'_{b,j}(\vartheta_j - \vartheta_{o,j})^3 , \tag{2}$$

where ϑ stands for a valence angle.

Various formulations have been used for $e_{nb,j}$, the interaction between a pair j of nonbonded atoms. They can all be written in terms of two parameters, ε and α. Label the two atoms by indices 1 and 2. ε_{12} is taken to be the depth of the potential well for atoms of the types to which 1 and 2 belong (C ... C, C ... H, etc.). α is r_{12}/r^*_{12}, where r^*_{12} is the optimal and r_{12} actual interatomic distance. r^*_{12} is a constant, r_{12} is a target of computation. The usual position is $\varepsilon_{12} = (\varepsilon_1\varepsilon_2)^{1/2}$ and $r^*_{12} = r^*_1 + r^*_2$ [41, 42]. Thus, the NB constants are made monatomic, with ε to measure hardness and r^* to measure size.

Out of the many conceivable variants, we shall cite here three formulations that have recently been compared [43]. For simplicity, the subscript j is omitted.

a) The so-called "Hill equation" [41]:

$$e_{nb} = \varepsilon\{-C/\alpha^6 + A \exp[-B\alpha]\} . \tag{3}$$

This is, no doubt, the expression used most commonly. It is incorporated in molecular mechanics programs of the MM-series, that is, programs devolving from the MM1 [44] and MM2 [45] force fields. With the original constants ($A = 2.9 \times 10^5$, $B = 12.5$, $C = 2.25$), the minimum is at $\alpha = 1$, whereat $e_{nb} = -1.2693\varepsilon$. A novel set of constants is incorporated in new versions of MM2 ($A = 1.84 \times 10^5$, $B = 12$, $C = 2.25$).

b) The Engler-Andose-Schleyer formula [46], which has found its way into other force fields [47]:

$$e_{nb} = \varepsilon\{-2/\alpha^6 + \exp[12(1 - \alpha)]\} . \tag{4}$$

c) The Ermer-Lifson formula [48], now incorporated in the "Consistent Force Field" [29]:

$$e_{nb} = \varepsilon\{-3/\alpha^6 + 2/\alpha^9\} . \tag{5}$$

In Eqs. (4) and (5), $\alpha_{min} = 1$ and $e_{nb,min} = -\varepsilon$.

Coming now to E_{ES}, its components are calculated — if calculated — either as a sum of dipole-dipole interactions or as a sum of charge-charge interactions [49]. Since questions of outlook are involved, and a definite historical evolution can be traced, the entire topic is best deferred to a section apart (Sect. 5).

As to E_{TOR}, its sub-components $e_{tor,j}$ refer to four-atom-sequences (diheders) j in the molecule, the internal coordinate being the dihedral angle ω_j. In principle, a cosine-series is a reasonable choice, since $e_{tor,j}$ is periodic. In practice, all fields truncate the series after the third term, and some limit themselves to the second or the third, depending on the diheder's nature [45, 50, 51]. In MM2 [45], the position is

$$e_{tor,j} = \frac{1}{2} V_{1,j}(1 + \cos \omega_j) + \frac{1}{2} V_{2,j}(1 - \cos 2\omega_j) + \frac{1}{2} V_{3,j}(1 - \cos 3\omega_j) \quad (6)$$

$V_{1,j}$, etc., are the force field constants for the diheder of appropriate atom types, and ω_j is the target of computation. The choice in Eq. (6) is somewhat confusing, since $e_{tor,j}$ vanishes at $\omega_j = 180°$ rather than at 0°. The definition of the reference state is affected accordingly.

Readers familiar with spectroscopic force fields [52] may wonder at the occurrence of E_{TOR}. It was incorporated quite early [53] with the apology that, although its origins are not even qualitatively understood, its absence is liable to incur serious error. This step followed the realization that E_{NB} alone could not account for the barrier to internal rotation in ethane [41]. The torsional contribution E_{TOR} to E_T is primarily regarded as a corrective term [19], inserted in the energy equation so as to take care of factors other than the readily interpretable strains. Fortunately, it has a quantum-chemical counterpart in the overlap repulsion (exchange) between vicinal bonds, which is considered to constitute the principal contributor to rotational barriers [54]. A typical molecular-mechanical calculation (MM2 [45]) gives the barrier height in ethane as 11.4 kJ mol^{-1}. The experimental value is about 12 kJ mol^{-1}: the force field is intentionally made to compute barrier heights somewhat lower than their measured counterparts [44]. Of the total, $\Delta E_{NB} = 1.9$ and $\Delta E_{TOR} = 8.9$ kJ mol^{-1}, that is, 17 and 78%. The rest is due to skeletal strains — ΔE_S, ΔE_B and ΔE_{SB}.

The constants V in $e_{tor,j}$ should not be confused with constants (say U_k) that characterize the periodic change of E_T along the course of internal rotation:

$$E_T = \frac{1}{2} \Sigma(k)\, U_k(1 - \cos k\omega)\,. \quad (7)$$

Here ω is the torsional angle relative to some chosen origin, *not* the dihedral angle characterizing any of the four-atom-sequences. A molecule of the general type $A^1A^2A^3M{-}NB^1B^2B^3$ has nine deheders (A^1MNB^1, A^1MNB^2, etc.), and computation of rotamer energies requires 27 constants V of the field as *input*. Once the energies are known, they may be used to derive the potential energy constants U of the particular molecule, which constitute a possible *output* of a series of computations (for an example from the quantum-chemical literature, see [55]). The spectroscopic literature abounds in experimental data of this type (for an example at random, with U_1–U_6, see [56]). Constants of type U [Eq. (7)], but not of type V [Eq. (6)], can be interpreted in terms of fundamental physical effects [28, p 6 and p 29].

The above suffices for saturated organic molecules and for molecules with localized unsaturation, e.g. aldehydes and ketones [57]. In conjugated systems, the effect of delocalization on l_o (in E_s) and on V's [in E_{TOR}, Eq. (6)] has to be taken into account. This is one occurrence where a quantum-chemical sub-calculation is let invade the molecular-mechanical process. Several research groups have dealt with the topic, but it has persistently evaded reviewers. Even the monograph [2, pp 52–55] avoids going into detail. Since the original literature presents the material piecemeal, it is helpful to outline here one of the approaches [58].

If the conjugated portion of the molecule is coplanar, it is subjected to a π-electronic SCF calculation. This furnishes, for each bond j, a bond-order p_j and a function β_j

of the overlap between π-orbitals on the bond-terminals [59]. For C—C bonds and C—C—C—C diheders, the missing constants are then expressed as

$$l_{o,j} = 1.512 - 0.179p_j$$

$$k_{s,j} = 5.0 + 4.6p_j$$

$$V_{2,j} = 16.25p_jf_j\beta_j$$

(the numerical constants are cited from the original literature and computer-program, and refer to bondlength in A units and energy in kcal mol^{-1}). The third expression applies to the diheder in which j is the central bond where, also, $V_{1,j} = V_{3,j} = 0$. At this stage, the constant f_j is put equal to 1.

If the conjugated moiety is not coplanar, two SCF calculations are performed: one for the actual geometry, and one for an hypothetical coplanar system. The duplicity is required to estimate the energy lost by the disruption of conjugation, due to deviation from coplanarity. Quantum-chemically, the loss amounts to

$$\Sigma\ (\pi\text{-bonds j})\ \{p_j\beta_j(\text{coplanar}) - p_j\beta_j)\}\ .$$

The corresponding mechanical expression is

$$\Sigma\ (\pi\text{-bonds j})\ \{p_j\beta_j(\text{coplanar})\ (1 - \cos\omega_j)\}\ ,$$

where ω_j is the dihedral angle between π-orbitals at the terminals of bond j. The ratio of the former expression to the latter is the factor f_j in $V_{2,j}$. The approach has been extended to hetero-compounds [60].

It may be added that dissatisfaction seems to be growing as regards the parameter f_j. Reasons have been advanced both in favor of replacing it by a better function [33] and in favor of eliminating it altogether [61]. To date, the new approaches have been checked only for unfunctionalized hydrocarbons.

Such are, *grosso modo*, the terms in the energy equation. For further details, the reader is referred to the reports cited in this and in the previous section.

4 Molecular Mechanics at Work

Computational processes in molecular mechanics comport two modes: development and application. In developing a force field, the main task is to assign suitable numerical values to the constants. These should not only reproduce the available data, but also comply with the physical model. For example, the stretching constant k_s for a given type of bond should not only lead to realistic lengths, but also resemble the corresponding spectroscopic quantity. In applying a force field, the task is to calculate the geometry of a molecule, its E_T [Eq. (1)], and derived quantities, e.g., heat of formation. One also wishes that light be shed on factors that determine details of structure.

Research articles can usually be categorized according to one or the other of the two modes. In reporting upon the development of force fields, authors frequently delineate their reasoning and deliberations. A recent article on ketones [57] may serve as an example. A typical description of application is a report on monochlorocineols [62], where a standard computer-program was used to identify products and interpret NMR spectra. In one of the isomers (I), a remarkable downfield shift of methyl-protons could be related to a long-range Cl ··· Me repulsion and the conse-

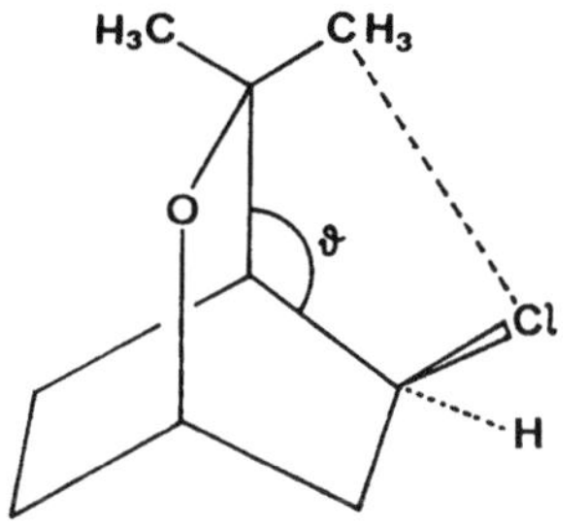

I

quent widening of a valence angle (ϑ in formula I). Articles of a mixed nature are also encountered, like a report on neutral molecule complexation with crown ethers [37]. Here, the authors start by modifying an extant field to their needs, then go on and apply it.

4.1 Application and Development

In application, the user starts by guessing a trial geometry. The computer-program sets up the force field for the given species. It then optimizes E_T by varying the internal coordinates. Details on current optimization techniques may be found elsewhere [2, pp. 64–76]. Note that both minima and saddle-points can be located.

The process of developing a force field is much more complex, and calls for a sequence of decisions. First, obviously, one decides on the range of applicability of the projected field. Second, one chooses whether to make the field independent of existing fields, or, rather, an extension to a given field. In the second option, formulation of the field — apart from numerical values for the new constants — is predetermined. Otherwise, the formulation has to be decided upon, e.g. whether to limit e_{tor} to the term in V_3 [cf. eqn. (6)]. Third, one chooses an „*original set*" of molecules, for which structural and energetic information is available. Molecules in this set are to serve as testing grounds. Trial values are now selected for the force field constants, E_T is optimized for all molecules in the original set, and the outcome examined. The last stage is re-iterated until the results of computation are deemed satisfactory. This done, the set of molecules has to be extended, in order to ensure that the derived constants are of wide applicability. If they are not, part of the process has to be recycled. Years may elapse between cycles, because fresh data are sometimes slow to forthcome. The process is summarized in Fig. 1.

The optimization of constants is usually performed by trial-and-error, but full or partial mathematical optimizations have been resorted to [48, 63, 64]. It is not evident which of the options is to be preferred. In some cases, the number of parameters to be assigned is almost as large as the number of traits to be reproduced. Mathematical fitting will then satisfy the original set, but not guarantee success with other molecules. Manual fitting, by contrast, leaves the investigator an opening for using his intuition and knowledge of background. There is then a better chance that results will not deteriorate appreciably when going from the original to an extended set.

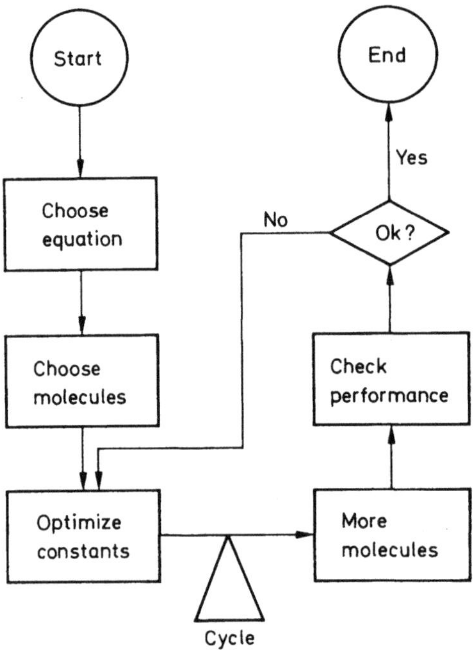

Fig. 1. Steps taken in developing a force field

Sometimes one has no choice. This happened, for example, when trans-chrysanthemic acid (II, 3-isobutenyl-2,2-dimethylcyclopropane-1-carboxylic acid [64]) was studied. An estimate was then required of the intramolecular dielectric constant (Sect. 5). An automated optimization, subject to the chosen constraints, led to 1.96 — a value that could not be foreseen.

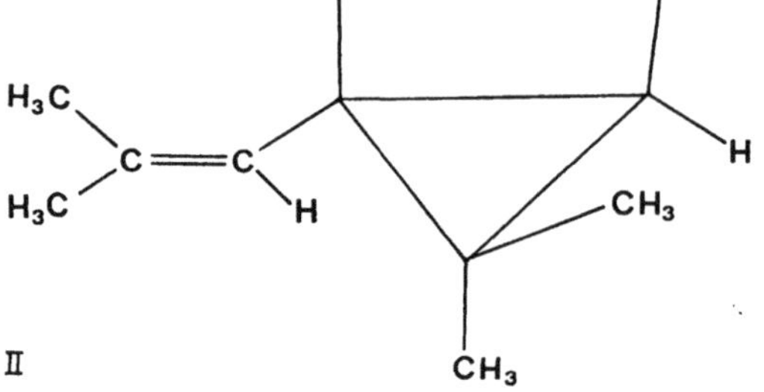

II

4.2 Organic Dibromides

To get a feeling for the two modes, consider organic bromine compounds. The force field [65] was projected as extension to an existing field for chlorides [66] and fluorides [67] (later, it was further extended to encompass iodides [68]). Numerical values were therefore required only for constants pertaining to bromine-containing environments, 20 in all. The original set comported a dozen or so of compounds, for which some geometrical data, and/or some conformational energies, were available from measurements.

Two of the compounds included in the original set were 1,2-dibromoethane and 1,2-*trans*-dibromocyclohexane. Dibromoethane exists as a mixture of conformers: *anti* (III) and *gauche* (IV). *Anti* is much more stable. By the experimental evidence (cited in [65]), the enthalpy difference in the vapor is 6.7 to 9.2 kJ mol^{-1}. The Br—C—C—Br dihedral angle in *gauche* (ω in formula IV) has been estimated as 73°, and the dipole moment of the *gauche*-form as 2.23 D. *trans*-Dibromocyclohexane

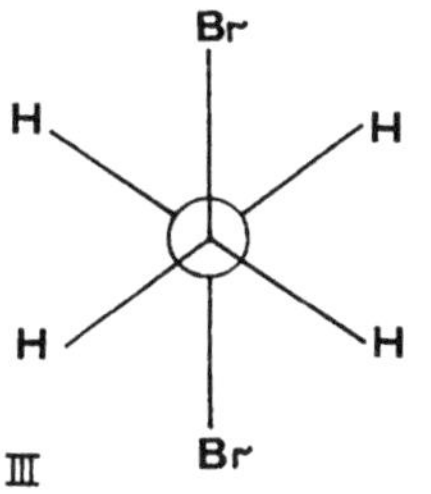

III

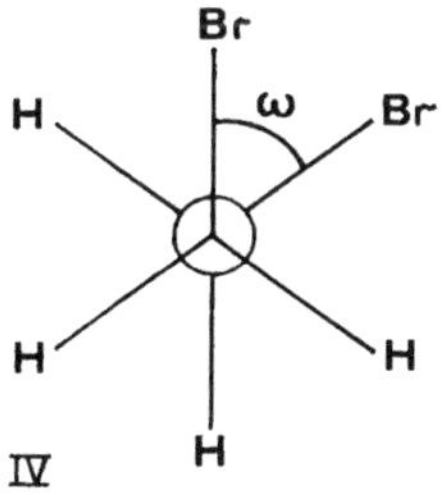

IV

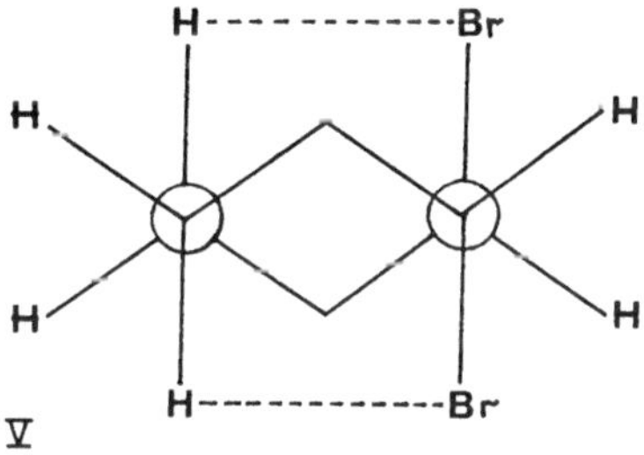

V

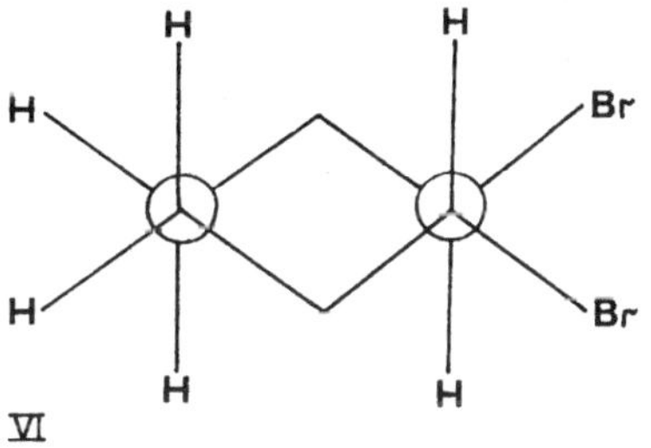

VI

exists also as a mixture of conformers: diaxial (*anti*, V, corresponding to III) and diequatorial (*gauche*, VI, corresponding to IV). Here also the *anti*-form is more stable, but the enthalpy difference is low. Judging by data on solutions [69], it would lie somewhere in the range 3.3–4.2 kJ mol^{-1}. The dipole moment of the vapor at 176 °C has been reported as 2.00 D [70].

The field was now fitted to the available data, including those cited above. Table 1 (entries a and b) lists the results. One thing to note is that the background information is indeed reproduced: for 1,2-dibromoethane one has ΔE_T and ω (*gauche*); for 1,2-dibromocyclohexane one has ΔE_T and the dipole moments which, when appropriately weighted [Sect. 5, Eqn. (13)], lead to 1.97 D at 176 °C. Another thing to note is that the results are faithful to the underlying physics; in particular, ΔE_{NB} does not contribute much to ΔE_T (see Sect. 3). The heaviest contributor to ΔE_T is ΔE_{TOR},

Table 1. Energy breakdown for three vicinal dibromides

	Anti	*Gauche*	ΔE(g — a)
a. 1,2-Dibromoethane			
ω	180°	73°	
χ	180°	83°	
μ	0	2.49	
E_S	0.15	0.21	0.06
E_B	1.96	2.79	0.83
E_{NB}	1.56	2.64	1.08
E_{TOR}	0.11	5.12	5.01
E_{ES}	—0.25	0.49	0.75
E_T	3.5	11.2	7.7
b. 1,2-*trans*-Dibromocyclohexane			
ω	161°	63°	
χ	162°	74°	
μ	1.09	3.24	
E_S	3.55	3.63	0.09
E_B	7.14	4.42	—2.72
E_{NB}	19.30	19.39	0.09
E_{TOR}	0.00	4.86	4.86
E_{ES}	—1.18	—0.07	1.10
E_T	28.8	32.3	3.4
c. 2,3-*trans*-Dibromodecalin			
ω	162°	62°	
χ	163°	74°	
μ	1.08	3.25	
E_S	7.10	6.95	—0.16
E_B	10.11	7.30	—2.81
E_{NB}	29.90	30.23	0.34
E_{TOR}	—0.04	5.12	5.16
E_{ES}	1.75	2.86	1.11
E_T	48.8	52.5	3.6

Energies in kJ mol^{-1}. E_B includes E_{SB}, E_{TOR} includes E_{TB}. ω — Br—C—C—Br dihedral angle; χ — C—Br/C—Br dipole angle (defined in Sect. 5 and Fig. 2); μ — dipole moment in Debye units

and this component does not behave wildly. A third thing to note is that the calculation actually adds to our understanding of structural factors. For example, we see that in open-chain fragments (entry a) the bending strain E_B is higher in *gauche* than in *anti*, whereas in cyclic fragments (entry b) it is the other way round.

Passing now to the mode of application, Table 1 (entry c) reports results on compounds not included in the original set. Little is known about 2,3-*trans*-dibromodecalin. There are two stereoisomers, *anti* (VII) and *gauche* (VIII). Unlike the con-

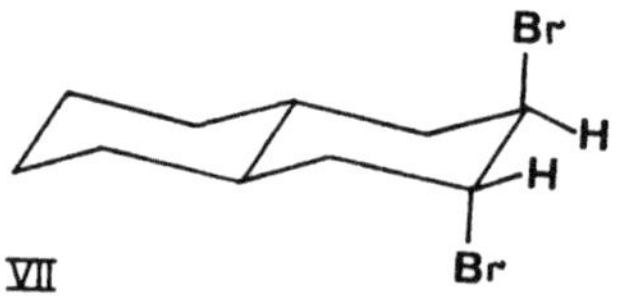

VII

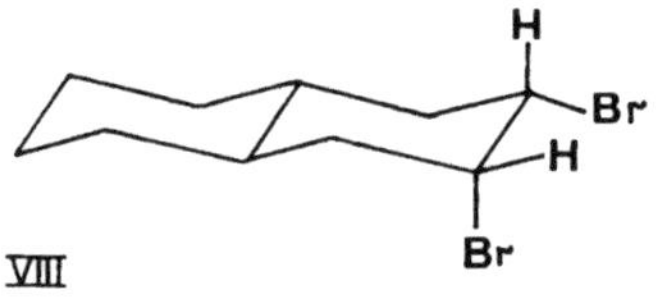

VIII

formers of dibromocyclohexane (V, VI), they cannot interconvert under ordinary conditions. Both have been isolated [71–73], and their dipole moments measured. These are: VII, 1.1 in cyclohexane [71] and 1.15 D in CCl_4 [74]; VIII, 3.55 D in cyclohexane [72]. The only check one can have on the calculation is the closeness of the computed moments to the experimental values. If this is deemed satisfactory, the rest is to be considered as prediction.

4.3 The Problem of Phase

The enthalpy-difference between stereoisomers changes on going from the vapor (v) to solution (s):

$$\Delta H^s = \Delta H^v + \Delta H_s ,$$

where ΔH_s is the difference between the solvation enthalpies of the stereoisomers. The need to estimate ΔH_s arises in two types of context, corresponding to the two modes of operation in molecular mechanics. When developing a force field, parameters are needed for isolated molecules (ΔH^v) while some of the required data may come from measurement on solutions. When applying a force field, results are for isolated molecules whereas predictions are frequently required for dissolved compounds. Users of molecular mechanics tend to estimate ΔH_s by Abraham's formalism, who also reviewed the topic in 1974 [75].

For dilute solutions of uncharged molecules that are not too polar, in solvents that are not too polar, Abraham breaks the solvation energy E_s into two components: the dipole term E_D and the quadrupole term E_Q:

$$E_D = \frac{\mu^2}{a^3} \cdot \frac{X}{1 - lX}$$

$$E_Q = \frac{3}{2a^5} \cdot \frac{3X}{5 - X} q^2 .$$

In these equations, μ is the molecular dipole moment, and q is the quadrupole moment, placed at the center. When q is written out, it becomes a rather complicated expression in the bond moments [76, 77]. In E_D and E_Q, X is a function of the bulk dielectric constant ε, $X = (\varepsilon - 1)/(2\varepsilon + 1)$, and $l = 2(n_D^2 - 1)/(n_D^2 + 2)$. The effective radius "a" is the radius of the solute cavity, assumed spherical (Sect. 6.2). Multiply E_D and E_Q by 14.3942 to obtain energies in kcal mol^{-1} from distances in A, bond moments and dipole moment in Debye units.

Recent developments in other aspects of molecular mechanics necessitated revisiting of the solvation formulas. Work has been concerned mainly with the positioning of the molecule in the coordinate system [78], the estimation of bond moments [79] or their replacement by atomic charges [77], and the estimation of effective molecular radii [79]. Further details on bond moments and effective radii will be given below, in Sect. 5.3 and 6.2.

4.4 Ancillary Remarks

When comparing the stability of stereoisomeric dibromides (Table 1), we encountered two types of trend in bending strains. In dibromoethane, the less stable conformer (IV) had the higher E_B; in dibromocyclohexane the more stable conformer (V) had the higher E_B. The latter eventuality is not unreasonable. A small sacrifice in bending strain can alleviate a sizeable amount of nonbonded strain. In conformer V, some distortion of valence and dihedral angles, which costs little energywise, relieves the severe nonbonded repulsions (dotted in formula V).

The phenomenon was discerned and first analyzed by Westheimer [4], when studying the syn-coplanar rotamer of 2,2'-dibromobiphenyl-4,4'-dicarboxylic acid (IX). He concludes as follows [4, p 550]. "Whenever two groups are forced into close proximity, the valence angles are considerably modified. This bending cannot be said

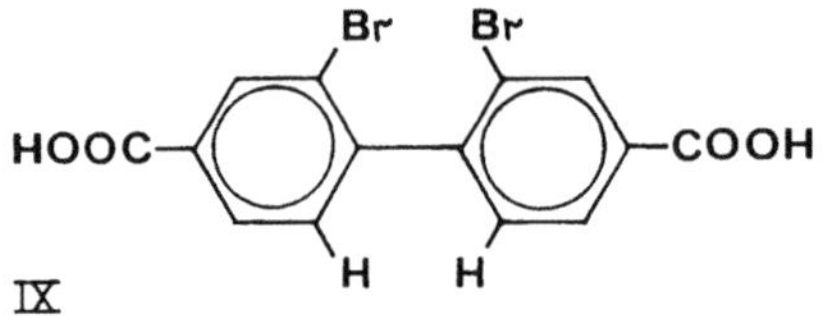

IX

to cause steric strain; rather, the bending occurs to mitigate or relieve the steric strain. But the major result of steric strain is the deformation of bond angles".

We are now ready to illustrate two remarks made in Sect. 1. One was that the molecular mechanist acts only above a certain threshold of complexity. Thus, dibromoethane and dibromocyclohexane are just simple enough to be included in an original set; chlorocineols, chrysanthemic acid and dibromodecalin are sufficiently complex to serve as candidates for application. The second remark was that molecular mechanics is best resorted to when delicate variations along a series are to be discerned. If one is interested, say, in the general nature of electronic interactions in —CHX—CHX— groupings, quantum chemistry should be his tool. As a matter of fact, the literature on this topic is enormous [80]. Molecular mechanics comes in, say, if it is the response of diheder X—C—C—X to cyclization that is to be studied.

5 Electrostatics

Almost every user of molecular mechanics has his pet way or ways of estimating E_{ES}, the contribution of intramolecular electrostatic interactions to E_T [Eq. (1)]. Interest in E_{ES} does not subside: existing pathways are constantly elaborated, and new pathways announced. One reason is that atomic charges and bond dipoles are involved, and there is a host of conceivable ways to estimate these. Another reason is that E_{ES} is not just one component in a somewhat artificial breakdown of E_T. Choices regarding the evaluation of E_{ES} affect the calculated dipole moment, which is an observable on its own. Such choices affect also the calculated energy of solvation (Sect. 4.3), again a separate observable. Two of the approaches to E_{ES} will now be described. The account is preceded by an overview and followed by an update. Note that the topic was covered a decade or so ago [28], and that a recent research article [81] cites many of the later developments.

5.1 Overview

When molecular mechanics was started, electrostatics was deliberately neglected [4, p 551]. Early applications concentrated on alkanes [44, 53, 82] and there was no urgent need to make amends: the small dipole moments of some alkanes [28, p 10] were not of concern, and mutual interaction of polarized C—H bonds was deemed small and, anyhow, implicitly included in E_{NB}. This is because r* and ε [Eqs. (3–5)] are fitted to properties of molecules in the original set and, if E_{ES} is absent, electrostatics is automatically absorbed into E_{NB}. The practice of ignoring E_{ES} was so convenient that it was carried over, perhaps with some lip service [22, p 48], to monofunctionalized hydrocarbons. Indeed, if only one moiety in the molecule is considered polar, it can have no partner to interact with [83]. Authors who did care to estimate E_{ES} noted that its presence did not affect the issue significantly [48]. This is frequently true, but neglect of E_{ES} precludes the opportunity of predicting dipole moments and solvation energies, and of pinpointing unexpected electronic effects. It certainly mars the physical interpretation of intramolecular effects.

When molecular mechanics ripened to deal with polyfunctional molecules, the realization that E_{ES} was indispensable fell on confused grounds: the term had now to be appended to fields that were already accounting, albeit implicity, for electrostatics. An example is provided by a pilot study of organic halogen compounds [84] wherein dipole-dipole interactions were superposed on the older formulations of E_{NB}. This, however, just preceded a period of transition, during which a new generation of force fields was being developed. At its term, E_{ES} was in, parameters were more commensurate with each other, and functionalized hydrocarbons could be tackled in a way more consistent than before [66, 85]. Hydrocarbon skeletons were still treated as if devoid of partial atomic charges, despite strong admonitions to the contrary [86, 87].

When different force fields are compared, one is struck by the range of values attributed to r* and ε. For example, r*(H) has been estimated as small as 143 pm [88] and as large as 182 pm [48]. A few years ago, it was suggested that the divergence had to do with the neglect of E_{ES} in treating hydrocarbons [43]. Since an essential component is missing from the list of strains, the set of optimized parameters is by necessity biased by the accidental choice of molecules in the original sets. Inclusion of E_{ES} for hydrocarbons, it was claimed, would help to make different force fields more similar and, even more important, capable each of tackling wider domains. By now, two research groups have taken C—H polarization into account [7, 89], though in a limited context. As anticipated [43], this incurs re-assignment of r* and ε for C and H (7). The situation actually risks to get out of hand, for some authors threaten to ignore all components in E_T but E_{ES} [90, 91]!

There are two ways to formulate a component $e_{es,ij}$ in E_{ES}: either

$$e_{es,ij} = q_i q_j / D r_{ij} \tag{8}$$

or

$$e_{es,ij} = \mu_i \mu_j (\cos \chi_{ij} - 3 \cos \alpha_i \cos \alpha_j) / D r_{ij}^3 , \tag{9}$$

where

$$E_{ES} = \sum_i \sum_j{}' e_{es,ij} \tag{10}$$

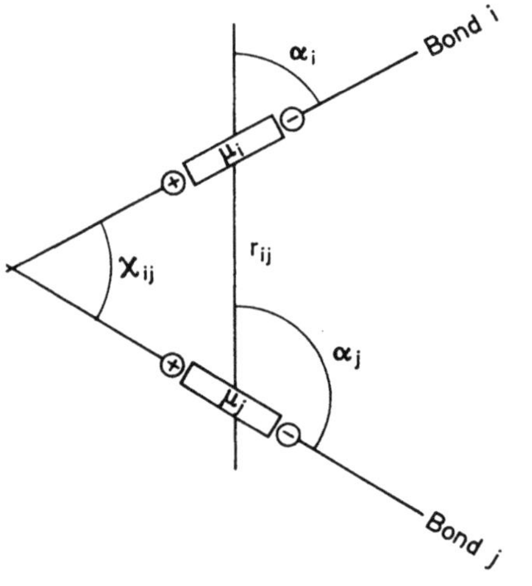

Fig. 2. Parameters in the calculation of dipole-dipole interaction: bond moments μ_i and μ_j, distance r_{ij}, mutual angles α_i and α_j, and the dipole angle χ_{ij}

Equation (8) is the "monopole approximation". In it, q_i and q_j are point-charges assumed to reside on atoms i and j, and r_{ij} is the distance between the atoms. Equation (9) is the scalar version of the "dipole approximation" [92], see Fig. 2. μ_i and μ_j are point-dipoles assumed to reside at the middle of bonds i and j, r_{ij} is their distance, α_i and α_j are the angles they make with the bonds, and χ_{ij}, the angle they make with each other, is the *dipole angle*. For four-atom-sequences X—(i)—Y—Z—(j)—W with tetrahedral valence angles X—Y—Z and Y—Z—W, the dihedral and dipole angles are approximately related through [93]

$$\cos \chi = 0.82 \cos \omega - 0.17 \, . \tag{11}$$

When using Eq. (8), the summation in Eq. (10) does not include bonded atoms nor atoms bonded to a common atom. This is done in order not to duplicate interactions covered, respectively, by e_s and e_b. In using Eq. (9), the summation does not include bonds emanating from the same atom. To obtain energies in kcal mol^{-1} from charges in eu, distances in Å and bond-dipoles in Debye units, multiply Eq. (8) by 332.05 and Eq. (9) by 14.39.

In examining computed geometries, Eq. (11) may be used to check how much, on the whole, do bond-angles deviate from tetrahedral values. For the gauche-forms cited in Table 1, the difference $\cos \chi - (0.82 \cos \omega - 0.17)$ is in the range 0.05 to 0.07. Dibromoethane is the least distorted and dibromocyclohexane is the most.

In Eqs. (8) and (9), the factor D is an intramolecular dielectric constant: the dielectric constant assumed to characterize the space between atoms i and j [Eq. (8)] or bonds i and j [Eq. (9)]. There is no reason for, and there is every reason against identifying it with the dielectric constant of the neat compound or of the medium. In molecular mechanics, D is more a parameter in a specific force field than an attribute of a specific molecule. Choices made by investigators in the previous decade have been tabulated [28, p 17]. They range from 1 to infinity, the latter obviously eliminating E_{ES}. Justifications given include theoretical considerations, fitting to an original set, and a wish to make the solvent field pervade the solute. Typical trends

Table 2. Intramolecular Dielectric Constant

Year	D	Context	Ref.	Justification given
1981	1	Oxamides (by monopoles)	[99]	
1983	1	Alcohols (by monopoles)	[100]	
1983	$1.6^{1/2}$	Organic halides (by monopoles)	[101]	Estimated charges are too high
1981	3	New force field	[102]	Within reasonable range
1983	4	Organic halides (by dipoles)	[68]	By fitting
1981	4r	AMBER force field	[103]	D is distance-dependent
1986	6r–15	Drug-receptor interaction	[104]	D is distance-depencent
1988	80	Water solutions	[105]	To simulate solvent

In the distance-dependent expressions, distance r is measured in A units

in the present decade may be gleaned from Table 2. In fields devolving from the MM-series [44, 45], when using the monopole approximation, D = 1 works quite well for various ways of estimating charges (first two entries and [94]). In the dipole approximation, a higher value seems more suitable (fifth entry). D = 4 is corroborated by certain measurements [95]. As can be seen, some authors make D distance-dependent [96] and others choose it such as to assimilate the solute with the solvent (last entry).

It is proper to add that the significance of intramolecular D is more fundamental than the value that happens to suit it in one force field or another [95]. This recognition underlies Cram's characterization of the hole that the six oxygens in spherand X delimit [97]. He says that the orbitals of the oxygen lone pairs "are in a microenvironment whose dielectric properties are between those of a vacuum and those of a

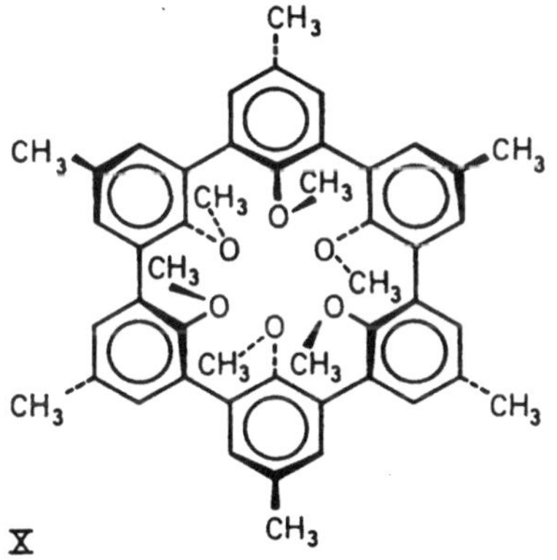

X

hydrocarbon". In vacuum, D = 1. For all we know, D may be as high as 5 for interaction of polar groups through an hydrocarbon skeleton [98].

The diversity of schools stems from the choice between Eqs. (8) and (9), and from the many options regarding the estimation of atomic charges and bond moments. Parameter-fitting is constrained by the requirement that the molecular dipole moment μ be fitted simultaneously:

$$\sum_i q_i \mathbf{r}_i = \boldsymbol{\mu} \tag{12}$$

in monopole calculations, or

$$\sum_i \mu_i = \mu$$

in dipole calculations. When fitting to the measured dipole moment of a conformational mixture, like 1,2-*trans*-dibromocyclohexane (V, VI), the weighting formula is

$$\mu^2 = \mu_I^2 X_I + \mu_{II}^2 X_{II} . \tag{13}$$

Here, μ_I and μ_{II} are the calculated dipole moments of the two conformers, X_I and X_{II} — their mole fractions, and μ is the observed dipole moment of the mixture. For a multi-component system, $\mu^2 = \Sigma\,(I)\,\mu_I^2 X_I$.

5.2 Del Re Methods

Del Re's computational method is the σ-system analog of the π-electronic Hückel method [106, 107]. It yields atomic charge distributions in saturated molecules and, through Eq. (12), the corresponding dipole moments. Note that Eq. (12) contains only one of the terms that enter the full expression of μ (the complete expectation value of the dipole moment operator is discussed in [108]). To cover for the omissions, extensive parametrization is unavoidable. Care has been taken [109] that π-electronic and σ-electronic parametrizations be consistent. Hence, there is a way to obtain the total charge distribution and dipole moment of heteroatomics by formulations based on "naive" Hamiltonians [79, 109].

Del Re's method decrees all σ-bonds of ordinary molecules to be fully localized, letting them interact with each other only via inductive effects [107]. Since the bonds are considered localized, a bond-orbital is atributable to each. This is an eigenvector, corresponding to the lower eigenvalue of the local secular equation,

$$\begin{pmatrix} H_{ii} - E & H_{ij} \\ H_{ij} & H_{jj} - E \end{pmatrix} \begin{pmatrix} C_i \\ C_j \end{pmatrix} = \bar{0} \tag{14}$$

where overlap has been neglected. The coefficients fulfil $C_i^2 + C_j^2 = 1$, the atomic charges are $2\,C_i^2$ and $2\,C_j^2$, and the net charges at the bond-terminals are $q_{i(j)} = 1 - 2\,C_i^2$ and $q_{j(i)} = 1 - 2\,C_j^2$. Since $q_{j(i)} = -q_{i(j)}$, a *bond charge* can be defined,

$$Q_{ij} = 1 - 2\,C_i{}^2 = C_j^2 - C_i^2 . \tag{15}$$

The total net charge on an atom is obtained by summing over all its bonds:

$$q_i = \sum_j Q_{ij} \tag{16}$$

In analogy with Hückel-type practices, matrix elements are formulated as $H_{ii} = \alpha + \delta_i\beta$ and $H_{ij} = \varepsilon_{ij}\beta$, where ε_{ij} is a constant of the bond. To introduce through-bond induction, δ_i is written as sum of an atomic constant δ_i^0 and contributions from linked atoms,

$$\delta_i = \delta_i^0 + \sum_j \gamma_{i(j)}\delta_j , \tag{17}$$

where $\gamma_{i(j)}$ and $\gamma_{j(i)}$ are additional constants of the bond. By fitting to experimental dipole moments, Del Re set up a table of values for ε, γ and δ^0. His values for C—H are: $\delta^0(H) = 0$ (the reference atom), $\varepsilon(C - H) = 1$ (the reference bond), $\delta^0(C) = 0.07$, $\gamma_{C(H)} = 0.3$, $\gamma_{H(C)} = 0.4$. Note that all values are positive. Later refinements and additions have been cited [81, 110].

We now show that bond charges [Eq. (15)] can be approximated without resolving a sequence of secular equations [Eq. (14)]. Divide the secular determinant by β, put $k = (\alpha - E)/\beta$, and solve for the lower root of

$$\begin{vmatrix} k + \delta_i & \varepsilon_{ij} \\ \varepsilon_{ji} & k + \delta_j \end{vmatrix} = 0 .$$

It is

$$k = -\frac{1}{2}\{(\delta_i + \delta_j) + [(-\Delta)^2 + 4\varepsilon_{ij}^2]^{1/2}\} ,$$

where $\Delta = \delta_j - \delta_i$. Since ε's cluster in the range 0.5–1.0, while Δ's are usually lower than 0.25, $(-\Delta)^2$ can here be neglected. Rearrangement yields $k + \delta_i = -\frac{1}{2}(\Delta + 2\varepsilon)$

Table 3. Dipole moment of alcohols

Alcohol	Calculated[a]	Measured[b]
CH_3OH	1.63[c]	1.69
CH_3CH_2OH		
exo	1.58	1.52
endo	1.69	1.62
$CH_3CH_2CH_2OH$	1.65[d]	1.65–1.68[e]
$CH_3CHOHCH_3$	1.67[d]	1.68[e]
exo	1.64	—
endo	1.74	1.58
$C_6H_{11}OH$	1.68[d]	1.6–1.8[e]
cis-4-*tert*-Butyl-1-cyclohexanol	1.70[d,f]	1.74[f]

Dipole moments in Debye units.
[a] Original Del Re method.
[b] See [100] for sources of information.
[c] The computed components are 1.36 and 0.89 D; the Stark-effect counterparts are 1.44 and 0.89 D.
[d] Average over conformers (when required, by extension of Eq. (11) to several contributions).
[e] In solvents.
[f] At 110 °C.

and $k + \delta_j = -\frac{1}{2}(2\varepsilon - \Delta)$. With these, Eq. (14) can be solved for C_i and C_j, and the bond charge [Eq. (15)] reformulated as

$$Q_{ij} = C_j^2 - C_i^2 = \frac{\Delta + 4\varepsilon}{\Delta^2 + 4\varepsilon\Delta + 8\varepsilon^2}\Delta$$

$$\sim \frac{4\varepsilon\Delta}{8\varepsilon^2} = \frac{\delta_j - \delta_i}{2\varepsilon}. \quad (18)$$

Thus, in a Del Re calculation, one starts by resolving the linear system [Eq. (17)], then goes on to evaluate bond charges [Eq. (18)] and net charges [Eq. (16)]. To illustrate how far the inductive model can go, Table 3 compares computed and experimental dipole moments for a few alcohols. The descriptor *endo* refers to the conformer wherein the hydroxylic H occupies the more encumbered environment; in alcohols with an H—O—C—H sequence, this is the anti-conformer (formula XI for CH_3CH_2OH and $CH_3CHOHCH_3$). *Exo* is the alternative (XII). The charges

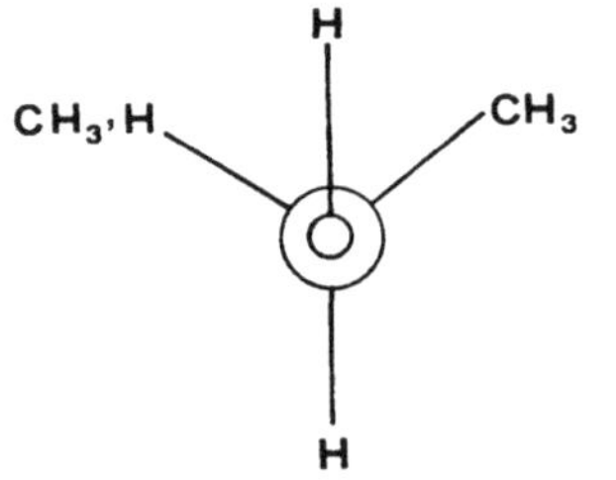

XI ENDO

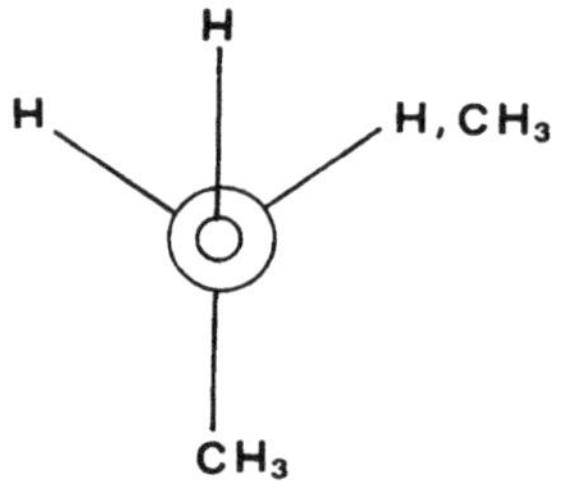

XII EXO

in Table 3 were by-products of a molecular mechanics study of alcohol conformations, using monopole contributions to E_{ES} in E_T [49, 100]. It is fair to note that new experimental information on alcohols [111] suggests that some of the reported steric constants [100] require updating. In the terms of Fig. 1, a new cycle is indicated.

Two situations are known where the original Del Re algorithm does not work well [110]. The first concerns molecules with geminal fluorine disubstitution, like CH_3CHF_2 and CH_3CF_3, where the calculated moment is frequently too low. The reason verges on the anecdotal, and so is the remedy. In geminal polyfluorides, the C—F bond is very short [112]. The moment is low because Del Re fitted fluorine constants to molecules with long C—F bonds. The way out, simply, is to use in this eventuality a reduced value for ε_{CF}. By Eqs. (18) and (16), this raises the fluorine charge. Incidentally, geminal fluorines must be attributed separate constants also in the steric components of E_T [69].

The second problem is more serious. Dipole moments calculated for non-geminal polyhalides are frequently too high, and so are the dipole moments of geminal polyhalides other than fluorides [110]. For example, the calculated value for 1,3-*trans*-dibromocyclohexane (XIII) is 2.30, while the experimental is only 2.19 D. This must be due to a field effect [95] that acts to reduce charges [113]. Two ways have been proposed to contend with this difficulty [110, 114]. One of these [110] is very simple: the field effect is introduced via the inductive mechanism by defining a

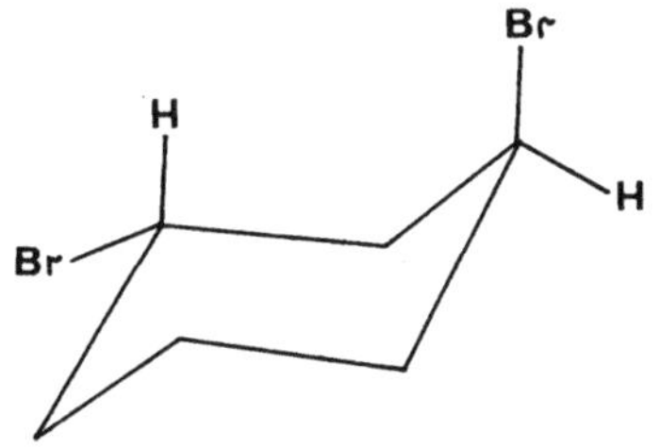

XIII

"pseudobond" of $\varepsilon = -0.2$ (negative!) between the interacting functions. A sample of results for non-geminal dihalides has been given in the original publication. Examples of geminal polysubstitution are CH_2Cl_2 and $CHCl_3$. For the former, one calculates 1.71 without and 1.53 D with the pseudobond; experimental values are in the range 1.54 to 1.63 D. For the latter one calculates 1.36 without and 1.06 D with three pseudobonds; the experimental range is 0.96 to 1.07 D.

What about unsaturated molecules? In principle, one has to perform a π-electron calculation and superpose the ensuing π-charges on the σ-charges [79]. In practice, pilot calculations show that the computed dipole moments are reasonable, but not necessarily their components along the inertial axes. For example, measurement on FCH_2COCH_3 led to $\mu_a = 0.907 \pm 0.020$ and $\mu_b = 0.76 \pm 0.05$, totalling at 1.18 D [115]. The computed $(\sigma + \pi)$ dipole moment is 1.13 D, seemingly in perfect agreement. Its components, however, are 1.11 and 0.19 D [79]. We shall report in Sect. 5.4 that another method has been proposed to meet unsaturation [116, 117]. That method, however, was not checked on inertial components.

5.3 Bond Moments

The notion of bond moments can be traced at least as far back as 1923, when Thompson provided formulas relating the dipole moment of a disubstituted benzene to those of its monosubstituted parents [117]. In these formulas, the dipole moments μ of C_6H_5X and C_6H_5Y behave mathematically as if residing, respectively, in C—X and C—Y: the dipole moment of C_6H_4XY is the vectorial sum of $\mu(C_6H_5X)$ and $\mu(C_6H_5Y)$. There follows the conception that the dipole moment of a molecule is the sum of moments attributable to its bonds. And more: each bond behaves as an electric dipole, being subject to dipole-dipole interaction and to dipole-dipole induction [119]. Hundreds of structural studies, based on these ideas, were reported. The vogue subsided at the advent of nuclear magnetic resonance, a tool of wider scope, much more rapid in practice and more direct as interpretations go. In molecular mechanics, interest in dipole moments survives. This is due to their tie with E_{ES}.

Quantitative estimates of electrostatic bond-bond interactions [Eq. (9)] can be traced back to Corey's work in 1953 [120]. He may have mixed up [121] the definition of angles α (Fig. 2), perhaps because of a literature ambiguity [93], but it is his work that heralds the inclusion of E_{ES} in E_T.

The problem with bond moments is that the decomposition of a molecular dipole moment into bond-tied components, even if physically sound, is not unique. Different modes of decomposition lead to different bond moments, and hence to different estimates of E_{ES} via Eq. (9). Therefore, any mode is by necessity limited in scope and meaningful only in conjunction with other parameters of the force field. A very simple example will illustrate the point. The dipole moment of CH_3CH_2Cl is 2.04 D. One can attribute this value to bond C—Cl, putting $\mu(C-H) = \mu(C-C) = 0$ [83]. With this choice, E_{ES} vanishes for all rotamers. Alternatively, one can reconstruct 2.04 D by attributing 1.60 to +CCl—, 0.33 to +HC— and 0.28 D to +C(β)C(α)— [66]. With this choice, E_{ES} is non-nul and varies along the course of internal rotation. Clearly, if one wishes to reproduce the barrier to internal rotation, parameters in E_{NB} have to differ in the two options.

We now mention briefly two ways to estimate bond moments.

The first, again, is Del Re. It is not widely realized that this method can be worked for bond moments as easily as it is worked for atomic charges [79]. What the algorithm primarily does is furnish bond charges [Eq. (17)], *not* atomic charges. The bond charges can be either summed for atomic charges [Eq. (16)], or, which is even simpler, multiplied by bondlengths to produce bond moments:

$$\mu_{ij} = Q_{ij} l_{ij} \tag{19}$$

There are two advantages. First, the technique provides moments that are consistent with charges, which is not the usual occurrence [122]. Second, it can be run either way, according to needs. In calculating solvation energies, for example, bond moments are more convenient to use (Sect. 4.3), while charges are more convenient, say, in analyzing drug-receptor interactions [123].

Other ways to obtain bond moments from charges have been developed [37].

The second approach was developed for force field calculations on organic halides [49, 65–68]. Numerical constants, as fitted to experimental dipole moments, are listed in Table 4. Here, one started by assigning a value, 0.33 D, to all —C—H+ bonds [136]. The choice is almost arbitrary, but in the correct range (as regards the sign, see [108]). Dipole moments of mono-halogenated alkanes can be reproduced by summ-

Table 4. Bond moments in organic halides

Bond	$\mu°$	$\mu[+C(\beta)-C(\alpha)-]$	k
—C—H+	0.33		
+C—F—	1.51	0.31	0.17
	1.35 (—CF_2—)	0.51	0.17
	1.21 (—CF_3)	0.51	0.17
+C—Cl—	1.60	0.28	0.28
+C—Br—	1.50	0.35	0.26
+C—I—	1.33	0.27	0.74

Bond moments in Debye units. Figures from [65—68]

ing vectorially $\mu(-C-H+)$ for C—H bonds, the intrinsic moment μ^0 of the +C—X— bond, and the moment induced in +C(β)—C(α)—, where C(α) is the carbon bonded to X. In polyhalogenated alkanes, if C—X bonds are close to each other, this construction does not suffice. The reason is that the strong C—X dipoles, polarizing and polarizable, act to reduce each other:

$$\mu_i = \mu_i^0 - \Delta\mu_i \,. \tag{20}$$

To see how $\Delta\mu$ can be estimated, rewrite Eq. (9) in the form

$$e_{es,\,ij} = \mu_i\mu_j F_{ij} \,,$$

where the spatial factor is

$$F_{ij} = (\cos\chi_{ij} - 3\cos\alpha_i \cos\alpha_j)/Dr_{ij}^3 \,.$$

Suppose bond i is affected by one sole bond j. If the interaction is calculated at the correct geometry, but with the intrinsic moments, one has

$$e^0_{es,\,ij} = \mu_i^0 u_j^0 F_{ij} \,.$$

Also [113],

$$\Delta\mu_i = \lambda_i \mu_j F_{ij} \,,$$

where λ_i is the polarizability of bond i. Now,

$$\Delta\mu_i \sim \lambda_i \mu_j^0 F_{ij}$$
$$= \left(\frac{\lambda_i}{\mu_i^0}\right) \mu_i^0 \mu_j^0 F_{ij} = k_i e^0_{es,\,ij} \,.$$

k_i is a new constant of the bond, which can again be assigned by fitting. If bond i is affected by several dipoles j, Eq. (20) is replaced by a sum,

$$\mu_i = \mu_i^0 - \sum_j \Delta\mu_{i(j)} \,.$$

Examples of application have been cited in the original publications.

Other ways to meet induction have been developed [124].

5.4 Current Trends

Developers of force fields prefer nowadays atomic charges to bond moments. As in the past [28], organic halides are the main testing grounds in checking new algorithms. Three items will illustrate present activities.

a) Two new schemes have been announced to evaluate atomic charges for force fields of the MM-series. One is Došen-Mićović's "Induced Dipole Moment and Energy Method" [114], which was tested on an extensive list of organic halides, ketones and cyclic ethers [78]. The other is Gasteiger's "Partial Equalization of Orbital Electronegativity" [125], applied to date to organic halides [81].

b) Halogen compounds, of types not tackled before, have been calculated by Stőlevik's force field which incorporates a technique of "excessive charges" [126]. By running calculations with and without electrostatics, it was found that neglection of E_{ES} does not affect ΔE_T markedly. The effect on torsional angles is erratic. Among the many applications reported, one may cite a variety of halogenated olefins [51], halogenated disilanes [127], halogenated enones and alkenoic acids [128].

c) An entirely new technique for evaluating charges is being developed by Abraham [129]. The atomic charge is written as an intrinsic value plus contributions due to interaction with other atoms: one-bond (A with B in A—B—C—D), two-bond (A with C) and three-bond effects (A with D). As of now, the method handles the dipole moments of variegated carbon and silicon compounds [130], including substituted olefins [116] and conjugated systems [117]. To date, the calculated charges have not been used in molecular mechanical work.

6 Atomic Radii

Among other constants in the force field, there appear the atomic nonbonded radii r^* (Sect. 3). Their fitted or optimized values look strange and unexpected: they are much larger than the better known "van der Waals atomic radii" r_w (subscript w for van der Waals). Thus, for hydrogen, $r_w(H) \sim 120$ pm, while fitted values of $r^*(H)$ lie in the range 140 to 180 pm (Sect. 5). Numbers for some other atoms are listed in Table 5. These high values have always been of concern and a cause for apology [22, 44] as if they are accidental issues of the arbitrary choices made in developing force fields.

Notions of atomic and molecular size are somewhat mixed up. This is because various measures are conceivable, and different contexts are suited by different measures. Since the molecular mechanist touches frequently on questions of size, some of the options will now be examined. A preliminary outline of the argument has been given [11].

The discussion pivots about two questions, one qualitative and one quantitative: Why are r^*'s so large? Why are they numerically as they are?

6.1 Qualitative

The notion of r_w is usually presented by aid of drawings such as Fig. 3. Two distant atoms (large r/R) are imagined to happen into mutual influence; attraction is dominant, they approach each other, the energy goes down. At a certain distance

Table 5. Intermolecular and intramolecular atomic radii[a]

		r_w	r_{rm}	r^*
C	(saturated)	175	—	190
C	(unsaturated)	182[b]	—	194, 196[c]
H		117	170	150[d], 162[c]
—O—		140	—	174
=O	(carbonyl)	150	213	174
=O	(carboxyl)	150	195	174
L[e]	(neat alcohol)	0[f]	0	120
	(dissolved alcohol[g])	0	185	120
	(ether)	0	228	120
F		130	185	165
Cl		177	209	203
Br		195	219	218
I		210	243	232

[a] In pm units. Unless noted otherwise, van der Waals radii r_w are from Gavezzotti's list [131] and intramolecular nonbonded radii r* are as incorporated in the MM2 program [45]. [b] Estimated by raising Bondi's value (173 pm versus 169 for saturated C [132]) to the height of Gavezzotti's scale. [c] Very recent readjustment [7]. [d] For H bonded to O, MM2 has r* = 120 pm. [e] L denotes each of the two lone electron-pairs in —O—. Molecular mechanics usually disregards lone pairs on =O. [f] Assumed, given that $r_w(X) < r_{rm}(X)$ and $r_{rm}(L) = 0$ by actual fitting to densities. [g] Infinite dilution in non-polar solvents (partial molar volumes at 20 °C estimated from data in [133] at 25 °C).

($r = R$, $e_{nb} = -\varepsilon$), repulsion becomes significant. Further approach ($r < R$, $e_{nb} < 0$) is attractive on the whole, but now the energy goes up. In Fig. 3, R stands by definition for the sum of the van der Waals radii of the two atoms. Let us call it the *contact distance*, sum of the atomic *contact radii*.

In a molecule, each of any two interacting atoms in covalently bonded to some residue. Its domain of influence is not spherical. Questions of orientation [134] and of the effect of residues may come in. Not necessarily, though. Consider, for example, fluoroacetone [85, 115]. As marked in Fig. 4, the distance $F^1 \ldots H^5$ is 252 pm. This is *exactly* the contact distance [$r_w(H) + r_w(F) = 117 + 135 = 252$]. It is only when the angle C—F ... X is acute, and to the extent that it diminishes, that F approaches X to less than the contact distance. Thus, distances $F^1 \ldots C^4$ and $F^1 \ldots C^3$ are 271 and 239 pm, while the contact distance is 310 [$r_w(F) + r_w(C)$]. A bonded atom looks indeed more like a teardrop than like a sphere. A mathematical representation of its shape was attempted many years ago [135, p 178].

Let us do away quickly with 1,3-interactions (like $F^1 \ldots C^3$ in Fig. 4) and with 1,4-interactions ($F^1 \ldots C^4$). In molecular mechanics, the former are covered by angle-bending strain [Sect. 3, Eqn. (2)], and not included among the nonbonded. For completeness, it should be mentioned that atoms have distinctive "sideways contact radii" [136], and these determine the bond angles [137]. Carbon, for example, is attributed an edge-on r_w of 175 pm but a lateral r_w of only 125 pm. As to 1,4-approaches, these are determined by five skeletal strains — stretchings 1—2, 2—3, 3—4 and bendings 1—2—3, 2—3—4 — on top of the nonbonded potential. Hence, the revealing distances are 1,5 at the closest.

What will happen if the molecular mechanist used contact radii r_w in his formulas for the nonbonded potential [e.g. Eqs. (3–5) of Sect. 3]? A few pairs, like atoms A and B in Fig. 5, are so positioned as to come into close contact. However, a host of other pairs, like C ... B and A ... D, cannot approach that close. In these pairs, distances correspond to the attractive branch (Fig. 3), and the process of optimization tends to bring the partners closer. The combined attractions would push A and B to less than the proper contact distance, and force them to interpenetrate ($r < R$ in Fig. 3).

This is why r*-values in nonbonded potential functions have to be large. They are such that penetration of r*(A) into r*(B), as brought about by interactions other than A ... B, just places A and B at the contact distance. Conversely,

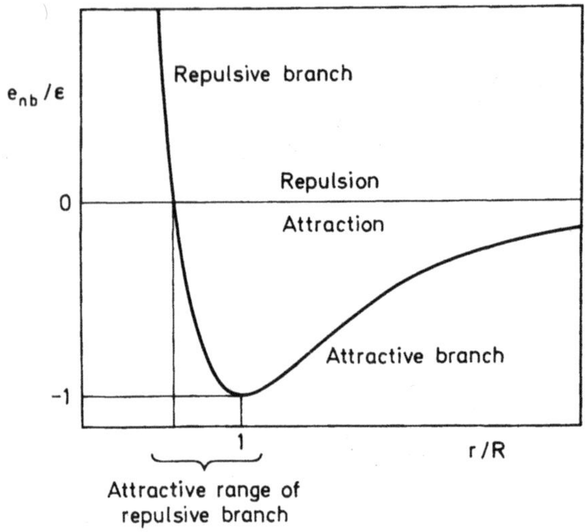

Fig. 3. Explicating the notions of "contact distance" ($r/R = 1$) and the "attractive range of the repulsive branch"

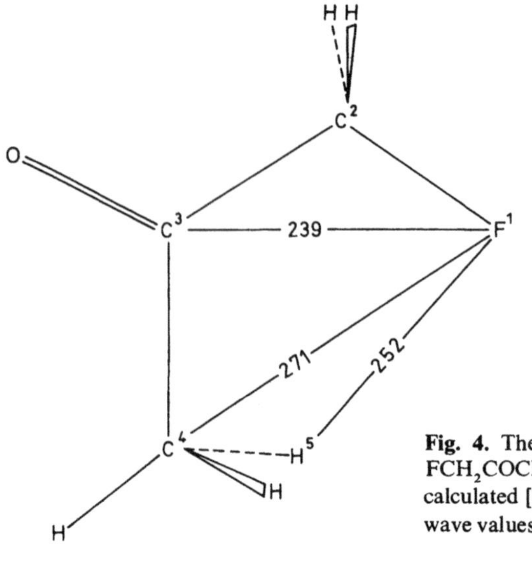

Fig. 4. The more stable conformer of fluoroacetone, FCH_2COCH_3. Interatomic distances in pm units are as calculated [85]. They should be very close to the microwave values [115]

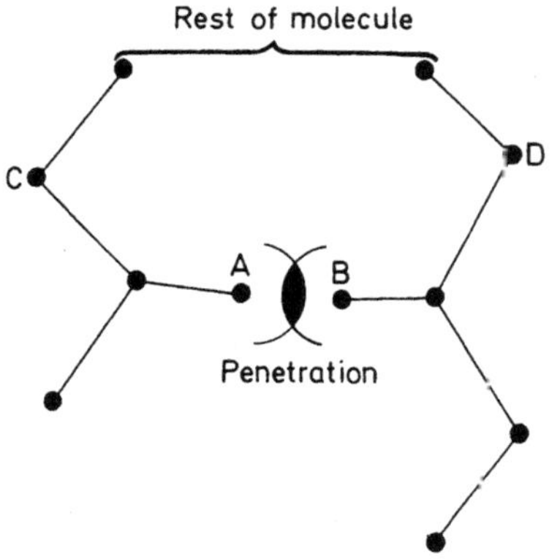

Fig. 5. Nonbonded interactions in a large molecule (schematic). Distances C ... B and A ... D correspond to the attractive branch of the nonbonded potential. The ensuing attractions force distance A ... B onto the repulsive branch, but not necessarily out of the range of attraction

molecules with very close nonbonded distances offer good testing grounds for nonbonded potentials [47, 88].

Obviously, it may happen that attractions are so numerous as to force a real van der Waals penetration at some point [138]. Figure 3 serves to remind us that a small penetration does not imply repulsion, and that the overall effect could be beneficial energywise [38].

6.2 Quantitative

It is common to envisage atoms in a molecule as spheres of radius r_w. This is, in fact, how molecular graphics [139] and space-filling mechanical models [18] work. The molecule becomes then a geometrical solid in space, made up of interlocking

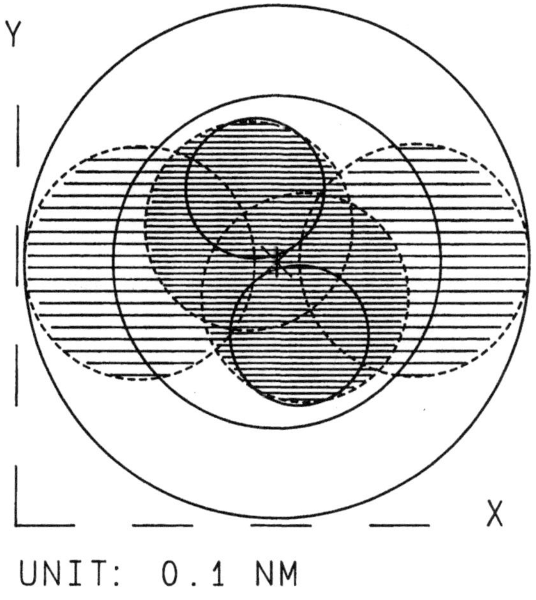

Fig. 6. Cut through the van der Waals body of anti-1,2-dibromoethane. Sparse hatching marks the part of bromine spheres that does not overlap with the body of the alkyl residue. Outer circle delimits the circumscribing sphere; inner circle delimits a sphere of volume equal to V_w. Center of molecule and spheres indicated by a star

spheres. Let us call this body the "w-body" (w for van der Waals). Figure 6 is a cut through the w-body of *anti*-1,2-dibromoethane.

There are ways — incremental [132], numerical [140, 141] and analytical [142] — to evaluate the volume of w-bodies (V_w), their surface area (S_w) and other attributes [11]. No chemical procedure directly measures V_w and S_w [143], but some molecular properties correlate with them [11, 144]. In Fig. 6, the inner circle represents a sphere of volume V_w.

If the molecular shape is approximated as spherical, the sphere that circumscribes the molecule becomes of interest. The sphere is also of interest when there are reasons to consider as spherical the molecular domain of influence — which is quite common in the study of liquids [145]. In Fig. 6, the circumscribing sphere is represented by the outer circle. Its volume V_c is another measure of volume. Correspondingly, there are two measures of molecular radius: the radius R_w of a sphere of volume V_w,

$$R_\omega = \left(\frac{3V_\omega}{4\pi}\right)^{1/3}$$

and the radius R_c of the circumscribing sphere. Obviously, $V_c > V_w$ and $R_c > R_w$.

One can arrive at intermediate estimates of molecular size [146]. To this end, one partitions the molar or partial molar volume $\bar{V}$ of a compound among N_0 molecules (N_0 — Avogadro constant). The ensuing "molecular volume" V_m of the "*m*-body" is the average volume allotted to one molecule in a condensed phase. It includes the molecule's average share in the empty spaces that run between molecules [147], and therefore is temperature-dependent [148]. Again, with a spherical approximation, a molecular radius R_m can be defined:

$$R_m = \left(\frac{3V_m}{4\pi}\right)^{1/3} .$$

Unlike the quantities mentioned heretofore, V_m is measurable and R_m therefore is an observable. The effective radius "a" of solvation theory (Sect. 4.3) is usually identified with R_m [75, 79].

As a numerical illustration, here are the data for 1,2-dibromoethane (Fig. 6). The units are pm for radii and cubic pm for volumes (recall that 10^2 pm $\equiv$ 1 Å, and 10^6 cubic pm $\equiv$ 1 Å^3).

w-body:	$V_w = 89 \times 10^6$,	$R_w = 277$;
m-body at 20 °C:	$V_m = 143 \times 10^6$,	$R_m = 325$;
sphere:	$V_c = 328 \times 10^6$,	$R_c = 428$.

The ratio of V_w to V_m is the "packing density" P [147, 149]:

$$P = \frac{V_w}{V_m} = \frac{N_0 V_w}{N_0 V_m} = \frac{V_I}{\bar{V}} \tag{21}$$

V_I has been labelled "the *intrinsic volume*" [150]. Theoretically, the upper limit on P is 0.74 [151]. Random packing of metallic spheres [152] and ellipsoids [153] leads to P in the range 0.63 ~ 0.59. In actual molecules [154] the range is sensitive to structural features. A representative series is: $(CH_3)_3CCH_2CH_3$ (branched, P = 0.51); C_6H_{14} (straight acyclic, 0.52); C_6H_{12} (cyclic, 0.56); $C_6H_{11}Br$ (functionalized cyclic, 0.61); $BrCH_2CH_2Br$ (bifunctionalized acyclic, 0.62); $C_6H_{10}Br_2$ (bifunctionalized cyclic, 0.64). The numbers above were calculated by Eq. (21), from calculated V_w's and measured densities.

Since V_m-values are observables, there is reason to look for a computational pathway that approximates them. In doing this, one would like to conserve an analogy with V_w's, that is, adhere to the depiction of a molecule as a system of interlocking spheres. In the technique proposed [146], atoms in a molecule are classified in two categories: core-atoms and mantle-atoms. Core-atoms are those that form several links ("connected atoms", like C in hydrocarbons, C and O in alcohols), and mantle-atoms are those that form one link ("attached atoms", like H in hydrocarbons, H and =O in ketones). Core-atoms are assigned the usual radii r_w: since they are well buried in the molecular body (like the carbons in Fig. 6), the actual value chosen is almost immaterial. Radii for mantle-atoms can be obtained by fitting the computed volumes of molecules to observed V_m's. Let us denote the fitted atomic radii by r_{rm}, and the molecular volumes they lead to by V_{rm} (rm for "reduced molecular").

As is to be expected, one sole value for hydrogen, $r_{rm}(H)$, cannot fit satisfactorily all saturated hydrocarbons. This is because packing densities span a wide range. Branched alkanes (low P) require a larger $r_{rm}(H)$ than straight-chain alkanes (intermediate P) and cycloalkanes (high P). Fitting to cyclohexane leads to $r_{rm}(H)$ = 170 pm at 20 °C [146]. This value, which also fits the hydroxylic hydrogen of neat

Table 6. Molecular volumes and "reduced molecular" volumes for molecules with hydrogen mantles

Compound	V_{rm}	V_m	Δ (%)
$Me_3CCH_2CHMe_2$	250.1	274.1	−8.8
Et_4C	273.2	282.6	−3.3
C_6H_{14}	202.8	217.0	−6.5
C_9H_{20}	289.1	296.8	−2.6
C_5H_{10}	152.2	156.2	−2.6
C_6H_{12}	177.3	179.5	−1.2
C_7H_{14}	204.3	201.3	+1.5
1,2-*cis*-$C_6H_{10}Me_2$	230.9	234.0	−1.3
1,2-*trans*-$C_6H_{10}Me_2$	232.5	240.1	−3.2
MeOH	65.1	67.2	−3.1
EtOH	95.4	96.9	−1.6
PrOH	124.6	124.2	+0.3
Me_2CHOH	123.7	127.1	−2.7
Me_3COH	150.8	156.1	−3.4
Me_3CCH_2OH	179.4	180.3	−0.5
$C_6H_{11}OH$	186.4	172.8	+7.9
$CH_3(CH_2)_8CH_2OH$	326.8	316.8	+3.2

To obtain volumes in cubic pm, multiply numbers by 10^6

alcohols [155], was adopted for all alkanes and for the hydroxylic hydrogen of neat carboxylic acids [156]. Table 6 compares the computed V_{rm} with the experimental V_m for several compounds. Agreement is somewhat erratic, due to the wide range of P. It is perfect for Me_3CCH_2OH (−0.5%) but off the mark for $C_6H_{11}OH$ (+7.9%). Because of the spread in packing densities, liquid densities of organic compounds can be predicted to 10% at most.

With a value chosen for $r_{rm}(H)$, r_{rm}-values for other atoms can be estimated. For an atom X other than H, one examines a group of liquid compounds that contain C, H and X, and seeks $r_{rm}(X)$ by fitting V_{rm}'s to V_m's. All values derived to date were listed in Table 5.

The rm-body of anti-1,2-dibromoethane is shown in Fig. 7.

At this point, one is ready to compare the molecular-mechanical r*'s with the set of r_{rm}'s. Recall that elements in the two sets are obtained independently. The former are fit to molecular geometries via the intermediacy of a chosen force field, the latter are fit directly to measured densities. As Table 5 shows, numbers in the two sets are very close. Eventual elaboration of force fields could bring them even closer. The molecular mechanist's r* is, then, a "quasi-observable".

6.3 More on Atomic Volumes

Reasoning by non-van der Waals radii can clarify phenomena which hitherto remain unexplained. Refer, for example, to the "striking fact [...] that halogen atoms behave as if their volumes were smaller than indicated by their w-volumes. [...] Bromine has about the same van der Waals volume as methyl, and [...] would be expected to have

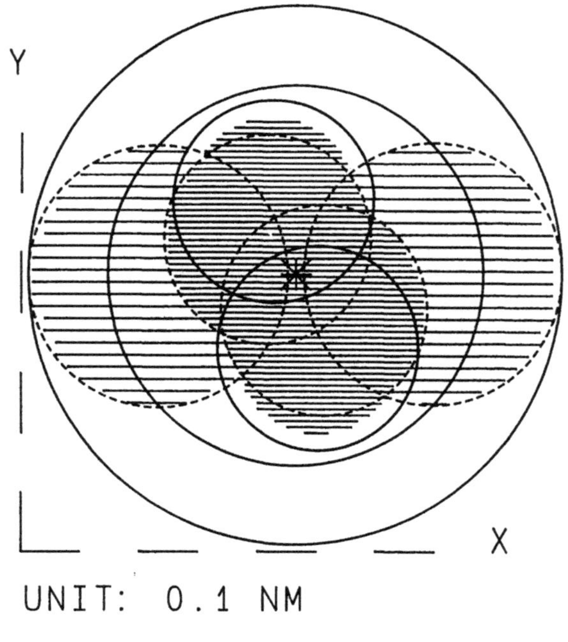

Fig. 7. Cut through the calculated rm-body of anti-1,2-dibromoethane. Internal circle delimits a sphere of volume equal to V_{rm}, which is quite close to the experimental V_m (R_{rm} = 318, R_m = 325 pm). Other details as in Fig. 6

about the same volume increment. In fact, [its increment] is considerably smaller" [157]. Usage of rm-radii, rather than of w-radii, reveals that Br is indeed smaller than CH_3: the rm-volume of a methyl-group is 48.7×10^6, while the portion of the rm-sphere of bromine which does not overlap with an alkyl residue has a smaller volume, 31.6×10^6 cubic pm. This portion is depicted by sparse hatching in Fig. 7. The corresponding numbers for the other halogens are 13.8 (F), 25.8 (Cl) and 46.0×10^6 cubic pm (I). For comparison, the w-volumes are: 26.8 (CH_3), 4.1 (F), 16.2 (Cl), 23.5 (Br), and 31.4×10^6 (I).

Numerical corroboration comes from a recent study of hydration enthalpies [158]. Its authors start by defining the radius of a solute as the radius of a sphere that contains a negligible electron density contribution from the surrounding solvent. For poly-atomic species, the boundary of the cavity is demarked by the union of envelopes of spheres centered on peripheral atoms. The radii of atomic spheres are estimated through considerations of liquid structure and from data on molecular association. The radius they derive for non-hydrogen-bonding hydrogen is 171 pm. This is virtually identical with $r_{rm}(H) = 170$ (Table 5).

In viewing the definition of P [Eq. (21)], one may ask whether the spread of packing densities is an artifact of the way V_w is defined. If the spread is real, correlation of size-dependent molecular properties with V_I should be statistically more meaningful than with $\bar{V}$. This is because $\bar{V}$-values contain varying amounts of empty space, and do not represent properly the volume occupied by the molecules themselves. Fortunately, material that sheds light on this question has recently become available.

Leahy et al. [150] regressed water-solubility, water-octanol partition coefficients, and other properties of organic compounds, against $\bar{V}$ and other parameters. In order to bring alicyclic and aromatic compounds into line, the investigators were forced to add 10 $cm^3\ mol^{-1}$ to the measured $\bar{V}$ of such compounds. We interpret the need for empirical adjustment as follows. What the increment does is render the V_m-values of alicyclic and aromatic molecules larger than they really are. Consequently, their P-values are made smaller [cf. Eq. (21)]: they are brought from the range of $0.56 \sim 0.62$ to the range of $0.50 \sim 0.55$. In this way, V_m for all types of compound becomes roughly proportional to V_w: $V_m \sim kV_w$, with k in the narrow range of $0.50 \sim 0.55$, rather than in the wide range of $0.50 \sim 0.64$. For a numerical demonstration, transform $\Delta = 10\ cm^3\ mol^{-1}$ into 16.61×10^6 cubic pm per molecule. The volume (in 10^{-6} cubic pm) and packing densities become:

	V_w	V_m	P
hexane (acyclic, unadjusted)	112.8	217.03	0.52
cyclohexane (cyclic, adjusted)	100.6	179.52 + 16.61	0.51
benzene (aromatic, adjusted)	90.9	147.56 + 16.61	0.55

Hence, in correlations with physical properties, the better parameter of size is V_w even though the better descriptor of volume is V_m. As a matter of fact, the cited investigators soon realized that intrinsic volumes V_I furnish better correlations than molar volumes.

Even McGowan's "characteristic volumes" (V_x [159]) correlate better than $\bar{V}$ with physical properties [160]. V_x-values are calculated in a simple way from the structural formula: there is an increment for each atom of a given type (C, 16.35; H, 8.71 cm^3

mol^{-1}), and a decrement (—6.56) for each bond, *irrespective of its multiplicity*. What the decrement does, of course, is lower the characteristic volumes of saturated and acyclic molecules more appreciably than of cyclic and unsaturated. The procedure makes the trend in V_x/N_o, and the spread of values, more akin to those of V_w than to those of V_m. In fact, the authors report an excellent correlation between V_x and the intrinsic volume:

$$V_I = 0.6823\, V_x + 0.597$$

(n = 208, sd = 1.24, r = 0.9998, standard errors 0.0023 in V_x and 0.003 in the free term).

7 Relation with Other Doctrines

7.1 Existing Relations

Molecular mechanics bears the unmistakable imprint of the organic chemist. This is incontestible, even if it may be hard to state what makes organic chemistry so individual. In part, no doubt, it is the shamelessness with which the microscopic is interpreted in terms of the macroscopic. It is "too mechanistic", to quote Walden's reaction to Bischoff's models [1]. From Bischoff's "dynamic theory" (1890, Fig. 8 [161]), through Mislow's analysis (1965 [39]) to modern textbooks, molecular structure is interpreted as a balance of mechanical strains. Inorganic and physical chemists have a very different philosophy [162], which resembles molecular mechanics in no way

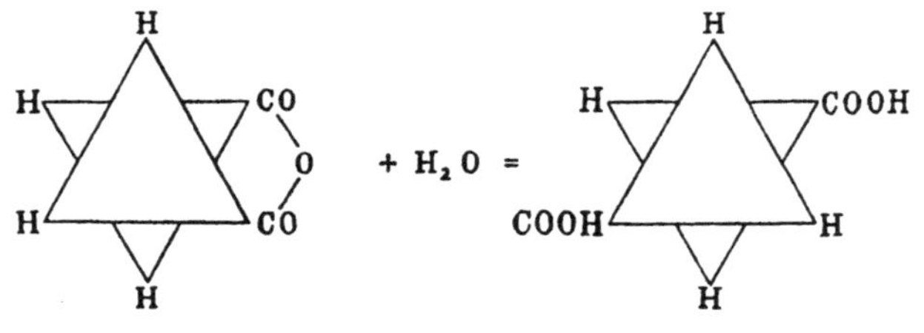

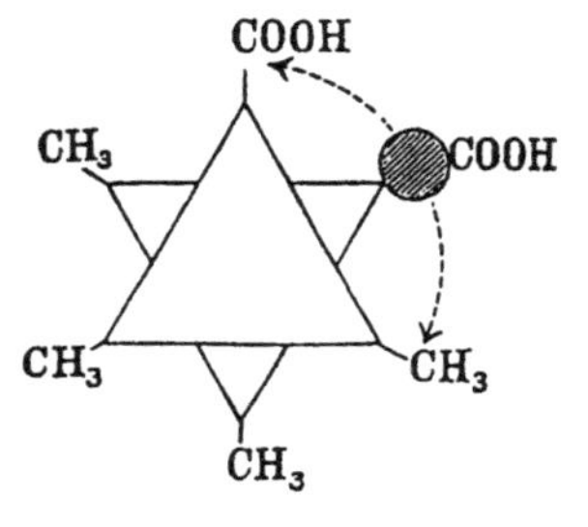

Fig. 8. Bischoff's molecular models, illustrating aspects of his "dynamic theory". The theory and models are now one hundred years old. Above: hydrolysis relieves strain by permitting bulky groups to get away from each other. Below: internal rotation is hindered because bulky groups cannot pass each other

and is in no way resembled by it. How similar are ethane ($H_3C—CH_3$) and dithionate ($O_3S—SO_3^{-2}$, [163])! How utterly unrelated they seem, just because they "belong" in separate departments!

Molecular mechanics descends from spectroscopic force fields [2, 21]. Westheimer's exposition (1956 [4]) is certainly "vibrational"; Hendrickson (1961 [53]) cares to apologize when forced to depart from spectroscopic practices; Wiberg (1965 [82]) mentions, but just in passing, the Urey-Bradley force field; but when Allinger comes in (1967 [164]) the main concern is already with long-range nonbonded interactions. One must construe that these and later workers exploit the spectroscopic machinery without borrowing its spirit. The exception, let us hurry to add, is when vibrational frequencies constitute the declared target of computation [29]. Even so, when vibrational theory itself is traced back, one falls again on "ingenious molecular models . . . which duplicate many details of the vibrational frequencies of molecules" [165]. In 1925, Dennison wrote [166]: "It is obvious that there must exist a very intimate connection between the mechanical structure and properties of a molecule and the frequencies of radiation which it may absorb". The qualifier "mechanical" survives in molecular mechanics.

By contrast, the relations with quantum chemistry are strong in spite of the antithetic machineries. There are six ways in which molecular mechanics and quantum chemistry interact.

a) *Estimation of atomic charges*. Techniques range from usage of naive Hamiltonians (Sect. 5.2) to the ab initio evaluation of charges contained within spheres that inscribe nuclei [90]. CNDO/2 is still a popular source of atomic charges [94, 99, 117, 167]. This pathway, however, suffers from some internal inconsistency, since the "atomic charges" of CNDO/2 are diagonal elements of the bond-order matrix and not meant to fulfil Eq. (12) [49].

b) *Estimation of bond orders in conjugated systems*. This facet was discussed in Sect. 3.

c) *Formulation of force field in terms of "quantum-chemical particles"*. This exception to the propensity of ignoring electrons is realized in the EPEN force field ("Empirical Potential using Electrons and Nuclei" [168]). Here, the interacting units are nuclei and electron-pairs. The components in E_T are Coulombic interactions between all units, overlap repulsion between electrons, and dispersive attraction between nuclei in molecular fragments. The method, in a version EPEN/2 [169], is finding application in studies of hydrogen-bonded clusters [170].

d) *Geometries from molecular mechanics for energies by quantum chemistry* [171]. It is recognized that the reliability of molecular mechanics is generally greater in geometries than in energy differences. Since the process of optimization is infinitely more rapid in molecular mechanics than by ab initio methods, it is logical to start an investigation by using the former. The optimized geometry is then used as input for obtention of quantum-chemical energies, ab initio or other. Care should be taken, though. In the "butterfly compound" XIV, for example, re-optimization by MNDO/2 changed not only structural details but also the symmetry [172].

e) *Joint attack*. Sometimes one hesitates to rely on either of the approaches. The molecule may contain a structural feature for which the force field is not adequately tested, or economic considerations preclude a high-level ab initio study. A way out is to perform, as well as one can, calculations of both types, and identify common

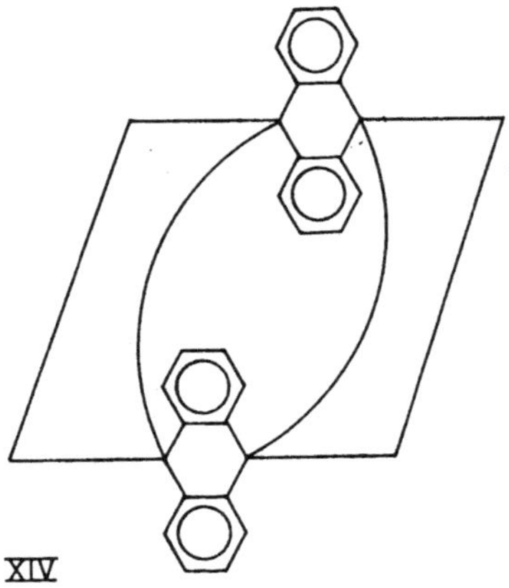

XIV

features in the results. One example is a study of cyclic nucleotides in which, on the one hand, many parameters were new to the force field and, on the other, the molecular dimensions limited the scope of ab initio to the STO-3G basis set [173]. Another example is provided by a study of prismanes, where quantum chemistry had to be limited to MNDO and AM1, and the presence of strained rings called for care in relying on force fields [174].

f) *Quantum-chemical data for construction of force fields*. By far, this is the most common meeting place. As said previously (Sect. 4), an original set of data is required for assigning constants. In the past, such data were expected to come from measurements. Nowadays, one frequently obtains the numbers from high-level ab initio calculations. This also permits the usage, as model compounds, of species that are hardly accessible to synthesis or measurement.

Typical courses of action can be illustrated by two rival constructions of a force field for organosilicon compounds [32, 175]. Both research groups started from MM2, which contained a dated set of parameters for silanes, and both relied on 3-21G* ab initio to update it. One group [175] used the program to obtain approximate geometries for conformers of certain compounds, which were then quantum-chemically re-optimized. By successive corrections in MM2, the force field was amended. The other group [32] started by quantum-chemical optimizations, then fitted MM2 parameters de novo. Other examples are the development of force fields for molecules with short H ... H distances [7] and for conjugated and aromatic hydrocarbons [33], and the improvement of torsional parameters for carbonyl compounds [57].

7.2 Non-existing Relations

It is a pity that, despite the joint usage of molecular mechanics and quantum chemistry, links have not been established between the corresponding formalisms. Some fifteen years ago, Maksić and others noted that the quantum-chemist's hybridization index could constitute such a link [176]. This is because the index measures the s-content in a bond, and the s-content is related to the molecular mechanist's reference values of bond angles and bondlengths [177, 178]. Among other things, bonds rich in s-character are shorter and partake of wider angles.

With the view to increase awareness of the possibilities, we now recapitulate some of the background. The discussion is limited to molecules that contain quadrivalent carbon atoms.

Denote by sp^n the hybridization of carbon in a given bond, where n is the *index of hybridization*. Let $n = n_i$ for bond i and $n = n_j$ for bond j (formula XV). Then:

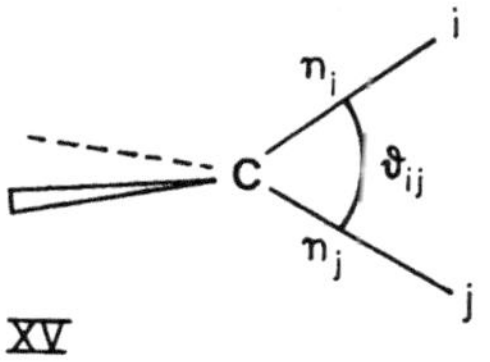

XV

$$\text{s-character in bond i} = \frac{1}{1 + n_i}$$

$$\text{p-character in bond i} = \frac{n_i}{1 + n_i}$$

$$n_i = \frac{\text{p-character}}{\text{s-character}}$$

and

$$\sum_{i=1}^{4} \frac{1}{1 + n_i} = 1 \,. \tag{22}$$

If the four hybrids can be considered orthogonal (which they can [176]), we also have [179]

$$\cos \theta_{ij} = -\left(\frac{1}{n_i n_j}\right)^{1/2} \tag{23}$$

Thus, Eqs. (22) and (23) establish relations between the six valence angles, and between the angles and the s-content of the bonds. For example [180], suppose the molecule is of type X_2CY_2 and let $S_1 = \cos \vartheta_1$ and $S_2 = \cos \vartheta_2$ (XVI). One derives

$$S_2 = \frac{S_1 + 1}{3\,S_1 - 1} \,.$$

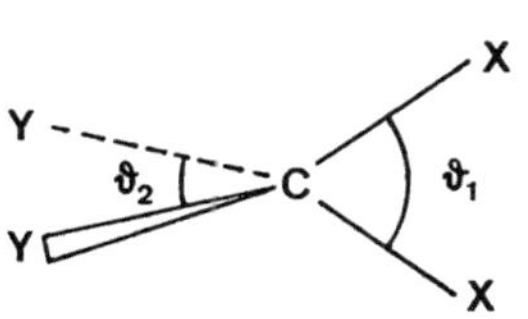

XVI

Thus, if ϑ_1 is tetrahedral ($S_1 = -\frac{1}{3}$) all bonds are sp^3 and all angles tetrahedral. Also, $(dS_2/dS_1) < 0$, and if ϑ_1 becomes larger than tetrahedral, ϑ_2 becomes smaller. In the latter case, $n_X < 3$, $n_Y > 3$, the s-character is high in C—X and low in C—Y.

Concurrently, hybridization indices can be derived from the bond orders of molecular orbital theory [181]. The sum of the squared bond orders P^2_{ru} between any one atomic orbital r and all other atomic orbitals u in a molecule fulfils [182]

$$\sum_u P^2_{ru} = \sum_u W_{ru} = 2P_{uu} - P^2_{uu} \tag{24}$$

W_{ru} is Wiberg's "bond index". In covalent single bonds, $P_{uu} \sim 1$ and hence $\Sigma\,(u)\; W_{ru} \sim 1$. We shall assume that this sum is mainly formed from neighboring-atom contributions.

Now, the amount of charge in the orbital 2s of C which is involved in a given C—H bond, is identified with the bond index between that orbital and the orbital 1s of H:

$$R_{CH} = P^2_{S-H} = W_{S-H}\,.$$

Likewise, the amount of charge in the orbital 2s of C which is involved in a given C—C′ bond, is identified with the sum

$$R_{cc'} = W_{s-s'} + W_{s-x'} + W_{s-y'} + W_{s-z'}\,.$$

In these expressions, s, x, y and z stand for the atomic orbitals 2s, $2p_x$, $2p_y$ and $2p_z$. Eq. (24), as interpreted above, leads to

$$\sum_r W_{r-H} = 1$$

$$\sum_r \sum_u W_{ru} = 1\,.$$

Here, r varies over s, x, y and z of C and u varies over s, x, y and z of C′. Consequently, the hybridization indices of C are

$$\left.\begin{aligned} n_H &= \frac{1 - R_{CH}}{R_{CH}} \quad \text{in bond} \quad C-H \\ n_{C'} &= \frac{1 - R_{CC'}}{R_{CC'}} \quad \text{in bond} \quad C-C'. \end{aligned}\right\} \tag{25}$$

Equations (25) provide the link between molecular-orbital bond orders, which are not too sensitive to the input geometry, and the hybridization indices. Eqs. (22) and (23) link the indices further with geometrical details and the nature of bonds (s-content, etc.). The nature of bonds is the information that molecular mechanics incorporates in constants of types k_s, k_b, l_o, ϑ_o.

The entire set of equations was used in the past [183] to estimate C—C—H bond angles. Starting with a trial geometry, bond orders were obtained by CNDO/2, from which hybridization indices were derived [Eqs. (25)]. These served to obtain better bond angles [Eq. (23)], and the process was recycled to convergence. Such procedures are definitely antiquated. But the underlying principles can now be exploited to new ends.

In sum, through the intermediacy of hybridization indices, the bond orders of quantum chemistry are potentially related to the skeletal constants of molecular mechanics. To the author's knowledge, no work has been reported along these lines.

8 Conclusion

Testa arranged the targets of theoretical stereochemistry by order of increasing sophistication [184]. His list may be abridged as follows.

Low dimensionality
Conceptual level: geometric
Properties considered:
 a. Atom connectivity
 b. Spatial structure
Higher dimensionality
Conceptual level: geometric and electronic
Properties considered:
 c. Spatio-temporal structure
 d. Electronic properties
High dimensionality
Conceptual level: geometric, electronic and interaction with environment
Properties considered:
 e. Solvation, hydration
 f. Intermolecular interactions

In summing up, let us recapitulate the extent to which molecular mechanics contends with these targets.

a) Connectivity. Molecular mechanics cannot identify bonded atoms, that is, atoms very close to each other [6]. On the contrary: the connectivity matrix is an essential input, and bonded atoms are treated differently than others.

b) Spatial structure, i.e. configuration. The comparison of configurational isomers is one of the domains in which molecular mechanics excels. Sometimes, as in the nine-species equilibrium of dissolved 3,3,5-trimethylcyclohexanol [49], one can hardly believe that such an unpresuming algorithm is so efficient in practice. This is particularly gratifying because, in many cases, configurational equilibria are not easily amenable to experimentation.

c) Spatio-temporal structure, i.e. molecular flexibility and conformational equilibria. It is to meet this context that molecular mechanics was created [4, 53, 82, 164], and this is the domain in which it works best.

d) Electronic properties. As emphasized above, mainstream molecular mechanics tends to ignore electrons. There are two exceptions: first, when bond orders are needed as input (Sect. 3); second, when bond charges are produced as output (Sect. 5.2 and 5.3).

e) Solvation. We have seen in Sect. 4.3 that molecular mechanics has its own way of dealing with solvation energies. This way bears no resemblance to the quantum-chemist's tool, which by necessity centers about an Hamiltonian (e.g. [185]).

f) Intermolecular interactions. The molecular mechanist's interest in "non-covalent bonding", e.g. in inclusion compounds [36, 37, 186], is quite recent. So is the interest in small intramolecular changes that accompany aggregation [38, 187] and in the geometry of voids between aggregated molecules [141]. As regards molecules that are polar but uncharged, the position now is that the aggregative geometry is determined predominantly by intermolecular terms in E_{NB}. Skeletal strains and electrostatic interactions play a minor role.

Finally, I cannot resist a paraphrase on the remark that Walden made in 1908 concerning molecular models (see motto to this Chapter and Fig. 8). Some may find molecular mechanics "too mechanistic". Few will deny that it is serving the chemist well.

9 References

1. Walden P (1908) Chemik. Zeitung 32: 1053
2. Burkert U, Allinger NL (1982) Molecular mechanics, ACS monograph no. 177. Amer. Chem. Soc., Washington
3. Frühbeis H, Klein R, Wallmeier H (1987) Angew. Chem. Intl. Ed. 26: 403
4. Westheimer H (1956) in: Newman MS (ed) Steric effects in organic chemistry. Wiley, New York p 524
5. Daudel R (1988) J. Mol. Struct. 165 (3/4): xiii
6. Daudel R (1971) Théorie quantique de la liaison chimique. Presses Universitaires de France, Paris. For the definition of the chemical bond see, in particular, p 55 and the epistemological note on p 175
7. Allinger NL, Lii JH (1987) J. Comput. Chem. 8: 1146
8. Walton A (1969) Progr. Stereochem. 4: 335; Walton A (1978) Molecular and crystal structure models. Ellis Horwood, Chichester
9. Houk KN, Moses SR, Wu YD, Rondan NG, Jäger V, Schohe R, Fronczek FR (1984) J. ACS 106: 3880
10. DeTar DF, Binzet S, Darba P (1987) J. Org. Chem. 52: 2074
11. Meyer AY (1986) Chem. Soc. Revs. 15: 449
12. Bernstein RB, Herschbach DR, Levine RD (1987) J. Phys. Chem. 91: 5365
13. Ōsawa E, Musso H (1982) Top. Stereochem. 13: 117
14. Condat M, Kahn O, Livage J (1972) Chimie théorique, concepts et problèmes. Hermann, Paris, p 20

15. Pilar FL (1968) Elementary quantum chemistry. McGraw-Hill, New York, p 92
16. Profeta S, Allinger NL (1985) J. ACS 107: 1907
17. Saunders M, Jarret RM (1986) J. Comput. Chem. 7: 578
18. Williams JE, Stang PJ, Schleyer PvR (1968) Ann. Rev. Phys. Chem. 19: 531
19. Altona C, Faber DH (1974) Top. Curr. Chem. 45: 1
20. Dunitz JD, Bürgi HB (1975) in: MTP international review of science, physical organic chemistry series 2. Butterworths, London. vol 11 chap 4
21. Ermer O (1976) Structure and bonding 27: 161
22. Allinger NL (1976) Adv. Phys. Org. Chem. 13: 1
23. DeTar DF (ed) (1977) Comp. and Chem. 1: 139
24. Mislow K, Dougherty DA, Hounshell WD (1978) Bull. Soc. Chim. Belges 87: 555
25. White DNJ, Beagley B (1979) in: Sutton LE, Truter MR (eds) Molecular structure by diffraction methods. Specialist periodical reports, Chem. Soc., London. vol 6 chaps 2 and 3
26. Boyd DB, Lipkowitz K (1982) J. Chem. Educ. 59: 269
27. Ōsawa E, Musso H (1983) Angew. Chem. Intl. Ed. 22: 1
28. Meyer AY (1983) in: Patai S, Rappoport Z (eds), The chemistry of the functional groups, Suppl. D. Wiley, Chichester
29. Rasmussen K (1985) Potential energy functions in conformational analysis. Lecture notes in chemistry no. 37. Springer, Berlin Heidelberg NewYork. The survey of research groups is on pp 17–37
30. Boeyens JCA (1985) Structure and Bonding 63: 65
31. Clark T (1985) A handbook of computational chemistry. Wiley, New York
32. Grigoras S, Lane TH (1988) J. Comput. Chem. 9: 25
33. Kao J (1987) J. ACS 109: 3817
34. Grunewald GL, Creese MW, Weintraub HJR (1988) J. Comput. Chem. 9: 315
35. Vedani A (1988) J. Comput. Chem. 9: 269
36. Uiterwijk JWHM, Harkema S, Feil D (1987) J. Chem. Soc., Perkin Trans. 2, p 721
37. Damewood JR, Anderson WP, Urban JJ (1988) J. Comput. Chem. 9: 111
38. Meyer AY (1988) J. Mol. Struct. 179: 83
39. Mislow K (1966) Introduction to stereochemistry. Benjamin, New York, p 33
40. Sadlej J (1985) Semi-empirical methods of quantum chemistry. Ellis Horwood, Chichester
41. Hill TL (1948) J. Chem. Phys. 16: 399
42. Mirskaya KV (1973) Tetrahedron 29: 679
43. Meyer AY, Forrest FRF (1985) J. Comput. Chem. 6: 1
44. Allinger NL, Tribble MT, Miller MA, Wertz DH (1971) J. ACS 93: 1637
45. Allinger NL (1977) J. ACS 99: 8127
46. Engler EM, Andose JD, Schleyer PvR (1973) J. ACS 95: 8005
47. White DNJ, Bovill MJ (1977) J. Chem. Soc., Perkin Trans. 2, p 1610
48. Ermer O, Lifson S (1973) J. ACS 94: 4121
49. Meyer AY (1984) J. Comput. Chem. 5: 299
50. Bartell LS (1977) J. ACS 99: 3279
51. Stavnebrekk PJ, Stölevik R (1987) J. Mol. Struct. 160: 143. Torsional expression retains only the term in V_3, with a minus sign preceding the trigonometric function
52. Wilson EB, Decius JC, Cross PC (1955) Molecular vibrations. McGraw-Hill, New York, chap 8
53. Hendrickson JB (1961) J. ACS 83: 4537
54. Pitzer RM (1983) Accts. Chem. Res. 16: 207
55. Radom L, Stiles PJ, Vincent MA (1978) J. Mol. Struct. 48: 259. This paper denotes by V what Eq. (7) denotes by U
56. Durig JR, Berry RJ, Groner P (1987) J. Chem. Phys. 87: 6303
57. Bowen JP, Pathiaseril A, Profeta S, Allinger NL (1987) J. Org. Chem. 52: 5162
58. Kao J, Allinger NL (1977) J. ACS 99: 975; Sprague JT, Tai JC, Yuh Y, Allinger NL (1987) J. Comput. Chem. 8: 581
59. Allinger NL, Tai JC (1965) J. ACS 87: 2081. The function β_j corresponds to the ratio β_{pq}/β_{rs} of this paper
60. Liljefors T, Allinger NL (1976) J. ACS 98: 2745 and (1978) 100: 1068
61. Liljefors T, Tai JC, Li S, Allinger NL (1987) J. Comput. Chem. 8: 1051
62. Carman RM, Fletcher MT (1983) Austral. J. Chem. 36: 1483

63. Lifson S, Warshel A (1968) J. Chem. Phys. 49: 5116
64. Castellani G, Scordamaglia R, Tosi C (1980) Gazz. Chim. Ital. 110: 457
65. Meyer AY, Ohmichi N (1981) J. Mol. Struct. 73: 145
66. Meyer AY (1977) J. Mol. Struct. 40: 127
67. Meyer AY (1978) J. Mol. Struct. 49: 383
68. Meyer AY (1983) J. Mol. Struct. 94: 95
69. Abraham RJ, Siverns TM (1972) J. Chem. Soc., Perkin Trans. 2, p 1587; Abraham RJ, Rossetti ZL (1973) J. Chem. Soc., Perkin Trans. 2, p 582
70. Rogers MT, Canon JM (1961) J. Phys. Chem. 65: 1417
71. Hückel W, Waiblinger H (1963) Lieb. Ann. 666: 17
72. Hückel W, Waiblinger H, Feltkamp H (1964) Lieb. Ann. 678: 24
73. Moura-Ramos JJ, Dumont L, Stien ML, Reisse J (1980) J. ACS 102: 4150
74. Altona C, Buys HR, Hageman HJ, Havinga E (1967) Tetrahedron 23: 2265
75. Abraham RJ, Bretschneider E (1974) in: Orville-Thomas (ed) Internal rotation in molecules. Wiley, London, p 481
76. Abraham RJ, Cooper MA (1967) J. Chem. Soc. B p 202
77. Došen-Mićović L (1986) J. Comput. Chem. 7: 523
78. Došen-Mićović L, Jeremić S, Allinger NL (1983) J. ACS 105: 1723
79. Meyer AY (1989) J. Mol. Struct. in press
80. Dixon DA, Smart BE (1988) J. Phys. Chem. 92: 2729
81. Hammarström LG, Liljefors T, Gasteiger J (1988) J. Comput. Chem. 9: 424
82. Wiberg KB (1965) J. ACS 87: 1070
83. Allinger NL, Hirsch JA, Miller MA, Tyminski (1969) J. ACS 91: 337
84. Allinger NL, Meyer AY (1975) Tetrahedron 31: 1807
85. Meyer AY, Allinger NL, Yuh Y (1980) Israel J. Chem. 20: 57. The parameters described in this publication are incorporated in the program MM2
86. Williams DE (1974) Acta Cryst. A 30: 71
87. Wiberg KB (1977) Comp. and Chem. 1: 221
88. Jaime C, Ōsawa E (1983) Tetrahedron 39: 2769
89. Petterson I, Liljefors T (1987) J. Comput. Chem. 8: 1139
90. Sackwild V, Richards WG (1982) J. Mol. Struct. 89: 269
91. Sacks LJ (1986) J. Chem. Educ. 63: 487
92. Alonso M, Finn EJ (1967) University physics. Addison-Wesley, Reading, M, vol 2 p 477
93. Smyth CP, Dornte RW, Wilson EB (1931) 53: 4242
94. Burkert U (1979) Tetrahedron 35: 209; and (1980) J. Comput. Chem. 1: 192
95. Reynolds WF (1980) J. Chem. Soc., Perkin Trans. 2: 985
96. Hopfinger AJ (1973) Conformational properties of macromolecules. Academic, New York, p 59
97. Cram DJ, Nobel Prize Lecture, 8 Dec 1987. Reprinted (1988) Science 240: 760
98. Hermann RB (1969) J. ACS 91: 3152
99. Isaksson, Liljefors T (1981) J. Chem. Soc., Perkin Trans. 2 p 1344
100. Meyer AY (1983) J. Mol. Struct. 105: 143
101. Fegerland S, Rydland T, Stölevik R, Seip R (1983) J. Mol. Struct. 96: 339
102. Oie T, Maggiora GM, Christoffersen RE, Duchamp DJ (1981) Intern. J. Quantum Chem., Quantum Biol. Symp. 8: 1
103. Weiner PK, Kollman PA (1981) J. Comput. Chem. 2: 287
104. Höltze HD, Zunker P (1986) J. Mol. Struct. 134: 429
105. Hruby VJ, Kao LF, Pettitt BM, Karplus M (1988) J. ACS 110: 3351
106. Del Re G (1958) J. Chem. Soc. p 4031
107. Del Re G (1964) in: Pullman B (ed) Electronic aspects of biochemistry. Academic, New York, p 221
108. Kutzelnigg W, Del Re G, Berthier G (1971) Top. Curr. Chem. 22: 1. For the complete expectation value of the dipole moment operator see p 105
109. Berthod H, Pullman A (1965) J. Chim. Phys. 62: 942; Berthod H, Giessner-Prettre C, Pullman A (1967) Theoret. Chim. Acta (Berl.) 8: 212
110. Meyer AY (1982) J. Chem. Soc., Perkin Trans. 2: 1199

111. Caballero JC, Bruynes C, van der Maas JH (1986) J. Mol. Struct. 147: 243, and preceding papers in the series
112. Yokozeki A, Bauer SH (1975) Top. Curr. Chem. 53: 71 p 82
113. Cantacuzene J (1962) J. Chim. Phys. 59: 186
114. Došen-Mićović L, Jeremić D, Allinger NL (1983) J. ACS 105: 1716
115. Saegebarth E, Krisher LC (1970) J. Chem. Phys. 52: 3555
116. Abraham RJ, Hudson B (1984) J. Comput. Chem. 5: 562
117. Abraham RJ, Smith PE (1988) J. Comput. Chem. 9: 288
118. Thompson JJ (1923) Phil. Mag. 46: 497 p 513
119. Smallwood HM, Herzfeld KF (1930) J. ACS 52: 1919
120. Corey EJ (1953) J. ACS 75: 2301
121. Lehn JM, Ourisson G (1963) Bull. Soc. Chim. Fr. p 1113
122. Meyer AY, Burkert U (1982) J. Mol. Struct. 88: 183
123. Nagy P, Angyán JG, Náray-Szabó G (1987) J. Mol. Struct. 149: 169, and preceding publications in the series
124. Boyd RH, Kesner L (1981) J. Polym. Sci., Polym. Phys. Ed. 19: 375
125. Gasteiger J, Marsili M (1980) Tetrahedron 36: 3219
126. Stőlevik R (1984) J. Mol. Struct. 109: 397
127. Stőlevik R, Bakken P (1985) J. Mol. Struct. 124: 133
128. Boström GO, Bakken P, Stőlevik R (1988) J. Mol. Struct. 172: 227
129. Abraham RJ, Griffiths L, Loftus P (1982) J. Comput. Chem. 3: 407
130. Abraham RJ, Grant GH (1988) J. Comput. Chem. 9: 244
131. Gavezzotti A (1983) J. ACS 105: 5220
132. Bondi A (1964) J. Phys. Chem. 68: 441; Bondi A (1968) Physical properties of molecular crystals, liquids and glasses. Wiley, New York
133. Kumaran MK, Benson GC (1983) J. Chem. Therm. 15: 245
134. Margenau H (1943) Phys. Rev. 63: 385
135. Dostrovsky J, Hughes ED, Ingold CK (1946) J. Chem. Soc. p 173. See Eq (1) on p 178
136. Bartell LS (1960) J. Chem. Phys. 32: 827
137. Glidewell C (1975) Inorg. Chim. Acta 12: 219, (1976) ibid 20: 113, and (1979) Educ. Chem. 16: 146
138. Ramasubbu N, Parthasarathy R, Murray-Rust P (1986) J. ACS 108: 4308
139. Max NL (1984) J. Mol. Graphics 2: 8
140. Connolly ML (1985) 107: 1118; Stouch TS, Jurs PC (1986) J. Chem. Inf. Comput. Sci. 26: 4
141. Wigderson E, Meyer AY (1988) Comp. and Chem. 12: 237
142. Gibson KD, Scheraga HA (1987) Mol. Phys. 62: 1247
143. Richards FM (1977) Ann. Rev. Biophys. Bioeng. 6: 151
144. Meyer AY (1988) J. Comput. Chem. 9: 18
145. Soffer N, Bloemendal M, Marcus Y (1988) J. Chem. Eng. Data 32: 43
146. Meyer AY (1985) J. Mol. Struct. 124: 93
147. King EJ (1970) 74: 4590
148. Terryn B, Bariol J (1981) J. Chim. Phys. 78: 207
149. Richards FM (1974) J. Mol. Biol. 82: 1
150. Leahy DE, Carr PW, Pearlman RS, Taft RW, Kamlet MJ (1986) Chromatographia 21: 473
151. Hill TL (1966) Lectures on matter and equilibrium. Benjamin, New York, p 96
152. Scott GD (1960) Nature 188: 908
153. Godbout G, Sicotte Y (1968) Canad. J. Chem. 46: 967
154. King EJ (1969) J. Phys. Chem. 73: 1220
155. Meyer AY (1986) J. Comput. Chem. 7: 144
156. Meyer AY, Farin D, Avnir D (1986) J. ACS 108: 7897
157. Shahidi F, Farrell PG, Edward JT (1979) J. Phys. Chem. 83: 419
158. Rashin AA, Namboodiri K (1987) J. Phys. Chem. 91: 6003
159. McGowan JC (1956) Rec. Trav. Chim. Pays Bas 75: 193
160. Abraham MH, McGowan JC (1987) Chromatographia 23: 243
161. Bykov GV (1975) in: Ramsay OB (ed) van't Hoff-Le Bel centennial. American Chemical Society, Washington, p 114
162. Burdett JK (1980) Molecular shapes, theoretical models of inorganic chemistry. Wiley, New York

163. Cline SJ, Scaringe RP, Hatfield WE, Hodgson DJ (1977) J. Chem. Soc., Dalton Trans. p 1662
164. Allinger NL, Miller MA, Van Catledge FA, Hirsch JA (1967) J. ACS 89: 4345
165. Urey HC, Bradley CA (1931) Phys. Rev. 38: 1969
166. Dennison DM (1925) Astrophys. J. 62: 84
167. Caballero JC, van der Maas H (1985) 127: 57
168. Shipman LL, Burgess AW, Scheraga HA (1945) 72: 543
169. Brink G, Glasser L (1982) S. Afr. J. Chem. 35: 105
170. Brink G, Glasser L (1987) J. Mol. Struct. 160: 357 and preceding papers in the series
171. Dougherty DA, Mislow K (1979) J. ACS 101: 1401
172. Dougherty DA, Choi CS, Buda AB, Rudziński JM, Ōsawa E (1986) J. Chem. Soc., Perkin Trans. 2, p 1063
173. Marsh FJ, Weiner P, Douglas JE, Kollman PA, Kenyon GL, Gerit JA (1980) J. ACS 102: 1660
174. Jemmis ED, Rudzińsky JM, Ōsawa E (1988) Chemistry Express (Japan) 3: 109
175. Profeta S, Unwalla RJ, Nguyen BT, Cartledge FK (1986) J. Comput. Chem. 7: 528
176. Kovačević K, Eckert-Maksić M, Maksić ZB (1974) Croat. Chem. Acta 46: 249
177. Mills IM (1963) Spectrochim. Acta 19: 1585
178. Randić M, Maksić ZB (1972) Chem. Revs. 72: 43
179. McWeeney R (1979) Coulson's valence, 3rd edn., Oxford University Press, Oxford, p 195
180. Glidewell C, Meyer AY (1981) J. Mol. Struct. 72: 209
181. Trindle C (1969) J. ACS 91: 219; Trindle C, Sinanoglu O (1969) J. ACS 91: 853
182. Wiberg KB (1968) Tetrahedron 24: 1083
183. Berthier G, Meyer AY, Praud L (1971) in: Bergmann ED, Pullman B (eds) Aromaticity, pseudo-aromaticity and anti-aromaticity: proceedings of an international symposium. The Israel Academy of Sciences and Humanities, Jerusalem, p 174
184. Testa B (1982) in: Tamm C (ed) Stereochemistry (vol 3 in: New comprehensive biochemistry). Elsevier Biomedical, Amsterdam, p 2
185. Bonaccorsi R, Hodoščék M, Tomasi J (1988) J. Mol. Struct. 164: 105
186. Matsui Y (1982) Bull. Chem. Soc. Japan 55: 1246
187. Hargittai I, Hargittai M (1987) in: Liebman JF, Greenberg A (eds) Molecular structures and energetics. VCH, Weinheim, vol 2

Atoms in Molecular Environments

Kenneth B. Wiberg

Department of Chemistry Yale University New Haven, Connecticut 06511/USA

Atoms are the most fundamental units in chemistry. Via the use of Bader's theory of atoms in molecules, it is now possible to define uniquely the volume element associated with an atom based on the molecular wave function. This allows properties of the atoms, such as their electron populations and energies, to be calculated via numerical integration of the proper function of the charge density. An analysis of these properties reveals a close connection to common concepts such as hybridization and electronegativity. In addition, the theory makes it possible to locate bent bonds and to relate them to changes in hybridization, structural deformations and steric interactions. The theory serves to recover many of the older concepts in chemistry and places them on a more quantitative basis.

1 Introduction

The formation of molecules from atoms is a basic concern of chemistry. How do two atoms interact with each other to form a bond, and how are atoms affected by bond formation? The following will explore some ways in which these and related questions may be examined via the use of theoretical methods based on ab initio molecular orbital theory.

The first question which must be addressed is concerned with the structure of a molecule. The theoretical calculations lead to a set of cartesian coordinates for the atomic nuclei which represent a minimum in energy based on the Born-Oppenheimer approximation for the basis set (atomic orbitals) which was chosen for the calculation. It does not necessarily represent the global minimum for these atoms, but often represent a local minimum which is kept from going to a lower energy arrangement by an energy barrier. Although the calculations define the geometry with good accuracy [1], they do not lead directly to a structure in the conventional sense of a set of bonds between nuclei.

One might infer the locations of the bonds by looking for pairs of atomic nuclei within conventional bonding distances. A more general approach is that of Bader [2] in which one first locates the *atomic interaction lines* which are paths of maximum charge density ($\varrho = \psi\psi^*$) connecting a given pair of nuclei in a molecule. The charge density will be a maximum at each of the two nuclei, and a minimum at some point along the interaction line. This point will be called a bond critical point if it represents a

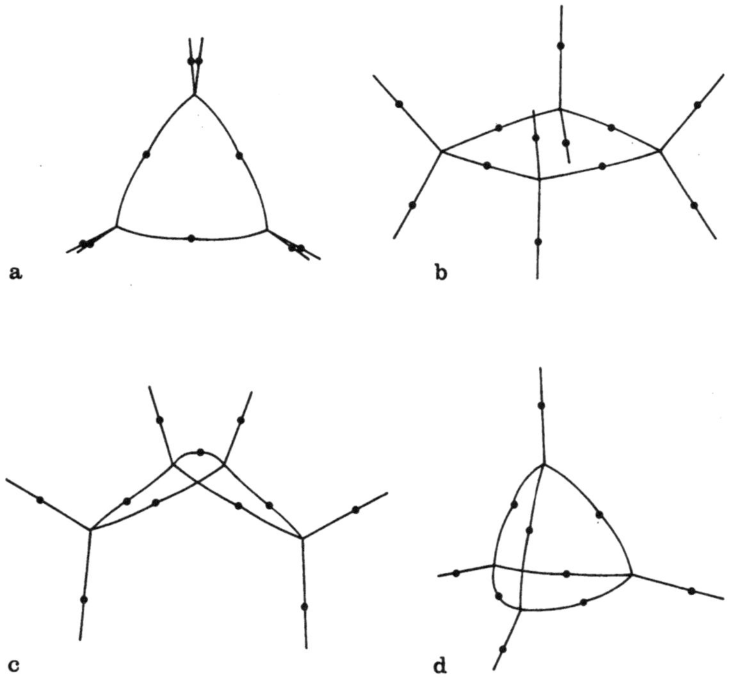

Fig. 1 a–d. Molecular graphs. The compounds are: a. cyclopropane; b. cyclobutane; c. bicyclo[1.1.0]-butane and d. tetrahedrane

maximum in charge density in directions perpendicular to the interaction line, and then the line will be called a bond path. The nature of these bond paths will be discussed in greater detail below, but for the present it is enough to state that the collection of bond paths for a molecule is known as its molecular graph. Some examples of molecular graphs are given in Fig. 1.

The cartesian coordinates (or their equivalent in terms of internal coordinates) that are derived by geometry optimization via ab initio calculations correspond to few if any of the molecules of the compound being considered. All molecules vibrate at or above the zero-point level for each of the 3n–6 normal vibrational modes, and some of these vibrational modes, especially low frequency bends and torsional motions, may have rather large amplitudes. Although the cartesian coordinates are constantly changing, the molecular graph for a stable compound remains unchanged. Thus, the structure in conventional chemical terms is given by the molecular graph. In this view, the often complex discussions [3] of the effect of the Born-Oppenheimer approximation on the structures of normal organic molecules become moot, and this question need not concern us.

The simplest chemical bond is that between two hydrogen atoms to form the hydrogen molecule. To what extent are the hydrogens perturbed in forming the bond? The energy of the hydrogen molecule is −1.1745 H (H = Hartree = 627.5 kcal/mol = 2625 kJ/mol [4], and that of two hydrogen atoms is −1.0000 H. Thus, the binding energy is 0.1745 H = 109.5 kcal/mol = 458.1 kJ/mol, and the energy of the molecule is 17% less than that of the two atoms. As far as the charge density is concerned, the difference between two hydrogen atoms and the hydrogen molecule is an increase in the region between the hydrogens and a depletion in the outer regions [5].

If one passed a plane through a hydrogen atom at the nucleus, and integrated ϱ from that plane to infinity, one would find 0.5 e. If one were to do the same starting at one of the hydrogens of H_2, the value would be reduced to 0.4848 e, corresponding to a 3% charge transfer from the wings to the interatomic region [6]. It can be seen that bond formation is a relatively small perturbation on the constituent atoms. With the first row or heavier atoms, the perturbation on the energy or charge density will be much smaller.

2 Bond Properties and Critical Points

In the introduction to a symposium on molecular quantum mechanics, Mulliken [7] reported a conversation with Löwdin in which they concluded, "that with old-fashioned chemical concepts, which at first seemed to find their counterparts in molecular quantum mechanics, the more accurate the calculations became, the more the concepts vanished into thin air." Many attempts have been made to recover these concepts, and the most comprehensive of these is Bader's theory of atoms in molecules [2], which is concerned with the charge distribution in molecules rather than with the molecular orbitals. It is important to note that the charge distribution is an experimentally measured quantity (via X-ray crystallography) [8] whereas the molecular orbitals are only a result of the way in which the quantum mechanical problem was set up. All of the properties of a molecule may directly be obtained from the charge density distribution [9].

The central quantities in the topology of the charge density are the critical points in a molecule. They are points at which the principal components of the derivative of ϱ with respect to the coordinates are zero (i.e., ϱ is either a maximum or a minimum in a given direction). The charge density in any molecule is dominated by the location of the nuclei. Whereas the charge density at the mid-point of a single bond is usually on the order of 0.1 to 0.3 e/au^3, it is about 0.43 e/au^3 at a hydrogen nucleus and on the order of 120 e/au^3 at a carbon nucleus. The nuclear positions represent critical points because the derivative of ϱ is zero in all directions. The value of ϱ decreases in all directions from the nuclei, and so the Laplacian of ϱ:

$$\nabla^2\varrho = (d^2/dx^2 + d^2/dy^2 + d^2/dz^2)\,\varrho$$

will have three negative components [10]. The critical points are usually designated by their rank (the dimensionality of the system which is 3 in all molecular systems) and their signature (the sum of +1 for each positive and −1 for each negative curvature in ϱ). Therefore the points at the nuclei are known as [3, −3] critical points. The nuclei are commonly referred to as attractors, and the region about a given nucleus is known as its basin. The definition of an atom in a molecule requires that basin regions be divided in an appropriate fashion among the attractors.

In this regard, one of the most important of the critical points is the *bond critical point* which was mentioned above. It is located along the bond path which is the path of maximum charge density connecting a pair of bonded nuclei. The critical point is that having the minimum value of ϱ. By its definition, it must be a maximum in ϱ in directions normal to the bond path. Therefore a bond critical point is designated as a [3, −1] critical point. Another type of critical point is found in rings. Consider a cyclopropane ring. At the center of the ring, ϱ will have a minimum value in the CCC plane, but it will be a maximum with respect to the three-fold rotational axis [11]. This is a *ring critical point* which is then designated as [3, +1]. In the middle of a cage, such as bicyclo[1.1.1]pentane, there is a point which is a minimum in ϱ in all directions. This is a *cage critical point*, designated as [3, +3].

Useful information may be derived from the locations of the bond critical points [12]. In the case of two equivalent atoms, such as the carbons of ethane, the bond critical point must be at the midpoint of the bond. If one of the methyl groups of ethane were replaced by an amino group forming methylamine, the bond critical point would move away from the nitrogen and closer to the carbon. This is a result of the greater electronegativity of nitrogen which tends to build up charge density at the nitrogen and push the critical point toward the less electronegative carbon. Some examples are shown in Table 1 where the location of the bond critical points for a number of methyl derivatives are given [13]. In all cases, the distance from the carbon to that point follows the commonly accepted order of electronegativity. The λ's in the table are the three curvatures in ϱ at the critical point with λ_3 lying along the bond path. Finally, $\nabla^2\varrho$, the Laplacian of ϱ, is the sum of the three λ's. With covalent bonds having relatively low ionic character, $\nabla^2\varrho$ will have a negative sign indicating that it is dominated by the curvatures in ϱ normal to the bond path. However, with bonds having a high degree of polar character, the curvature along the bond path becomes dominant and $\nabla^2\varrho$ takes a positive value. Methyllithium is an example, having low values of ϱ_c, λ_1 and λ_2 indicating little covalent character. The carbon-lithium "bond" might best be considered as a polarization of the methyl anion by the lithium cation.

The question as to which of the atomic interaction lines containing a [3, −1] critical point should be called a bond path has been raised [14]. There may be a [3, −1] critical point between two non-bonded atoms which are forced to be in close proximity, but this would not be a bond critical point. One possible criterion is that $\nabla^2\varrho$ at the critical point should be negative. It is true that this will be the case for bonds between atoms with similar electronegatives, as can be seen in the data given in Table 1.

Table 1. Bond properties of methyl derivatives[a]

X	Bond	r_A	ϱ_c	λ_1	λ_2	λ_3	$\nabla^2\varrho$
H	C—H	0.6605	0.2855	−0.7273	−0.7273	0.4010	−1.0536
Li	C—Li	1.2988	0.0422	−0.0607	−0.0607	0.3610	+0.2077
BeH	C—Be	1.1421	0.1030	−0.2075	−0.2075	0.7615	+0.3465
BH_2	C—B	1.0768	0.1866	−0.4456	−0.3546	0.6950	−0.1052
CH_3	C—C	0.7637	0.2528	−0.4808	−0.4808	0.2967	−0.6649
NH_2	C—N	0.5358	0.2776	−0.5668	−0.5555	0.1636	−0.9538
OH	C—O	0.4461	0.2643	−0.5213	−0.5182	0.8653	−0.1742
F	C—F	0.4316	0.2371	−0.4243	−0.4243	1.3335	+0.4849
Na	C—Na	1.3383	0.0347	−0.0357	−0.0357	0.2383	+0.1667
MgH	C—Mg	1.2439	0.0543	−0.0638	−0.0638	0.4071	+0.2796
AlH_2	C—Al	1.2005	0.0826	−0.1130	−0.1102	0.5777	+0.3545
SiH_3	C—Si	1.1799	0.1174	−0.1708	−0.1708	0.6564	+0.3148
PH_2	C—P	1.1732	0.1536	−0.2087	−0.1822	0.3695	−0.0214
SH	C—S	0.8536	0.1848	−0.2893	−0.2650	0.1941	−0.3603
Cl	C—Cl	0.7364	0.1840	−0.2936	−0.2936	0.2865	−0.3007

[a] r_A is the distance from the carbon to the bond critical point in Å, ϱ_c is the charge density (e/au^3) at the critical point, the λ's are the curvatures of ϱ at the critical point, and $\nabla^2\varrho$ is the Laplacian of ϱ at this point. The results are based on 6-31G** wave functions.

However, the C—F bond in methyl fluoride is certainly appropriately considered to be normal bond, and here $\nabla^2\varrho$ is positive, as it is for all bonds between atoms having large differences in electronegativity. It seems reasonable to state that all atomic interaction lines having a [3, −1] critical point with a negative $\nabla^2\varrho$ must be considered to be bond paths, and that those having a [3, −1] critical point with a positive $\nabla^2\varrho$, but a relatively large value of ϱ at the critical point also should be called bond paths. A more precise definition requires a calculation of the local energy density at, or better in the vicinity of the critical point along the atomic interaction line [14]. A negative value is indicative of a bond critical point.

In the series ethane, ethylene and acetylene the electronegativity of the carbon will increase with increasing s-character. This should lead to a shift in the bond critical point. The data in Table 2 show that this is the case. In these and in all other cases which have been examined, the shift in the bond critical point has been a useful indicator of changes in electronegativity. We shall return to this criterion subsequently when the effect of electronegativity on electron populations are considered.

The bond paths also are useful in studying structural effects in molecules. Consider methyl fluoride. The H—C—F bond angle is 109°, close to the tetrahedral value. One

Table 2. Distances to bond critical points[a]

Compound	Bond	r_A (Å)	r_B (Å)
CH_3-CH_3	C—H	0.6589	0.4267
$CH_2=CH_2$	C—H	0.6614	0.4149
HC≡CH	C—H	0.6804	0.3766

[a] r_A is the distance from the carbon to the critical point, and r_B is the distance from the hydrogen to that point. The values are based on 6-31G** wave functions.

might have expected a considerably smaller angle. Bent's rule [15] suggests that the strongly electronegative fluorine will prefer to bond to a carbon orbital having increased *p*-character, leading to increased *s*-character in the bonds to the hydrogens. This in turn should lead to an increase in the H—C—H angles and a decrease in the H—C—F angle. If the bond paths are examined, it is found that those for the C—H bonds are bent. The angle between the C—F and C—H bond paths at the carbon is 106.7°, in agreement with expectations [16]. It is clear that there must be a steric interaction between the fluorine and hydrogen which leads to a movement of the hydrogen away from the fluorine and the formation of a bent bond. Steric effects are generally associated with the formation of bent bonds. As will be noted below, bent bonds also result from bond angle distortion, as found in small ring compounds.

The bond path angles have been examined for a variety of methyl derivatives giving the data summarized in Table 3 [13, 16]. It is seen that the largest deviation from the conventional angles are found with the most electronegative atoms. Here, the electronegativity will lead to smaller H—C—X bond path angles and an increased H—X repulsion due to the decreased distance between them. The steric interaction leads to the bent bond. The strongly electropositive substituents such as BeH and Li give large H—C—X bond path angles as would be expected. Here, the deviation from the conventional angle is negative indicating that here the hydrogens of the methyl group have moved close enough together so that the dominant steric interaction is H—H nonbonded repulsion.

Table 3. Bond path angles for methyl derivatives[a]

Compound	Angle	Bond path	Convent	$\Delta\alpha$
MeN_2^+	HCN	101.17	104.98	3.81
$MeNH_3^+$	HCN	104.71	108.10	3.39
MeF	HCF	106.68	109.17	2.49
MeNC	HCN	107.42	109.62	2.20
MeCN	HCC	108.26	109.79	1.53
MeC≡CH	HCC	109.43	110.61	1.18
MeMe	HCC	110.25	111.20	0.95
$MeSiH_3$	HCSi	110.54	111.12	0.35
MeBeH	HCBe	112.77	112.07	−0.70
MeLi	HCLi	113.37	112.56	−1.81

[a] The conventional angles were calculated using the 6-31G* basis set, and are in all cases close to the experimental angles.

Another example of bent bonds is found with compounds having bond angle distortion which is forced by geometrical constraints. In the case of cyclopropane, for example, the bond paths do not coincide with the conventional bonds, but rather are bent, just as predicted by the Coulson-Moffitt model [17]. A convenient measure of the bond bending is given by the angle between bond paths at the central atom. Some of these data for small ring hydrocarbons are given in Table 4 [18].

Cyclopropane has a bond path angle of 78.8° which is 18.8° larger than the conventional angle [19], whereas cyclobutane has a bond path angle which is only 6.7° larger than the conventional angle as might be expected based on its lower strain energy per CH_2 group. With larger rings, the deviation between the two angles becomes fairly small. The larger deviations are generally found with the cyclopropanes, with 16 to 20° as normal values. The largest deviation is found with spiropentane. Whereas most cyclopropanes may improve C—C bonding by rehybridization, the central carbon in spiropentane is forced by symmetry to a hybridization similar to sp^3 and consequently a larger bond path angle. Not surprisingly, tetrahedrane also is calculated to have a large degree of bond bending.

In this regard, [1.1.1]propellane is anomalous. The bond path angle at the methylene is smaller than the conventional angle. A similar trend may be seen with bicyclobutane where the bond path angle at the methylene group is considerably smaller than that for cyclopropane. This may be related to the dramatic change (174 ppm) in the ^{13}C NMR shielding tensor component perpendicular to the C—C—C bonds of the methylene groups of cyclopropane, bicyclobutane and [1.1.1]propellane [20].

Table 4. Bond path angles for cycloalkanes

Molecule		Angle	Geometric angle α_e	Bond path angle α_b	$\Delta\alpha$ $\alpha_b - \alpha_e$	Strain energy, kcal
cyclopropane		1	60.0	78.84	18.84	27.5
cyclobutane		1	89.01	95.73	6.72	26.5
cyclopentane	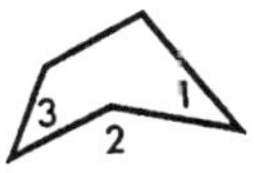	1	104.55	104.02	—0.53	6.2
		2	105.36	104.43	—0.93	
		3	106.50	105.03	—1.54	
cyclohexane	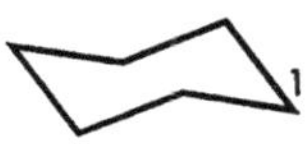	1	111.41	110.08	—1.34	0.0
bicyclo[1.1.0]butane	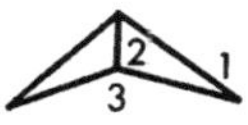	1	58.99	72.78	13.79	63.9
		2	60.50	76.62	16.12	
		3	97.91	105.07	7.16	
bicyclo[2.1.0]pentane	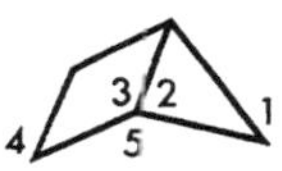	1	60.86	79.28	18.42	54.7
		2	59.57	78.30	18.72	
		3	90.84	96.76	5.92	
		4	89.16	95.77	6.61	
		5	110.03	109.72	—0.31	

Table 4 (continued)

Molecule		Angle	Geometric angle α_e	Bond path angle α_b	$\Delta\alpha$ $\alpha_b - \alpha_e$	Strain energy, kcal
bicyclo[2.2.0]hexane	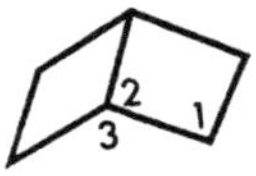	1	90.17	97.01	6.85	51.8
		2	89.91	96.59	6.67	
		3	114.43	112.64	−1.79	
bicyclo[1.1.1]pentane		1	74.44	84.72	10.27	68.0
		2	87.20	95.85	8.65	
bicyclo[2.1.1]hexane		1	99.20	100.69	1.48	37.0
		2	101.77	103.46	1.69	
		3	86.11	94.21	8.10	
		4	82.60	90.95	8.35	
bicyclo[2.2.1]heptane	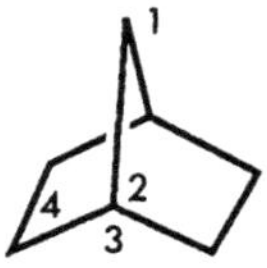	1	94.37	97.42	3.05	14.4
		2	101.51	102.90	1.39	
		3	108.43	108.69	0.26	
		4	103.13	103.57	0.44	
bicyclo[2.2.2]octane	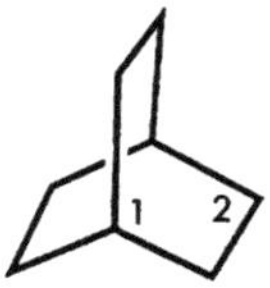	1	109.30	108.68	−0.62	7.4
		2	109.64	108.56	−1.08	
[1.1.1]propellane	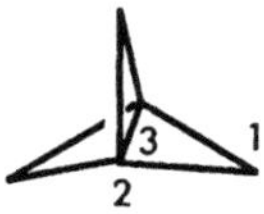	1	61.81	59.37	−2.44	98.0
		2	95.98	107.90	12.01	
		3	59.09	69.09	9.99	
[2.1.1]propellane	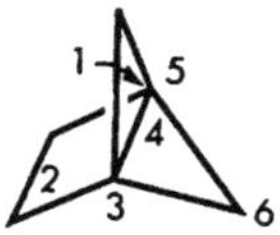	1	88.97	93.65	4.69	104.0
		2	91.03	97.91	6.88	
		3	112.30	116.3	4.00	
		4	57.72	67.96	10.25	
		5	97.27	111.45	14.18	
		6	64.57	65.27	0.71	
[2.2.1]propellane	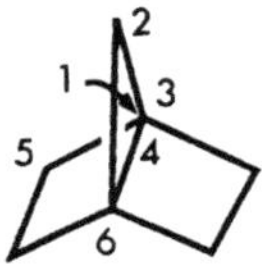	1	59.18	75.40	16.22	105.0
		2	61.64	74.32	12.68	
		3	112.38	116.43	4.06	
		4	90.86	92.97	2.11	
		5	89.14	96.43	7.30	
		6	128.49	126.57	−1.91	
[2.2.2]propellane	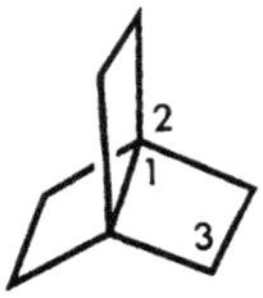	1	91.15	97.05	5.90	89.0
		2	119.96	118.51	−1.44	
		3	88.85	96.81	7.97	

Table 4 (continued)

Molecule		Angle	Geometric angle α_e	Bond path angle α_b	$\Delta\alpha$ $\alpha_b - \alpha_e$	Strain energy, kcal
tetrahedrane		1	60.00	81.36	21.36	140.0
cubane		1	90.00	97.43	7.43	154.7
spiropentane	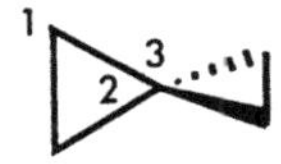	1	59.24	79.04	19.80	63.2
		2	61.51	84.84	23.33	
		3	137.6	123.02	−14.58	

[a] Angles in degrees.
(Reproduced from Ref. 18 with permission of the American Chemical Society).

Table 5. Electron populations and energies of hydrocarbon groups

Compound	Group	n	T = −E(au)
Ethane	CH_3	8.9993	39.6145
Propane	CH_3	9.0171	39.6311
	CH_2	7.9658	39.0014
Butane	CH_3	9.0163	39.6329
	CH_2	7.9837	39.0163
Pentane	CH_3	9.0154	39.6301
	CH_2	7.9845	39.0183
	CH_2	8.0002	39.0363
Hexane	CH_3	9.0166	39.6315
	CH_2	7.9844	39.0169
	CH_2	7.9990	39.0355

3 Electron Populations and Energies for Atoms and Groups

Chemists have been interested in assigning electron populations to atoms in order to see how changes in structure or substituents affect the atoms in a molecule. One of the first of the population analyses was suggested by Mulliken [21] who assigned the diagonal density matrix elements to the individual atoms, and split the off-diagonal elements evenly between the atoms involved. This procedure has been widely used, but has been recognized by Mulliken [22] and others [23] to be inadequate. Other more sophisticated procedures have been suggested, such as the use of natural orbitals [24]. However, any procedure which assigns atomic orbitals to atoms has an inherent problem- the orbitals have "tails" which might be more appropriately assigned to another atom.

Other procedures have been developed which are based on the charge density. For example, in the case of methane, one might separate the C—H bonds by dividing space, starting at the carbon, into four equal volume elements. If one starts at one of the hydrogens and integrates the 6-31G** charge density from that atom to infinity subject to the above boundary condition (in a fashion similar to that for the hydrogen molecule described above) one should find slightly less than 0.5 e if the C—H bond had no charge separation. The calculation gave a value of 0.516 e [6], showing that the sign of the bond dipole was C^+H^-. If the scaling factor found for the similar calculation for H_2 applied here (i.e. 1.00/0.4848 = 2.063), then the electron population at H would be 1.065 e.

Another approach makes use of the covalent radius of carbon, which may be taken as one-half of the C—C length in ethane. If this distance is used to give the boundary between C and H in methane, and ϱ is integrated from there to infinity, the electron population is found to be 1.05 e, in good agreement with the value given above [6]. Clearly, the sense of the bond dipole is $C^+—H^-$.

Procedures of the types described above may be extended to other bonds, but have the disadvantage of using arbitrarily defined volume elements [25]. A more general approach has been provided by Bader [2] which allows the atoms in a molecule to be uniquely defined, and allows the properties to be examined in detail. This approach gives the electron population at a hydrogen of methane as 1.062 e.

An essential part of the theory is the unique definition of the volume element which is to be assigned to an atom in a molecule. What is needed is a set of surfaces which separate pairs of bonded atoms, for a set of these surfaces will define volume elements. The appropriate surfaces can be shown to pass through the critical points [26]. The surfaces are developed by starting at a critical point and calculating a set of paths by which the charge density decreases most rapidly (i.e. the gradient paths). The set of paths formed by starting in different directions from the critical point will define a zero flux surface (i.e. a surface across which the rate of change of ϱ is zero), and the set of these surfaces will lead to volume elements which may be assigned to individual atoms in a molecule. An important characteristic of these regions is that they are quantum mechanical subspaces for which the hypervirial theorem is obeyed. Consequently, for each atom, Ω, $V_\Omega = -2T_\Omega$. Integration of ϱ over the volume element gives the electron population associated with the atom, and integration of G(r) over the volume gives the kinetic energy:

$$G(r) = (h^2/8m)\, \Sigma\, n_i \nabla\varrho_i \nabla\varrho_i/\varrho_i$$

$$T_\Omega = \int_\Omega dr\, G(r)$$

where ϱ_i is the orbital density. Since for each atom $E_\Omega = V_\Omega + T_\Omega$, a corollary of the virial theorem is that the total energy of an atom is given by the negative of its kinetic energy. As a result, the ability to obtain the kinetic energy directly leads to the total energy of each atom. Other properties of the atoms may be defined as well [23–25].

If a procedure of this type is to be genuinely useful, it must lead to transferability of properties. Thus, for example, methylene groups in similar environments should have similar calculated properties. This has been examined for some hydrocarbons. Table 5 gives the electron populations and energies for the CH_3 and CH_2 groups of the n-alkanes from ethane through hexane [18]. In this series, there is a constant increment in the experimental ΔH_f per methylene group, and there also is a constant increment in the calculated total energies. The energy of a methyl group adjacent to a methylene is constant throughout the series (i.e. from propane through hexane). However, there are some changes in the energy of the methylene groups. The central CH_2 groups in pentane and hexane have the same energies, but a methylene between a CH_3 and a CH_2 has a slightly lower energy.

These data tell us two things. First, when the methylenes are in comparable locations, the energies are the same. However, when placed adjacent to another group such as methyl, the energy is somewhat different. A methyl, having three electronegative hydrogens, is more electronegative than a methylene group, and it can be seen that there is charge transfer from the methylene to the methyl. The change in population leads to a change in energy, and the energy loss found for the methylene is equal and opposite to that for the methyl. The constant increment per methylene group is then due to compensating energy changes when some charge density is transferred from methylene to methyl. It is likely that this will be found with many cases where the charge density shift is small.

The ability to assign an energy to a group or atom in a "normal" environment makes it possible to examine the consequences of placing atoms in unusual environments. For hydrocarbons, an example of the latter is found with small ring compounds containing cyclopropane or cyclobutane rings. The electron populations and atom energies for some of these hydrocarbons are given in Table 6. It can be seen that the hydrogens of the cyclopropane ethylene group have a smaller electron population than those of a standard methylene group. This is a result of the rehybridization at carbon which places more *p*-character in the C—C bonds and more *s*-character in the C—H bonds [18]. In this way, the carbon becomes more electronegative, and withdraws charge density from the hydrogens. At the same time, the bond critical point moves toward the hydrogens. The combination of these two factors leads to a net charge transfer from hydrogen to carbon.

The energy of an atom usually follows its electron population since the larger populations normally lead to increased kinetic energies. This is found with the small ring hydrocarbons. The carbon of cyclopropane has a lower energy than that of a normal methylene carbon, and the hydrogen has a higher energy than that of a normal methylene hydrogen. The two energy changes do not balance, and the net energy of the CH_2 group has increased over that of the standard methylene by 9.6 kcal/mol. For the three methylene groups, the total increase in energy is 28.8 kcal/mol, in very good agreement with the observed strain energy of 27.5 kcal/mol [30].

These considerations lead to the surprising conclusion that the strain energy of cyclopropane is associated with the increase in energy of the hydrogens. The calculation of the energy changes at the groups present in a variety of small ring compounds lead to the same conclusions. In each case, the strained carbon decreases in energy as a result of its increased electronegativity, and the hydrogens increase in energy.

With cyclobutane, the electron populations at carbon and hydrogen are closer to those of cyclohexane. The net change in energy with respect to a standard methylene is reduced to 6.7 kcal/mol per methylene group, or a total of 26.8 kcal/mol, which may be compared with the experimental strain energy of 26.5 kcal/mol.

The bridgehead carbon of bicyclo[1.1.0]butane has the largest electron population and lowest energy of all the compounds in Table 6, and correspondingly, the bridgehead proton has one of the lowest electron populations. This is a result of the increased strain at the bridgehead position as compared to cyclopropane, which leads to increased *p*-character in all of the C—C bonds, and increased *s*-character in the C—H bond. The changes in population are in good accord with the large ^{13}C—H NMR coupling constant [31] and decreased pK_a for the bridgehead C—H bond [18]. Among known saturated hydrocarbons, the bridgehead C—H bonds of bicyclobutane and its derivatives are closest in *s*-character and properties to an acetylenic C—H bond.

Bicyclo[1.1.1]pentane and [1.1.1]propellane provide another pair of related compounds. The increase in strain at the bridgehead position of the latter compound leads to an increase in the electron population at the carbon. However, a standard quarternary carbon (i.e. the energy of neopentane less that of four standard methyl groups) has an energy of −37.8763 H (Hartrees), and in comparison to this, the bridgehead carbon of [1.1.1]propellane has a greater energy by 44 kcal/mol. The methylene group is similar to that of a standard methylene group and is raised in energy by only 5 kcal/mol. Thus of the total calculated strain energy of 104 kcal/mol (compared to 98 kcal/mol as the experimental value), 85% resides at the bridgehead position.

The most highly strained of the compounds listed in the table is tetrahedrane, and here there is an unusually large shift in charge from hydrogen to carbon. Since all of the carbons are equivalent, there is no possibility of charge transfer from one to another (as there was with bicyclobutane), and here the entire effect is seen at the hydrogens. Each CH group has an energy greater than that of a standard CH group (−38.4558 au) by 35 kcal/mol, giving a predicted strain energy of 140 kcal/mol. A much more detailed analysis of the populations and energies for cyclic hydrocarbons is available [18].

Having examined hydrocarbons in some detail, let us see how the carbons are affected by the introduction of substituents. A first indication of the effects to be noted was given in Table 1 which showed that the bond critical point moved away from the more electronegative atom. The latter then gains electron population in two ways. The higher electronegativity leads to an increase in charge density near the atom, and the movement of the bond critical point away from the atom will lead to an increased volume assigned to the atom. The effect of substitution on the total electron densities may conveniently be visualized via the projection density plots

Table 6. Electron population and energies for some cycloalkanes[a]

Compound	Atom	n	T = −E(au)
Propane	C1	5.937	37.7500
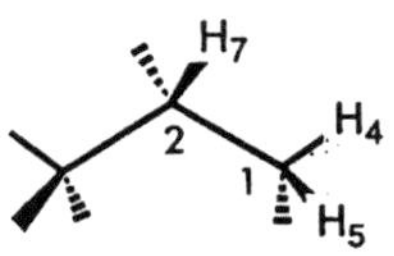	C2	5.892	37.7249
	H4	1.025	0.6264
	H5	1.028	0.6274
	H7	1.037	0.6383
Cyclopropane	C	6.010	37.7824
	H	0.995	0.6187

Table 6 (continued)

Compound	Atom	n	T = −E(au)
Cyclobutane 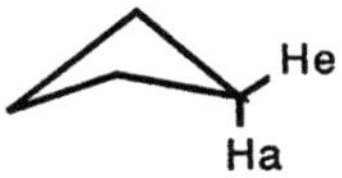	C	5.943	37.7593
	He	1.029	0.6322
	Ha	1.028	0.6325
Cyclohexane 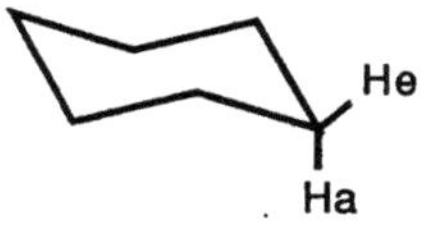	C	5.920	37.7580
	He	1.036	0.6371
	Ha	1.043	0.6397
Bicyclo[1.1.0]butane 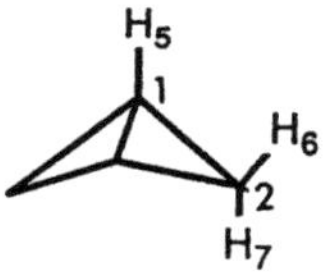	C1	6.121	37.8602
	C2	5.928	37.7415
	H5	0.946	0.5930
	H6	0.998	0.6188
	H7	1.007	0.6223
Bicyclo[1.1.1]pentane 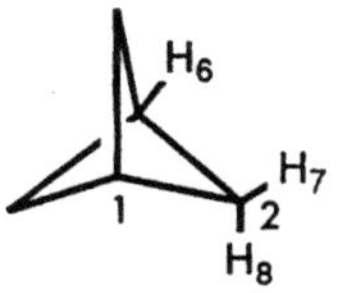	C1	5.964	37.7670
	C2	5.957	37.7787
	H6	1.011	0.6214
	H7	1.030	0.6322
[1.1.1]Propellane 	C1	6.108	37.8062
	C2	5.974	37.8000
	H	0.977	0.6131
Spiropentane 	C1	5.997	37.7821
	C2	6.072	37.8563
	H	0.993	0.6166
Cubane 	C	5.997	37.7821
	H	1.003	0.6148
Tetrahedrane 	C	6.111	37.8367
	H	0.889	0.5630

[a] These data were calculated using 6-31G* wave functions calculated at the 6-31G* geometries. The hydrogen populations are systematically 0.055 e less than those obtained using the 6-31G** basis. See Sect. 6 of the text.

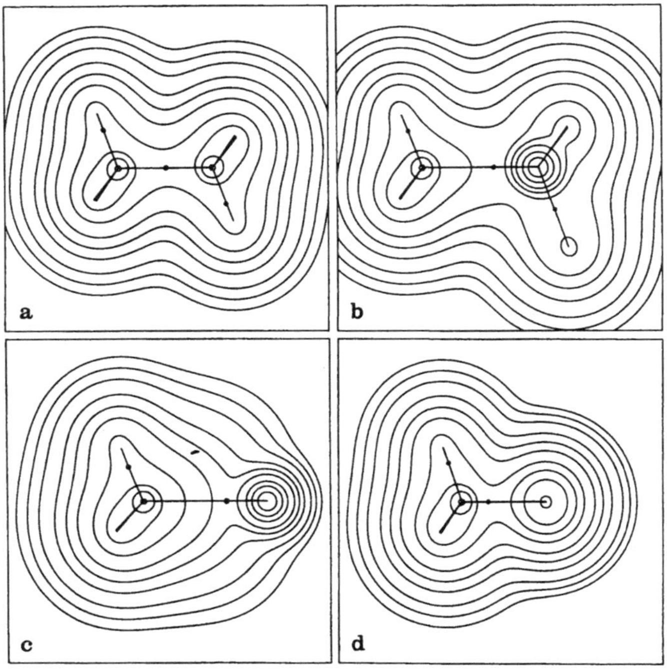

Fig. 2a–d. Projection plots for **a.** ethane; **b.** methylsilane; **c.** methyllithium; and **d.** methyl fluoride

developed by Streitwieser [32]. Here, all of the charge density above and below a given molecular plane is projected onto the plane. Some examples are shown in Fig. 2. In each case the location of the bond critical points are indicated by filled circles.

With ethane, symmetry requires that the two methyl groups be identical. When a methyl group of ethane is replaced by a lithium, it is quite evident that the volume element reasonably assigned to the methyl carbon is considerably increased in comparison to ethane. The opposite is found with methyl fluoride where the bond critical point has shifted toward the carbon, and as a result the volume assigned to the methyl carbon has considerably decreased.

The electron populations and atom (or group) energies for some of these compounds as determined by numerical integration of the ab initio wave functions are given in Table 7 [13]. It is sometimes easier to recognize the differences among compounds when the populations are converted to the equivalent charges by subtracting the atomic number from each. The results are shown in diagrammatic form in Fig. 3 for methyl groups attached to the set of hydrides for the first row and part of the second row of the periodic table.

At first glance, the magnitude of these charges may appear surprising. In the case of methyl fluoride, the methyl group bears a +0.75 charge and the fluorine a —0.75 e charge. We do not normally think of this compound as having a strongly polar bond. However, a methyl group and a hydrogen have almost equal electronegativities. We recognize hydrogen fluoride to be a polar molecule. Methyl fluoride should be similar to HF in this regard. As noted above, Fig. 2 shows the marked effect the fluorine has on the electron density distribution, and the reason for the strong polarization of the C—F bond.

Table 7. Electron populations and atom energies for methyl derivatives[a]

Compound	Atom	n_T	T = —E(au)	
CH_4	CH_3	8.9387	39.5539	
	C	5.7541	37.6105	
	H	1.0613	0.6478	
	sum	10.0000	40.2017	(—40.2017)
CH_3Li	CH_3	9.9021	39.6172	
	C	6.5059	37.6308	
	H	1.1320	0.6635	
	Li	2.0968	7.3995	
	sum	11.9987	47.0208	(—47.0210)
CH_3BeH	CH_3	9.8749	39.7853	
	C	6.6688	37.8589	
	H	1.0665	0.6424	
	Be	2.2536	14.2610	
	H	1.8723	0.7756	
	sum	13.9988	54.8219	(—54.8221)
CH_3BH_2	CH_3	9.7414	39.9291	
	C	6.5721	38.0024	
	Ha	1.0469	0.6315	
	Hb	1.0612	0.6476	
	B	2.8048	23.7760	
	H	1.7271	0.8917	
	sum	16.0004	65.4488	(—65.4487)
CH_3CH_3	CH_3	9.0000	39.6188	
	C	5.7627	37.6326	
	H	1.0791	0.6621	
	sum	18.0000	79.2376	(—76.2382)
CH_3NH_2	CH_3	8.5549	39.4137	
	C	5.3227	37.4095	
	H	1.1006	0.6776	
	Hb	1.0657	0.6636	
	N	8.1660	54.8382	
	H	0.6395	0.4856	
	sum	17.9999	95.2231	(—95.2219)
CH_3OH	CH_3	8.3523	39.2983	
	C	5.1519	37.2356	
	Ha	1.0438	0.6595	
	Hb	1.0780	0.6805	
	O	9.2594	75.3981	
	H	0.3880	0.3522	
	sum	17.9997	115.0486	(—115.0467)
CH_3F	CH_3	8.2573	39.2450	
	C	5.1333	37.2684	
	H	1.0413	0.6589	
	F	9.7426	99.7838	
	sum	17.9999	139.0297	(—139.0397)

Table 7 (continued)

Compound	Atom	n_T	T = −E(au)	
CH_3AlH_3	CH_3	9.7995	39.8522	
	C	6.6159	37.9310	
	Ha	1.0602	0.6441	
	Hb	1.0613	0.6514	
	Al	10.6051	214.3297*	
	H	1.7977	0.7441	
	sum	24.0000	282.6701	
CH_3SiH_3	CH_3	9.7463	39.8515	
	C	6.5955	37.9280	
	H	1.0484	0.6400	
	Si	10.9978	287.9651	
	H	1.7517	0.8231	
	sum	25.9992	330.2859	(−330.2819)
CH_3PH_2	CH_3	9.5762	39.8415	
	C	6.4459	37.9101	
	Ha	1.0442	0.6459	
	Hb	1.0430	0.6429	
	P	13.1575	339.9497*	
	H	1.6324	0.8521	
	sum	25.9985	381.4954	
CH_3SH	CH_3	8.9598	39.7138	
	C	5.8769	37.6686	
	Ha	1.0184	0.6393	
	Hb	1.0319	0.6480	
	S	15.7446	397.2470*	
	H	1.2956	0.7481	
	sum	26.0000	437.7089	
CH_3Cl	CH_3	8.6839	39.4803	
	C	5.6731	37.5688	
	H	1.0032	0.6373	
	Cl	17.3161	459.6176*	
	sum	26.0000	499.0979	

[a] The energies are given in Hartrees, and the populations are given in electrons. The results are based on 6-31G** wave functions obtained at the 6-31G* geometries. Ha is a unique methyl hydrogen, and Hb is one of a pair of hydrogens. The energies marked with an asterisk were obtained by difference because of the difficulty in integrating the kinetic energy with second row elements. The numbers in parentheses are the total energies calculated in the MO procedure.

Methyllithium is at the other end of the periodic table in Fig. 3, and again a strongly polar C—Li bond is found. There has been much discussion concerning the nature of C—Li bonds, but there now appears to be general agreement that they have little covalent character [33]. The projection density plot for methyllithium in Fig. 2 shows the strong polarization, and an examination of Table 1 shows that the charge density at the bond critical point is only 0.04 e/au^3, indicating little covalent character. Once

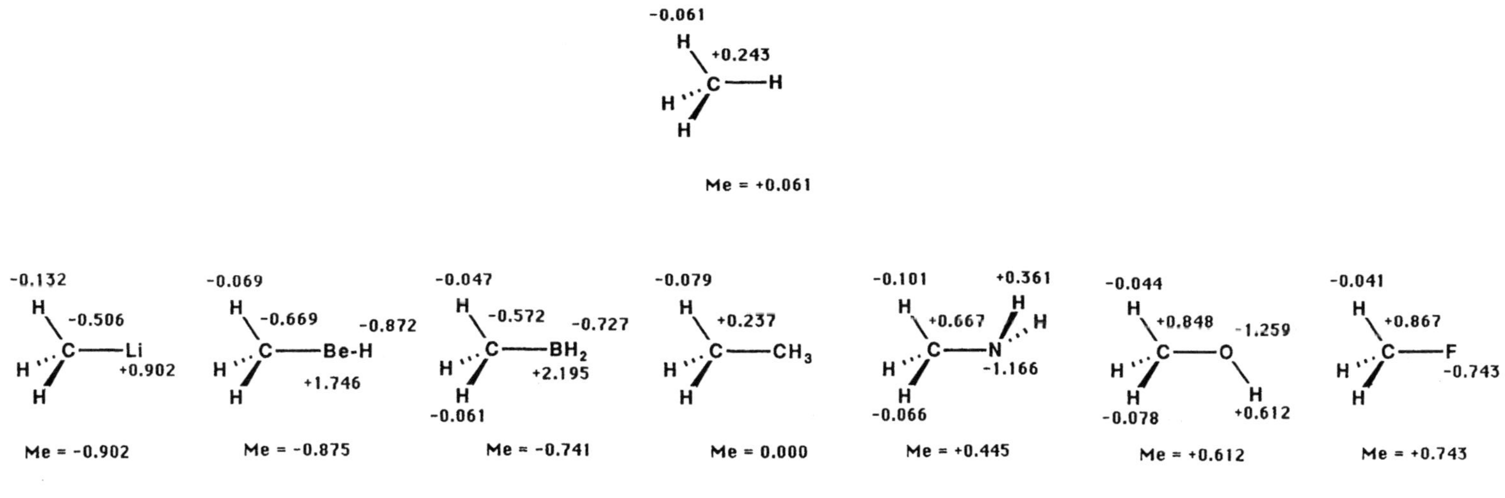

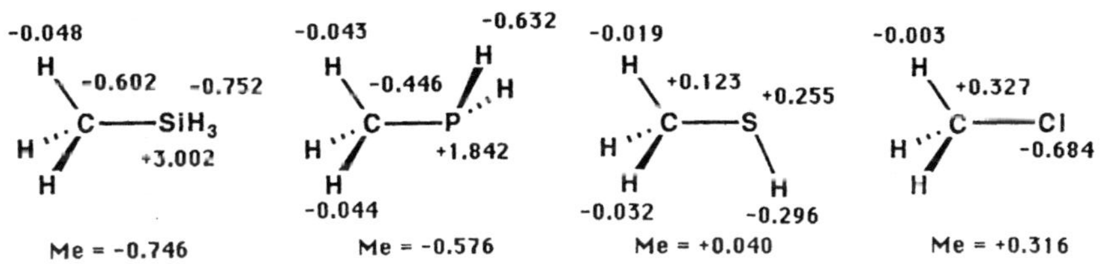

Fig. 3. Charges for methyl derivatives

one recognizes the origin of the polarization of the bonds in the extreme cases, it is easy to see how the other compounds in the first row fit in. The electron populations for beryllium and boron deserve comment. These atoms are attached to a relatively electronegative carbon, and also to equally electronegative hydrogens. As a result, the charge assigned to these atoms becomes rather large. The monomeric forms are not observed experimentally, and the dimers involve bridging between the electron deficient Be or B and the electron rich hydrogens.

It is generally recognized that the second row elements are less electronegative than those in the first row. This is reflected in the methyl and substituent charges. With methylsilane, for example, the charge assigned to the methyl group is -0.75 e, whereas that for the silicon is $+3.01$. Silicon is considerably less electronegative than carbon, and as a result loses electron population to the carbon. But, since hydrogen also has an electronegativity similar to that of carbon, the hydrogens also have a charge of -0.75. The relationship between the atoms in methylsilane may be seen in Fig. 2 where the bond critical point has moved from the carbon and hydrogens toward the silicon. The high susceptibility of silicon towards reaction with anions such as F^- or RO^- probably is related to the positive charge on silicon.

Again, the electronegativity of the atoms increases on going toward the right of the periodic table, with the methyl of methyl chloride being comparable to that of methylamine.

Charge distributions in unsaturated compounds which may lead to conjugated systems have been of considerable interest to chemists over the years. Formamide provides an interesting example. The barrier to rotation about the C—N bond has been measured by NMR spectroscopy, and was found to be 18 kcal/mol. The barrier is generally attributed to a resonance interaction of the type:

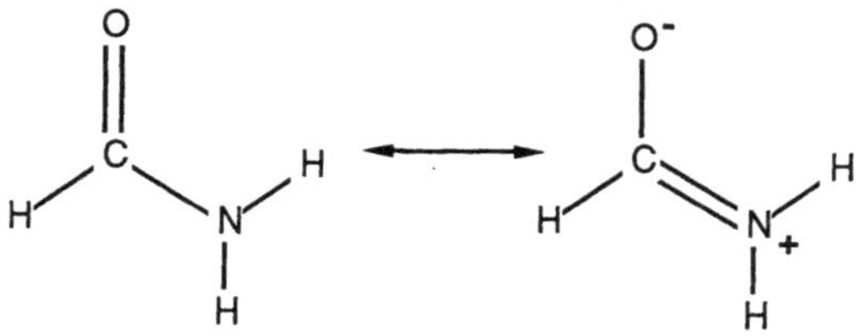

which introduces some double bond character into the C—N bond. The loss of this interaction on rotation would then lead to the barrier to rotation. An examination of the barrier using ab initio MO theory reproduced the observed rotational barrier, but there were unexpected changes in both the structural parameters and the electron populations on rotation [34]. The structures of the planar and rotated forms as well as the electron populations are shown in Fig. 4. The C—N bond is found to lengthen by 0.08 Å on rotation, but the C—O bond shortens by only 0.01 Å. The increase in C—N bond length is in accord with an expectation based on the resonance formulation, but the C—O bond is relatively unaffected. The same trend is found with the electron populations. Here, there is only a small change in the oxygen population, whereas both the carbon and nitrogen populations change by 0.2 e each. Further, the direction of the change is opposite to that anticipated by the resonance formulation. It is clear that this simple formalism cannot be correct and that the interactions are more complex.

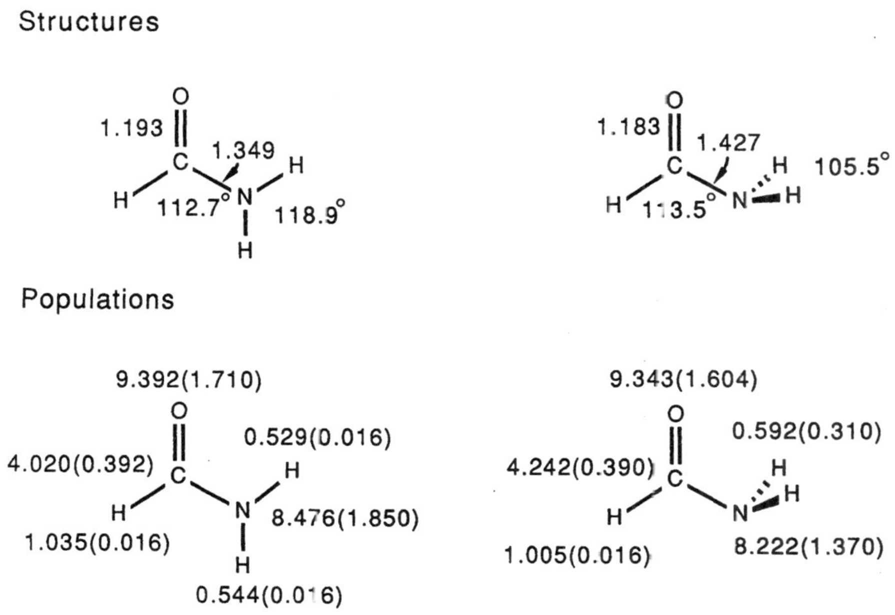

Fig. 4. Structures and electron populations for formamide

Fig. 5. Projection density plots ($\int \varrho \, dx$) for acetic acid (upper left = total density, upper right = π density) and acetate ion (lower left = total density, lower right = π density). The carbonyl group is aligned vertically and the methyl group is to the left. The contour values are: A = 2, B = 0.8, C = 0.4, D = 0.2 e/au^2, a = 0.4, b = 0,2, c = 0.08 and d = 0.04 e/au^2

The increased population at nitrogen in the planar form is readily explained on the basis of the calculated structures. In the lower energy form, the amino group is planar with approximately 120° bond angles. Here, the nitrogen forms bonds using sp^2 orbitals and places its lone pair in a p orbital. The amino group in the rotated (saddle point) conformer has 106° H—N—H bond angles. Here, the nitrogen uses orbitals with high p-character to form the bonds, and places the lone pair in an orbital with high s-character. The nitrogen is now much less electronegative, and as a result accepts less electron population from the carbon. The shorter bond in the planar form may now be attributed to the increased ionic character of the bond, resulting in internal electrostatic stabilization and to the sp^2 hybridization. A tendency for charge alternation as found with formamide has also been found for allyl type anions and related systems [35].

A related example is found with the carboxylic acids. For many years chemists have attributed the greater acidity of acetic acid as compared to ethanol to the resonance stabilization of the acetate ion [36]. However, the work of Thomas and Siggel [37] casts doubt on this interpretation, and rather suggests that the increased acidity is due to the polarization inherent in the carbonyl group. An examination of the projection density plots for acetic acid and acetate ion (Fig. 5) shows a remarkable similarity [38]. The main change which occurs on ionization is that the carbonyl carbon moves slightly to a position symmetrically related to the oxygens.

The electron populations for the atoms in acetic acid and in acetate ion are given in Table 8. The strong polarization for acetic acid seen in the projection density plot is found in the carbon and oxygen populations. The loss of the proton does not lead to a marked change in populations, rather both oxygens gain some. Similar observations have been made for other weak acids such as nitrous acid [39].

Table 8. Electron populations for acetic acid and acetate ion

Atom	Acetic acid	Acetate ion
C(Me)	5.849	5.908
H_a*	0.984	1.079
H_b	0.989	1.078
C(=O)	4.191	3.964
=O	9.346	9.446
—O—	9.298	9.451
H	0.353	
sum	31.999	32.004

* The unique methyl hydrogen

In retrospect, it is surprising that the difference between having a carbonyl and a methylene group next to a hydroxy was not generally recognized earlier. The carbonyl group is known to have a large bond dipole, and the projection density plots in Fig. 5 show how strongly polarized the charge distribution has become. A strongly polar group of this type next to a hydroxy could not help but stabilize a charge on an adjacent atom better than does a methylene group.

4 Other Properties: Atomic Volumes and Dipoles

The volume occupied by a molecule is reproduced quite well by the volume included by the 0.002 e/au^3 contour level derived from an ab initio calculation [40]. Thus, the shape of the molecule may be determined from these calculations. The constant charge density sheet which covers the molecule will intersect the zero-flux surfaces and convert them into closed volumes. It is now readily possible to calculate the volumes which may be assigned to the atoms in a molecule [40].

It was found that atomic and group volumes were transferable in the same fashion as noted above for their energies. The volume of an atom changed with a change in environment. Thus, the bridgehead carbons of bicyclo[1.1.0]butane and [1.1.1]-propellane had volumes 1.2 and 1.5 times that of a methyl carbon. Similarly, in the series with increasing *s* character, ethane, ethylene and acetylene, the electron population, stability and volume of a carbon increases in that order. The effect of hydrogen bonding on atomic volumes also has been studied.

A frequently used semiquantitative concept is that of a bond moment. With many simple compounds, it is possible to estimate the total dipole moment as the sum of individual bond moments, and this has sometimes been used to try to determine the conformation of a molecule. This simple model, however, ignores important interactions in molecules. In particular, it usually ignores atomic dipoles which can be quite large when lone pairs are involved.

The concept of bond dipoles has been widely used in interpreting the intensities of infrared bonds [41]. The intensity is related to the change in dipole moment with the change in coordinates which corresponds to a given vibration. From these data, it is possible to derive dipole moment derivatives with respect to internal or symmetry coordinates.

If one considers the antisymmetric stretching and bending modes of methane and silane, the signs of the dipole moment derivatives are:

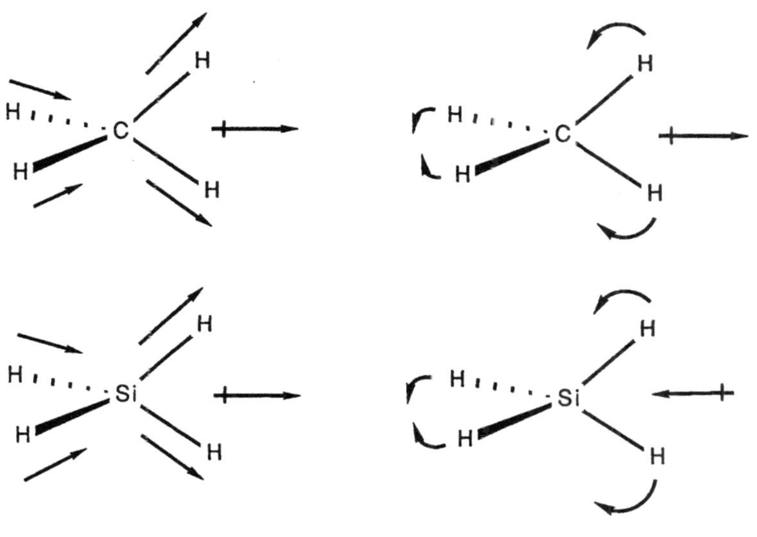

The signs for methane pose a problem: the stretching mode suggests that the CH bond dipole has the sense $C^+—H^-$ whereas the bending mode suggests $C^-—H^+$. Making the assumption that bending mode are simpler electrically than stretching modes, the latter sign has frequently been used, and the reversed apparent sign for the stretching mode was attributed to a charge flux accompanying stretching which overcame the moment expected using the bond dipole [42].

The above assumption can be shown to be incorrect. A calculation of the change in population during bond stretching showed that it was not sufficient to affect the sign of the dipole moment derivative [43]. Thus, the sense of the bond dipole must be $C^+—H^-$, and there now appears to be fairly general agreement as to the sign [24]. The bending mode has only limited orbital following, with the nucleus moving ahead of the charge distribution in the overlap region. This leads to the apparent motion of a positively charged hydrogen. This picture fits silane quite well. Here there is a large bond dipole in the sense $Si^+—H^-$, giving the same sign for the stretching mode as for methane. With the bending mode, the bond dipole is so large that even with limited orbital following, the sign corresponds to the movement of a negatively charged hydrogen.

In all cases which have been studied in this fashion, the sign of the dipole moment derivative for M—H bond stretching corresponds to the sign of the M—H bond dipole. Thus, with the ammonium ion it is reversed from that found with methane, and in the series ethane, ethylene and acetylene, the sign corresponds to $C^+—H^-$ with the first two, but with the much more electronegative carbon in acetylene which uses an sp orbital, the sign is reversed [43].

This simple picture cannot be readily generalized to the full range of molecules which might be studied. In particular, compounds with lone pairs cause difficulty because their electrical properties are stongly affected by the atomic dipoles.These dipoles are one of the properties which may readily be calculated from the wave functions [44], and it was found that atomic and group atomic dipoles represented another transferable quantity.

The molecular dipole moment is the sum of the charge transfer terms arising from the net charge on the atoms and a first moment contribution arising from the polarization of the atomic charge densities. Both terms are equally important. The two terms were calculated for stretching and bending modes of methane, silane, ethylene, cyclopropane and cyclobutene, and showed that the contribution from the polarization of the charge densities was important in each case [44].

5 Basis Set Dependence and Effect of Electron Correlation

The charge density is derived from the molecular orbital wave functions, and thus must be in some way dependent on the quality of the wave functions. It is clearly not possible to use a complete set of atomic orbitals at each center, and even if one were to do so, the result would still have to be corrected for the effects of electron correlation. It is therefore important to examine both the effect of basis set size, and of electron correlation on the calculated properties.

First, the importance of a balanced basis set should be noted. If one considers a C—H bond, it should be represented by approximately equal numbers of atomic orbitals at each center. A 6-31G basis set is reasonably well balanced. It uses eight valence functions (two each of *s*, *px*, *py* and *pz*) to represent the carbon, and two 1*s* functions to represent each hydrogen. So, each of the four C—H bonds may be though of as being represented by four functions, two from the carbon and two from the hydrogen. This basis set is however not adequate for compounds having significant dipole moments since the electric fields inside the molecule will cause the orbitals to deform. This deformation may be accommodated in the calculations if one introduces polarization functions (*d*-orbitals at carbon or *p*-orbitals at hydrogens). It is common to use the 6-31G* basis set containing *d*-functions on atoms other than hydrogen for geometry optimization on hydrocarbons and related compounds because polarization functions at hydrogen do not have much of an effect on the calculated structures. However, this is not a well balance basis set. It uses the eight valence functions noted above plus 6*d*-type functions giving a total of 14 per carbon, or 3.5 per bond as compared to two per hydrogen. The 6-31G** basis would be more satisfactory, using 5 functions per hydrogen (two 1*s*, and 3*p* type orbitals). It might appear overbalanced with respect to the hydrogens, but only one of the *p*-functions will be especially important in most cases.

The effect of using the 6-31G* and 6-31G** basis sets has been examined for a series of hydrocarbons [18]. The smaller basis set was found to shift the bond critical point toward the hydrogen by 0.002 Å and to give an electron population at hydrogen 0.055 e smaller than that obtained using the 6-31G** basis. No effect was found on the C—C bonds. Similar observations have been made concerning substitued compounds [45]. It can be seen that the effect of basis set on populations is fairly predictable, and is generally small.

The effect of electron correlation on the calculated bond properties and electron populations has only recently been examined. It is now possible to obtain the appropriate density matrix [46], and to carry out the numerical integrations in much the same fashion as for the Hartree-Fock wave functions. The effects of electron correlation including all single and double excitations (CISD) on charge densities for water and for ethylene are shown in Fig. 6. The effect of using the Møller-Plesset

Table 9. Effect of electron correlation on atomic charges[a]

Compound	Atom	RHF	MP2	CISD
Water	O	−1.226	−1.164	−1.148
	H	+0.613	+0.582	+0.574
Ammonia	N	−1.084	−1.062	−1.025
	H	+0.362	+0.354	+0.342
Phosphine	P	+1.872	+1.700	+1.730
	H	−0.625	−0.567	−0.577
Ethylene	C	+0.069	+0.015	+0.043
	H	−0.034	−0.007	−0.021
Ethane	C	+0.235	+0.115	+0.165
	m	−0.078	−0.038	−0.055

[a] Calculated using the 6-31G** basis set.

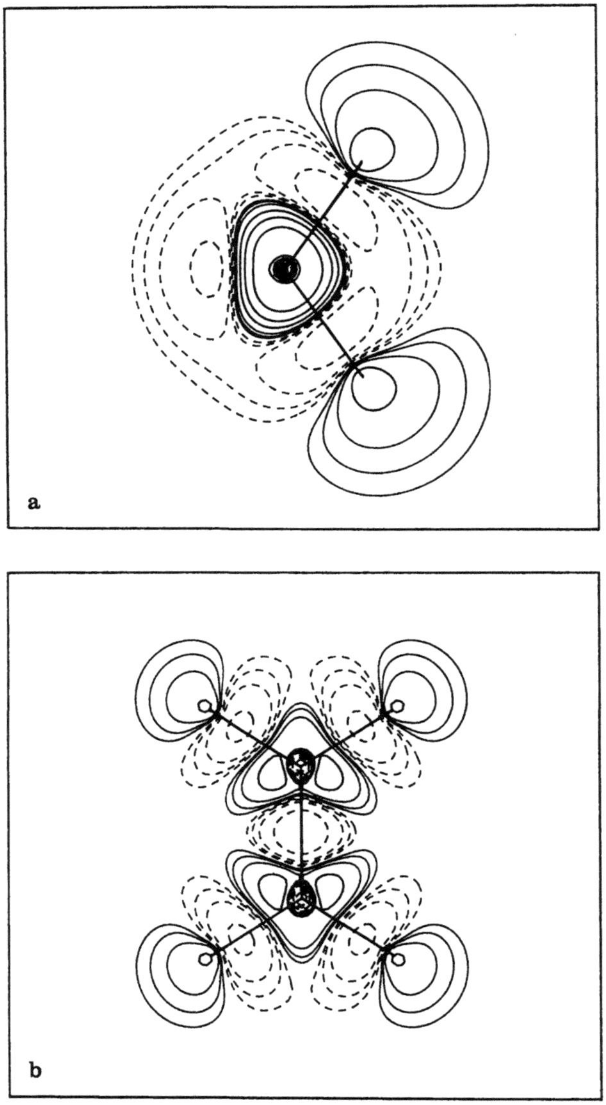

Fig. 6a, b. Change in charge density in the molecular planes of **a.** water, and **b.** ethylene between the CISD and RHF calculations. The outer contour level is 4×10^{-4} e/au^3. The positive contours indicate greater charge density in the CISD calculation, and the dashed lines indicate regions of decreased charge density

perturbation method (MP2) [47] was similar, but give even larger changes in charge density in accord with the general expectation that it will overcorrect. It can be seen that the charge density becomes more diffuse in the bonding region, both along the bond and normal to the bond. This results in an increase in calculated bond length giving structures which are usually in very good agreement with the calculated structures [48]. The charge density at the critical point generally decreases by 1 to 2%.

The effect on atom populations are shown in Table 9 which compares the RHF populations with those obtained by correction for electron correlation using MP2 and CISD. In each case, the calculated populations change so as to decrease the atomic charges when correlation is included. However, in most cases, the changes are relatively small. Data are not as yet available for a wide variety of compounds, but the relatively small changes usually found are encouraging, and suggest that the RHF populations will normally be satisfactory in comparing compounds. However, it remains to be determined how large an effect may be found with electron deficient species such as carbocations, free radicals and carbenes.

6 Changes in Molecular Graphs During Reactions

When a reaction occurs, the molecular graph of the reactants must be transformed into that of the products. As a result, the bond critical points must move, in some cases coalesce, and then move apart again. The points along the reaction coordinate where they coalesce are known as catastrophe points, and represent the points at which the transition from reactants to products occurs [49].

This idea has been applied to the isomerization of HNC to HCN and of CH_3NC to CH_3CN, reactions which have been well studied via molecular orbital procedures. The change in the molecular graph as found to be an abrupt process in the vicinity of the transition state. It also has been applied to the problem of the structure of the norbornyl cation [50].

7 The Laplacian of the Charge Density

An examination of the charge density distribution about an atom does not reveal any special characteristics which might have been expected from the shell structure of the atomic orbitals. However, if one examines the Laplacian of ϱ (the sum of the second derivatives of ϱ with respect to the coordinates), the shell structure becomes quite evident. It has been found that $\nabla^2\varrho$ is a generally valuable tool for examining the properties of atoms in molecules [51].

The integral of $\nabla^2\varrho$ over an entire molecule, or over an atom or group as defined by the theory of atoms in molecules is zero. However, $\nabla^2\varrho$ may have non-zero local values, and they provide useful information. A negative value of $\nabla^2\varrho$ indicates a region of local charge concentration. Similarly, a positive value indicates a region of local charge depletion. An examination of methane, ammonia and water found four local minima in $\Delta^2\varrho$ corresponding to the bonds and lone pairs [52]. The structures of many compounds have been examined using $\Delta^2\varrho$ and gave results in accord with the VSEPR model of molecular geometry [53].

The Laplacian of ϱ also has been related to the position of nucleophilic attack at a carbonyl group. The regions of local charge depletion corresponded to the angle of approach deduced by Burgi and Dunitz [54].

8 Conclusions

The theory of atoms in molecules provides a definition of molecular structure in terms of molecular graphs, along with a unique definition of atoms in molecules. This allows a number of properties of atoms such as their electron populations, kinetic energies, atomic volumes and atomic dipoles to be calculated. The properties thus derived are found to be transferable. Descriptions of molecular system in terms of the Laplacian of ϱ also has proven valuable in studying geometries and reactivity.

The theory has received many applications in addition to the ones mentioned above. These include descriptions of conjugation and hyperconjugation [55], of homoaromaticity [56], and properties of three membered rings [57], as well as other hydrocarbons [58]. It also has been applied to a study of electron distributions in vinyl groups [12], the properties of carbocations [59], the structures of dications [60], and the properties of carbon-lithium bonds [33].

9 References

1. Hehre JJ, Radom L, Schleyer PvR, Pople JA (1986) Ab initio molecular orbital theory, Wiley-Interscience, New York
2. a. Bader RFW, Nguyen-Dang TT (1981) Adv. Quantum Chem. 14: 63; b. Bader RFW, Nguyen-Dang TT, Tal, Y (1981) Rep. Prog. Phys. 44: 893; c. Bader RFW (1986) J. Chem. Phys. 85: 3133; d. Bader RFW (1985) Accts. Chem. Res. 9: 18
3. Wooley RG (1978) Chem. Phys. Lett. 55: 443; (1978) J. Am. Chem. Soc. 100: 1073; (1976) Adv. Phys. 25: 27
4. Pople JA, Frisch MJ, Luke BT, Binkley JS (1983) Int. J. Quantum Chem. Symp. 17: 307
5. This is not neccessarily the case for all bonds: Dunitz JD, Seiler PJ (1983) J. Am. Chem. Soc. 105: 7056. The apparent lack of accumulation of charge density appears, however, to be due to the use of spherically averaged atoms as the reference rather than atoms in appropriately prepared valence states: Kunze KL, Hall MB (1987) J. Am. Chem. Soc. 109: 7617
6. Wiberg KB, Wendoloski JJ (1981) J. Comput. Chem. 2: 53
7. Mulliken RS (1965) J. Chem. Phys. 43: S2
8. Coppens P (1982) in: Hall MB (ed) Electron distribution and the chemical bond, Plenum, NY
9. Barnzai AS, Deb BM (1981) Rev. Mod. Phys. 53: 95
10. The directions must be chosen so as to eliminate cross-terms. One normally constructs the Hessian of ϱ and diagonalizes it.
11. Bader RFW, Slee TS, Cremer D, Kraka E (1983) J. Am. Chem. Soc. 105: 5061
12. Slee TS (1986) J. Am. Chem. Soc. 108: 606, 7541
13. Wiberg KB, Breneman C (to be published)
14. a. Cremer D, Kraka E (1984) Croat. Chem. Acta 57: 1259; b. Koch W, Frenking G, Gauss J, Cremer D, Collins JR (1987) J. Am. Chem. Soc. 109: 5917
15. Bent HA (1961) Chem. Rev. 61: 275
16. Wiberg KB, Murcko MA (1988) J. Mol. Struct. 169: 355
17. Coulson CA, Moffitt WE (1949) Philos. Mag. 40: 1
18. Wiberg KB, Bader RFW, Lau CDH (1987) J. Am. Chem. Soc. 109: 985, 1001
19. One might wonder why the angle is not larger than 90°, the smallest angle which can be achieved using *p* orbitals. Although orbital following is normally quite limited (Nakatsuji H (1974) J. Am. Chem. Soc. 96: 24.), when large angular excursions occur, some orbital following will occur and the bond path angle will be smaller than expected. This is especially noticeable with spiropentane where the interorbital angle at the central carbon is 109.5° by symmetry, but the bond path angle is only 84.8°.

20. Orendt AM, Facelli JC, Grant DM, Michl J, Walker FH, Dailey WP, Waddell ST, Wiberg KB, Schindler M, Kutzelnigg W (1985) Theor. Chim. Acta 68: 421
21. Mulliken RS (1962) J. Chem. Phys. 36: 3428
22. Mulliken RS, Politzer P (1971) J. Chem. Phys. 55: 5135
23. Grier DL, Streitwieser A (1982) J. Am. Chem. Soc. 104: 3556 and references therein.
24. Reed AE, Weinhold F (1986) J. Chem. Phys. 84: 2428
25. Cf. Hirshfeld FL (1977) Theor. Chim. Acta 49: 127 for populations based on spherically averaged atom densities. However, note Figure 2 which shows that the atoms in molecules are not spherically symmetrical.
26. Bader RWF, Tal Y, Anderson SG, Nguyen-Dang TT (1980) Isr. J. Chem. 19: 8
27. a. Biegler-König FW, Bader RFW, Tang T-H. (1982) J. Comput Chem. 3: 317; b. Bader RFW, Tang T-H, Tal Y, Biegler-König FW (1982) J. Am. Chem. Soc. 104: 946
28. Bader RFW, Carroll MT, Cheeseman JR, Chang C (1987) J. Am. Chem. Soc. 109: 7968
29. Bader RFW, Larouche A, Gatti C, Carroll MT, MacDougall PJ, Wiberg KB (1987) J. Chem. Phys. 87: 1142
30. Wiberg KB (1986) Angew. Chem. Int. Ed. Eng. 25: 312
31. Wiberg KB, Lampman GM, Ciula RP, Connor DS, Schetler P, Lavanish J (1965) Tetrahedron 21: 2749
32. Streitwieser AJr, Collins JB, McKelvey JM, Grier D, Sender J, Toczko AG (1979) Proc. Natl Acad Sci. U.S.A. 76: 2499
33. Ritchie JP, Bachrach SM (1987) J. Am. Chem. Soc. 109: 5909 and references therein
34. Wiberg KB, Laidig K (1987) J. Am. Chem. Soc. 109: 5935
35. Wiberg KB, Breneman C, LePage TJ, Glaser R (to be published)
36. This argument is given in most textbooks. Cf. Morrison RT, Boyn RN (1983) Organic chemistry, 4th edn Allyn and Bacon, Boston, MA, p 793
37. Siggel MR, Thomas TD (1986) J. Am. Chem. Soc. 108: 4360
38. Wiberg KB, Laidig K (1988) J. Am. Chem. Soc. 110: (in press)
39. Wiberg KB (in press) Inorg. Chem.; Thomas TD (in press) Inorg. Chem.
40. Bader RFW, Carroll MT, Cheeseman JR, Chang C (1987) J. Am. Chem. Soc. 109: 7968
41. a. Person WB, Steele D (1974) Mol. Spectrosc. 2: 357; b. Steele D (1978) Mol. Spectrosc. 5: 106; c. Gussoni M (1980) Adv. Infrared Raman Spectrosc. 6: 61
42. a. Decius JC (1975) J. Mol. Spectrosc. 57: 348; b. Decius JC, Mast GB (1978) Mol. Spectrosc. 70: 294
43. Wiberg KB, Wendoloski JJ (1984) J. Phys. Chem. 88: 586
44. Bader RFW, Larouche A, Gatti C, Carroll MT, MacDougall PJ, Wiberg KB (1987) J. Chem. Phys. 87: 1142
45. Wiberg KB, Wendoloski JJ (1981) Proc. Natl. Acad. Sci, U.S.A. 78: 6561
46. Frisch MJ (to be published)
47. Moller C, Plesset MS (1934) Phys. Rev. 46: 618; b. Binkley JS, Pople JA (1975) Int. J. Quantum Chem. 9: 229; Pople JA, Binkley JS, Seeger R (1976) Int. J. Quantum Chem. 10: 1
48. Ref. 1, p 156ff.
49. Tal Y, Bader RFW, Nguyen-Dang TT, Ojha M, Anderson SG (1981) J. Chem. Phys. 74: 5162
50. Bader RFW (1985) Acc. Chem. Res. 18: 9
51. Bader RFW, Essen H (1984) J. Chem. Phys. 80: 1984
52. Bader RFW, MacDougall PJ, Lau CDH (1984) J. Am. Chem. Soc. 106: 1594
53. Gillespie RJ (1967) Angew. Chem. 6: 819
54. Burgi HB, Dunitz JD (1983) Acc. Chem. Res. 16: 153
55. Bader RFW, Slee TS, Cremer D, Kraka (1983) J. Am. Chem. Soc. 105: 5061
56. Cremer D, Kraka E, Slee TS, Bader RFW, Lau CDH, Nguyen-Dang TT, MacDougall PJ (1983) J. Am. Chem. Soc. 105: 5069
57. a. Cremer D, Kraka E (1985) J. Am. Chem. Soc. 107: 3800, 3811; b. Cremer D, Gauss J (1986) J. Am. Chem. Soc. 108: 7467
58. Bader RFW, Tang T-H, Tal Y, Biegler-Konig FW (1982) J. Am. Chem. Soc. 104: 940, 946
59. Bader RFW (1986) Can. J. Chem. 64: 1036
60. Koch W, Frenking G, Gauss J, Cremer D (1986) J. Am. Chem. Soc. 108: 5808

The Modelling of Molecules as Collections of Modified Atoms

Zvonimir B. Maksić

Theoretical Chemistry Group, Ruđer Bošković Institute 41001 Zagreb, Croatia, Yugoslavia and Faculty of Natural Sciences and Mathematics, University of Zagreb, Marulićev trg 19, 41000 Zagreb, Croatia, Yugoslavia

The notion of atoms in molecular systems is briefly discussed. The models of independent (IAM) and modified (MAM) atoms are described and applied in calculating a wide variety of molecular properties. It is concluded that they have a high cognitive value by providing a rationale for a number of features of molecular behavior. They represent an important link between the rigorous quantum theory and chemical phenomenology. The striking characteristics of the IAM and MAM models are their pictorial power and computational feasibility. Although they are designed to give qualitative information they sometimes yield semiquantitative results. It is worth pointing out that the IAM model provides data which can serve as a useful tool in detecting intrinsic drawbacks of the approximate treatments of the electronic structure of molecules.

Apparently there is colour, apparently sweetness, apparently bitterness; actually there are only atoms and the void.

Democritus, 420 B.C.

1 Introduction

The idea that matter consists of indivisible particles goes back to Democritus (420 B.C.). His penetrating imagination gave impetus to many thinkers and researchers for centuries. An important refinement of the idea of Democritus was made by Bošković by identifying atoms with centres of specific forces [1], which had a significant impact on the subsequent development of physics and chemistry [2]. The atomic hypothesis was definitely confirmed in the 19th century mainly through the work of Dalton in chemistry and Boltzman in physics. That was probably one of the most important discoveries in the history of natural sciences. The famous physicist R. Feynman in his well known series of books "The Feynman Lectures on Physics" writes: "If, in some cataclysm, all of scientific knowledge were to be destroyed, and only one sentence passed on to the next generation of creatures, what statement would contain the most information in the fewest words? I believe it is the *atomic hypothesis* (or the atomic fact, or whatever you wish to call it) *that all things are made of atoms — little particles that move around in perpetual motion, attracting each other when they are a small distance apart, but repelling upon being squeezed into one another*. In that one sentence there is an enormous amount of information about the world, if just a little imagination and thinking are applied [3]". There is nothing to be added to this statement.

The next crucial step in building up the theory of classical chemistry was the notion of atomic valency and the chemical bond as an interaction between two atoms at a time. Since Pasteur, Kekulè, van't Hoff and LeBel (1874) it has been known that chemical bonds have specific spatial arrangements which introduced the third dimension in chemistry and they laid down the foundation for the molecular structure concept. The latter, in its nongeneric geometric sense, is one of the most fruitful themes in chemistry. It mirrors a vast variety of (electronic) properties of molecules and gives and important clue to understanding their interactions. As an illustration we shall mention the application of structural principles in molecular biology like the "lock- and key" complementariness of Fischer or α-helix and β-sheet structural patterns revealed by Pauling.

The notions of atoms, valencies, chemical bonds and molecular structure provide the very basis of the impressive edifice of classical phenomenological chemistry. Rich experience has shown that properties of substancies could be interpreted at the microscopic molecular level. Molecules themselves, in turn, exhibit features which can be identified as originating from particular atoms and/or atomic groupings forming molecular fragments. In other words, experimental knowledge strongly supports a conjucture that a molecule has a "memory" and "remembers" atoms which took part in its formation. Atoms retain their identity within a molecule, but they are perturbed and distorted to a different extent depending on the immediate neighbourhood. It is an important corrollary that perturbed atoms can be further roughly classified according to coordination numbers or the number of the nearest neighbours. For example, one can distinguish in organic chemistry sp^1, sp^2 and sp^3 carbon atoms which exhibit different chemical and physical properties. This holds in general and one can identify characteristic moieties exhibiting specific features which vary in a narrow range within a family of related molecules. There are a number of atomic "fingerprints" to mention only NMR chemical shifts.

It appears also that atoms have localized electronic substructures e.g. inner core and valence shell lone pairs of electrons giving rise to characteristic ESCA chemical shifts and PES ionization potentials, respectively. It should be mentioned in this context that the so-called substituent effect also assumes the existence of atoms in molecules because the influence of atoms A and B on the fragment M are compared in compounds MA and MB.

It is remarkable that (localized) chemical bonds can be also classified according to types of modified atoms. For instance, $C(sp^3)$—$C(sp^3)$ bond differs from the $C(sp^3)$—$C(sp^1)$ bond, the latter being shorter, stronger etc. But each of these bonds has remarkably stable properties in various compounds. This finding can be rationalized in terms of the Lewis concept of coupled electron pairs [4] in its refined quantum mechanical version [5, 6] and characteristics of "*sp*3 and *sp*1" carbon atoms. This is the reason behind characteristic stretching vibration frequencies, force constants, bond distances and some other properties. Classification of modified atoms and the corresponding bonds offers the simplest rationalization of various group additivity rules. We mention a well known additivity scheme for enthalpies of formation in hydrocarbons which involves empirically adjusted but constant values of constituent molecular fragments —CH, —CH_2, —CH_3 etc. Further, a practicising chemist often discusses molecular properties in terms of electron-donating and electron-withdrawing groups (of atoms) which can be traced down to properties of constituent atoms.

It follows that a large body of the experimental findings in chemistry can be reduced to the behaviour of perturbed and modified atoms in chemical environments. The possibility of classification of distorted atoms and their chemical bonds offers a selection of relatively few simple building blocks which can describe and rationalize properties of myriads of complex molecular systems at the phenomenological level of sophistication.

Molecules are of course quantum objects and consequently the existence of molecular subunits (atoms, fragments) should be extracted from quantum mechanics. This is not an easy task. It is very difficult to define rigorously both molecular structure [7, 8] and atomic border-surfaces by quantum theory. A breaktrough was recently made by Bader's theory of atoms which are encompassed by zero-flux surfaces of the electronic charge density [9–11]. The atomic domain is defined by all paths of steepest ascent through the electron charge distribution which terminate at the corresponding nucleus. The electron density within the domain and the nucleus in question form an atom immersed in a molecular environment. Remarkably, the virial theorem is satisfied for each atomic basin separately. Since the decomposition of the molecular volume to exclusive domains is obtained by using the observable ϱ (electron density), all molecular properties including the total energy are rigorously defined and are directly additive in terms of atomic contributions. We note in passing that the chemical bond is defined in Bader's theory by a ridge of the electron density connecting bonded nuclei. This bond path has a minimum at the point of intersection with the interatomic surface. It is called a bond critical point and it has a property of a saddle. Many other concepts of classical chemistry can be described by this approach [12–15]. Analyses of the Laplacian $\nabla^2\varrho$ of the electron density gave additional insight into empirical ideas provided by experimental research [16–18]. Finally, it should be pointed out that generic quantum (topological) molecular structure is easily defined by the theory of zero-flux density atoms.

Bader's theory of topological atoms in chemical moieties is theoretically sound and attractive but requires intricate calculations. One should recall Pauli's remark that surfaces were invented by the devil. Zero-flux surfaces are complicated indeed. Extended basis sets are required too if meaningful quantitative results are desired. It would be, therefore, useful to have at hand a simple model which by sacrifying some accuracy gains in feasibility being applicable in large systems let's say of DNA size. On the other hand a conventional many-electron wavefunction gives more information than we need to know. Simplified models are better tailored to identify trends and dominant effects. We feel that the model of modified atoms (MAM) could serve the purpose rather well as evidenced by abundance of useful results [18–20]. The MAM model yields a transparent qualitative interpretation of some molecular properties observed experimentally or calculated at the ab initio level. It is interesting that despite its utmost simplicity the MAM model sometimes gives results of semiquantitative quality. It can be also used in a predictive manner serving as a guide in directing the execution of intricate and accurate calculations. Surprisingly, the model of independent atoms (IAM) consisting of neutral spherical atoms placed at the equilibrium positions in molecules frequently gives a good performance. We shall discuss applications of IAM and MAM models in calculating some molecular properties in some detail. The following section deals with the IAM model.

2 Independent Atom Model (IAM)

It has already been mentioned that the nongeneric geometric structure of molecules provides the most fundamental sort of information on which much of the rest of chemical understanding is based. The striking feature of the IAM model is the use of the Born-Oppenheimer clamped nuclei approximation and equilibrium bond distances. Neutral atoms are then placed at the equilibrium positions and their mutual interactions are set up to zero. Since the constituents are considered as completely independent, their electronic charge distribution is spherical. The effect of nuclear vibrations on molecular properties is later taken into account as a correction if necessary. This simple model is useful in discussing some magnetic properties and provides a treshold of quality in quantitative appraisal of approximate wavefunctions.

2.1 Description of some Molecular Magnetic Properties by the IAM model

2.1.1 Diamagnetic Shielding of the Nuclei

The spherically average magnetic shielding of nuclei can be split into two contributions:

$$\sigma_{av}(A) = \sigma^{d}_{av}(A) + \sigma^{p}_{av}(A) \qquad (1)$$

where the superscripts d and p refer to diamagnetic and paramagnetic contributions, respectively [21]. We shall concentrate on the former, which is known as the Lamb shift. It explicitly reads as follows:

$$\sigma^{d}_{av}(A) = (e^2/3mc^2)\,\langle 0|\ 1/r_A\ |0\rangle \qquad (2)$$

Table 1. Average diamagnetic shielding in some small molecules as obtained by IAM model and ab initio procedures (in ppm)

Molecule	Gauge origin	IAM	Ab initio	Molecule	Gauge origin	IAM	Ab initio
HCl	Cl	1150.4	1150.2	OF_2	O	514.9	512.3
	H	143.0	141.9		F	562.5	561.8
H_2S	S	1064.1	1064.9	NO_2	N	451.9	446.5[j]
	H	135.8	136.1		O	484.3	483.6
PH_3	P	980.8	980.9	O(2)O(1)O(2)	O(1)	512.6	508.4
	H	126.2	127.6		O(2)	488.4	487.4
Cl_2	Cl	1223.3	1222.5	OLi_2	O	429.0	434.4
	Mn	1947.3	1948.9		Li	155.9	157.3
	H	150.2	153.4	SO_2	S	1154.9	1151.7
NO	N	391.3	388.9		O	530.3	531.0
	O	452.1	451.5	HBO	H	90.3	90.6
MgH_2	Mg	720.0	751.2		B	272.6	272.4
AlH_3	Al	807.1	804.7		O	438.1	439.3
ZnF_2	Zn	2618.3	2616.6	HOB	H	119.2	116.8
	F	655.6	657.8		O	441.9	442.5
FeF_3	Fe	2186.5	2180.2		B	265.2	265.3
	F	658.8	652.2	SiH_4	Si	899.4	899.9
NaH	Na	634.0	627.2	SiF	Si	1091.3	1088.8
ClF	Cl	1194.9	1193.5	SiO_4^{4-}	Si	1074.8	1068.2
	F	569.1	569.0	SiF_6^{2-}	Si	1169.5	1169.0
BeF_2	Be	268.2	266.4	PF_3	P	1122.5	1119.6
	F	533.4	528.2	PF_5	P	1233.5	1227.7
BF_2	B	332.0	330.1	PF_6^-	P	1287.9	1281.0
	F	544.7	544.9	PF_3O	P	1179.6	1175.2
CF_2	C	391.0	387.8	PO_4^{3-}	P	1178.7	1170.9
	F	555.4	554.9	PO_3F^{2-}	P	1175.9	1169.5
NF_2	N	449.4	445.3	$PO_2F_2^-$	P	1177.5	1172.3
	F	558.1	557.8	PF_4^+	P	1188.5	1183.2

According to the early suggestion of Ramsey [22] $\sigma^d_{av}(A)$ can be expressed in a transparent form if the intramolecular charge transfer is neglected and only a monopole of the electron density of atoms $B \neq A$ is taken into account:

$$\sigma^d_{av}(A) = \sigma^d_{av}(FA) + (e^2/3mc^2) \sum_B{}' Z_B/R_{AB} \tag{3}$$

where Z_B stands actually for a number of electrons of the neutral atom B. The first term $\sigma^d_{av}(FA)$ corresponds to the value of free atom A. Obviously, Eq. (3) is developed within the independent atom model although without explicit reference to it. We give some representative results in Table 1. Perusal of the numerical data reveals a surprisingly good performance of the IAM model. Information is usually at the semiquantitative level of quality in neutral molecules. The largest discrepancies against ab initio results are 5–6 ppm, being very small percentagewise (~1%). The studied molecules cover widely different bonding situations. In order to test the model on charged species a number of anions are considered. It is assumed that the central atom is electroneutral and that the charge is evenly distributed over ligands. This is applied to highly symmetric species like SiO_4^{4-} (tetrahedron) and SiF_6^{2-} (octahedron) possesing equivalent ligands as well as to less symmetric compounds like PO_3F^{2-} and $PO_2F_2^-$ involving different peripheral atoms. Thus, fluorine and each oxygen atom in PO_3F^{2-} carry an effective charge of $(-1/2)\,|e|$, in $PO_2F_2^{2-}$ they have a charge of $(-1/4)\,|e|$ etc. This is a very crude picture which represents the simplest generalization of the independent atom approach. Nevertheless, the results are not unreasonable although the errors are naturally larger than in closed shell neutral molecules. These results illustrate rather nicely that the diamagnetic shielding is extremely insensitive to the finer details of the electronic charge distributions in molecules. The IAM results will be compared with some MAM values later in Sect. 2.3.

2.1.2 Diamagnetic Susceptibility of Molecules

The temperature independent part of the magnetic susceptibility has two terms:

$$\chi_{aa} = \chi^d_{aa} + \chi^p_{aa} \tag{4}$$

where the first part designated by χ^d_{aa} is the Langevin's diamagnetic term and χ^p_{aa} is the Van Vleck's paramagnetic contribution [23]. Langevin's term written more specifically reads:

$$\chi^d_{aa} = -K[\langle b^2\rangle_e + \langle c^2\rangle_e] \tag{5}$$

where a, b and c stand for the inertial coordinates. The abbreviation K denotes $Ne^2/4mc^2$ where constants have their usual meaning. It appears that the diagonal elements of the diamagnetic susceptibility tensor are given by the corresponding second moments $\langle b^2\rangle_e$ and $\langle c^2\rangle_e$ of the electronic charge distribution. We have shown that the second moment is easily expressed in terms of two contribution within the one-electron MO-LCAO picture and IAM model [18–20]:

$$\langle r_b^2\rangle \cong \sum_A Z_A R_{Ab}^2 + \sum_p n_p k_p \tag{6}$$

where R_{Ab} denotes the b-th coordinate of atom A measured from the center of mass. The first contribution represents the monopole term of the electronic densities. It obviously dominates in Eq. (6). The second contributions is isotropic and relatively small. It describes the spatial extension of the atomic orbitals. Remarkably enough, k_p constants are the same for all atoms belonging to the same row of the Mendeleev system of elements. They correspond very closely to the ab initio values of (1/3) $\langle 0| r^2 |0\rangle$ for free atoms averaged over the period in question. Hence the approximate Eq. (6) can be given in the parameter-free form. Of course, one could consider a more flexible formula:

$$\langle r_b^2 \rangle \cong \sum_A Z_A R_{Ab}^2 + \sum_A n_A k_A \tag{7}$$

where each atom possesses its own empirically adjustable parameter. Extensive calculations have shown that Eq. (7) yields only a marginal improvement of the Eq. (6) [24]. Comparison of empirical constants k_p and k_A is given in Table 2. The striking feature of both Eqs. (6) and (7) is that the second moment is determined solely by the equilibrium distances and electron densities of neutral atoms. In what follows, Eq. (6) will be employed, if it is not otherwise stated.

We commence discussion of results obtained by the IAM model by comparing the estimated diamagnetic susceptibilities in hydrocarbons [25] with ab initio and/or experimental data (Table 3). The IAM is very successful in providing reliable second moments and χ^d values as evidenced by excellent accordance between the corresponding numbers. However, hydrocarbons do not offer a critical test of the IAM model because of their very low polarity. Therefore, we checked the model against the more accurate ab initio results and observed values in some molecules involving highly electronegative atoms like fluorine and oxygen. Discrepancies (Table 4) are somewhat increased but the overall agreement is still very good. Performance of the IAM model is so high that it can easily detect errors in ab initio

Table 2. Empirical k_p and k_A constants describing isotropic part of the second moments

Atom	H	B	C	N	O	F
Parametrization						
Original IAM	0.2	1.0	1.0	1.0	1.0	1.0
Revised IAM	0.1[a]; 0.2[b]	1.5	1.0	1.1	1.3	1.1

Atom	Si	P	S	Cl	Br	I
Parametrization						
Original IAM	2.5	2.5	2.5	2.5	3.5	5.5
Revised IAM	3.3	2.9	2.7	2.5	3.7	5.7

[a] For H-atom attached to the sp^3 center
[b] For H-atom bound to the sp^2 center

Table 3. Second moments of the electron charge density distribution and diamagnetic susceptibilities in hydrocarbons estimated by the IAM model, ab initio methods and experimental measurements

Molecule	Second moments of charge (in 10^{-16} cm^2)			Diamagnetic susceptibilities (in 10^{-6} cm^3/mole)	
	Axis	Calcd.	Exp. or ab initio	Calcd.	Exp. or ab initio
Methane	*a*	3.28	3.28*	− 27.8	− 27.8*
Ethane	*a*	18.0	18.07*	− 53.5	− 52.60*
	b	6.3	6.20	− 103.1	− 103.00
	c	6.3	6.20	− 103.1	− 103.00
Propane	*a*	57.6	57.9*	− 95.9	− 94.3
	b	13.9	13.9	− 281.3	− 280.8
	c	8.7	8.3	− 303.3	− 304.7
Ethylene	*a*	14.1	13.819*	− 37.3	− 39.64*
	b	6.1	5.996	− 71.3	− 72.83
	c	2.7	3.348	− 85.7	− 84.06
Methylacetylene	*a*	39.4	38.72 ± 0.3	− 44.1	− 46.1 ± 2.1
	b	5.2	5.44 ± 0.25	− 189.2	− 187.3 ± 2.1
	c	5.2	5.44 ± 0.25	− 189.2	− 187.3 ± 2.1
Propene	*a*	37.5	37.5 ± 0.5	− 72.5	− 75.3 ± 1.3
	b	11.6	11.2 ± 0.5	− 182.4	− 184.8 ± 1.6
	c	5.5	6.1 ± 0.5	− 208.3	− 206.6 ± 1.7
Cyclopropene	*a*	17.5	17.8 ± 0.2	− 80.6	− 82.3 ± 0.3
	b	13.7	13.5 ± 0.2	− 96.7	− 100.4 ± 0.3
	c	5.3	5.8 ± 0.2	− 132.4	− 133.3 ± 0.3
Cyclobutene	*a*	29.1	29.4	− 153.1	− 154.1
	b	27.8	27.8	− 158.7	− 160.8
	c	8.3	8.5	− 241.4	− 242.8
Cyclopropane	*a*	17.2	17.60*	− 111.6	− 112.42*
	b	17.2	17.60	− 111.6	− 112.42
	c	9.1	8.90	− 145.9	− 149.33
Methylene-cyclopropane	*a*	47.9	48.2	− 115.0	− 115.8
	b	18.8	18.8	− 238.4	− 240.6
	c	8.3	8.5	− 283.0	− 284.3
Cyclopentadiene	*a*	42.3	42.3	− 210.8	− 213.0 ± 5.8
	b	42.2	41.8	− 211.3	− 214.8 ± 5.0
	c	7.5	8.4	− 358.5	− 356.9 ± 5.1
Benzene	*a*	60.5	60.1 ± 1.5	− 285.5	− 286 ± 10
	b	60.5	60.1 ± 1.5	− 285.5	− 286 ± 10
	c	6.8	7.7 ± 1.2	− 513.3	− 508 ± 20
Naphthalene	*a*	112.0	111.5*	−1019.5	−1031.3
	b	229.3	229.2	− 521.9	− 532.0
	c	11.0	13.9	−1447.9	−1445.4

* Ab initio results

calculations or experimental measurements. This is illustrated by SOF_2 where experimental $\langle x^2 \rangle$ and $\langle y^2 \rangle$ values [26] are obviously wrong. On the other hand IAM and ab initio results for this molecule are compatible. Trimethyleneoxide (C_3H_6O) is another interesting example which documents the usefulness of the IAM model in recognizing errors. By using the geometric structural parameters given in the theoretical work of the Brouckere et al. [27] we found a good agreement with the ab initio results, but there was a somewhat disturbing discrepancy with ex-

Table 4. Comparison of the second moments in some heteroatomic molecules as obtained by the IAM model, ab initio calculations and experimental data

Molecule	IAM	Ab initio	Exptl.
CS	$\langle x^2\rangle = \langle y^2\rangle = 3.5$	3.9	
	$\langle z^2\rangle = 13.8$	14.6	
H_2S	$\langle x^2\rangle = 2.9$	3.5	
	$\langle y^2\rangle = 4.8$	4.4	
	$\langle z^2\rangle = 4.5$	4.4	
SOF_2	$\langle x^2\rangle = 13.2$	11.9	6.9
	$\langle y^2\rangle = 27.5$	28.4	36.6
	$\langle z^2\rangle = 29.2$	29.6	28.9
SO_2F_2	$\langle x^2\rangle = 31.1$	32.3	31.4
	$\langle y^2\rangle = 29.8$	30.0	30.5
	$\langle z^2\rangle = 31.0$	29.2	30.7
	$\langle x^2\rangle = 10.0$	9.8	10.0
	$\langle y^2\rangle = 29.2$	27.8	29.2
	$\langle z^2\rangle = 27.3$	28.1	28.3
	$\langle x^2\rangle = 12.2$	12.5	
	$\langle y^2\rangle = 46.2$	46.9	
	$\langle z^2\rangle = 35.0$	34.7	
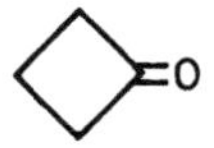	$\langle x^2\rangle = 11.1$	10.8	10.8 ± 1.8
	$\langle y^2\rangle = 33.1$	33.0	32.8 ± 1.8
	$\langle z^2\rangle = 61.5$	63.2	63.2 ± 1.8
	$\langle x^2\rangle = 7.8$	8.6	8.2 ± 2.2
	$\langle y^2\rangle = 42.6$	42.5	42.3 ± 2.2
	$\langle z^2\rangle = 42.5$	42.1	41.8 ± 2.2
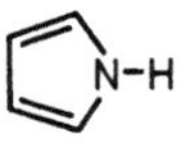	$\langle x^2\rangle = 6.0$	7.2	7.4 ± 0.6
	$\langle y^2\rangle = 39.7$	39.2	38.6 ± 0.6
	$\langle z^2\rangle = 40.2$	39.2	39.1 ± 0.6
	$\langle x^2\rangle = 5.8$	6.7	6.8 ± 0.7
	$\langle y^2\rangle = 38.4$	37.4	37.8 ± 0.7
	$\langle z^2\rangle = 35.8$	36.2	36.2 ± 0.7
	$\langle x^2\rangle = 7.3$	8.4	8.5 ± 1.2
	$\langle y^2\rangle = 41.7$	41.2	44.6 ± 1.2
	$\langle z^2\rangle = 58.7$	58.7	58.6 ± 1.2
	$\langle x^2\rangle = 10.9$	11.6	
	$\langle y^2\rangle = 53.9$	53.2	
	$\langle z^2\rangle = 63.7$	64.3	

In 10^{-10} cm^2. All values should be multiplied by -1

periment [26] for the $\langle y^2\rangle$ component. Careful examination of data revealed that de Brouckere et al. [27] did not quite correctly employ the experimental geometry. The IAM model gave a result with excellent accordance with the measured value for the $\langle y^2\rangle$ second moment (Table 4) if appropriate experimental structure was utilized. Similar simple calculations have shown that measured second moments of FClCO and PF_3 were erroneous [28]. Additional experimental investigation of the former molecule has led to corrected values which are in good agreement with the IAM results [29]. In this connection it should be pointed out that the diagonal elements of the molecular g-tensor are experimentally determined up to the sign and that the IAM second moments can help in choosing the proper sign.

The ab initio diamagnetic susceptibilities in diatomics H_2, HF, HCl, F_2, ClF, BrF, HBr, HI, IF, Cl_2, BrCl, Br_2, IBr and I_2 [30] were an additional challenge for the IAM model. The approximate χ^d values are in very good accordance with rigorous ab initio results over the large range of data (Fig. 1).

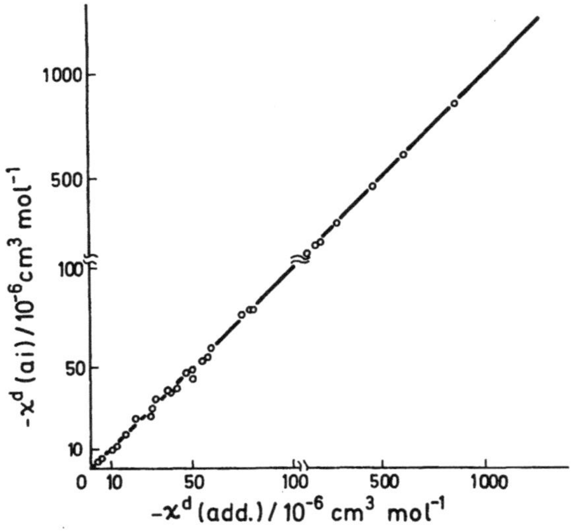

Fig. 1. Diamagnetic susceptibility in some heterodiatomics as calculated by the IAM model and ab initio DZ method

Recently, considerable research interest has been focused on the structure and properties of van der Waals complexes. Unfortunately, the experimental data for the second moments and diamagnetic susceptibilities are scarce. The available measured values for some simple vdW systems are compared with the IAM results (Table 5). Both sets of data are in fine agreement indicating that the IAM model might prove very useful in predicting second moments and χ^d values in vdW aggregates.

2.1.3 Pascal Rules

Pascal observed as early as 1910 that the temperature independent molecular magnetic susceptibility exhibits additivity rules which can be expressed by a simple mathematical formula [31]:

$$\chi_{av} = \sum_{A} \lambda_A + \sum_{S} \lambda_S \tag{8}$$

where λ_A are relatively large atomic contributions whilst λ_S are structural corrections necessary for molecules possessing specific structural features. Since Pascal's empirical increments λ are capable of reproducing a plethora of experimental data

Table 5. Comparison between IAM estimated second moments and diamagnetic susceptibilities and available experimental data or ab initio results in some simple van der Waals complexes

Molecule	Second moments		Diamagnetic susceptibilities	
	Calc.	Exp.	Calc.	Exp.
OC ... HF	$\langle x^2 \rangle = 3.2$	3.16	$\chi_\parallel^d = 27.2$	26.81
	$\langle z^2 \rangle = 84.2$	85.26	$\chi_\perp^d = 370.8$	375.10
	$\langle r^2 \rangle = 90.6$	91.58		
OC ... HCl	$\langle x^2 \rangle = 4.7$	4.84	$\chi_\parallel^d = 39.9$	41.06
	$\langle z^2 \rangle = 154.5$	254.65	$\chi_\perp^d = 675.4$	676.59
	$\langle r^2 \rangle = 163.9$	164.33		
N_2 ... HF	$\langle x^2 \rangle = 3.2$	3.18	$\chi_\parallel^d = 27.2$	26.98
	$\langle z^2 \rangle = 80.8$	81.90	$\chi_\perp^d = 356.3$	360.93
	$\langle r^2 \rangle = 87.2$	88.27		
OCO ... HF	$\langle x^2 \rangle = 4.2$	4.34 ± 0.65	$\chi_\parallel^d = 35.6$	36.8 ± 5.5
	$\langle y^2 \rangle = 132.0$	133.34 ± 0.65	$\chi_\perp^d = 577.8$	584.1 ± 5.5
	$\langle r^2 \rangle = 140.4$	137.7 ± 1.95		
OCO ... HCl	$\langle x^2 \rangle = 5.7$	6.06 ± 0.81	$\chi_\parallel^d = 48.4$	51.4 ± 6.9
	$\langle z^2 \rangle = 231.5$	231.86 ± 0.81	$\chi_\perp^d = 1006.3$	1009.3 ± 6.9
	$\langle r^2 \rangle = 242.9$	—		
SCO ... HF	$\langle x^2 \rangle = 5.7$	6.04 ± 0.8	$\chi_\parallel^d = 48.4$	51.2 ± 6.9
	$\langle z^2 \rangle = 196.6$	197.47 ± 0.8	$\chi_\perp^d = 858.2$	863.3 ± 6.9
	$\langle r^2 \rangle = 208.0$	203.51 ± 2.4		
$\triangleleft$ ... HCl	$\langle x^2 \rangle = 12.0$	11.4[e]	$\chi_{xx}^d = 856.9$	855.7[e]
	$\langle y^2 \rangle = 20.1$	19.8	$\chi_{yy}^d = 822.6$	820.0
	$\langle z^2 \rangle = 181.9$	181.9	$\chi_{zz}^d = 136.2$	132.4

* Second moments in 10^{-16} cm^2 and susceptibilities in 10^{-6} cm^3/mole units. In axially symmetric compounds z-axis coincides with C_∞

[32], they call for theoretical rationalization. Atomic additivity of the diamagnetic part was explained by us resulting (vide supra) in the simple expression:

$$\chi_{cc}^{d} \cong -K\left[\sum_{A} Z_A(R_{Aa}^2 + R_{Ab}^2) + 2\sum_{A} n_A k_A\right] \tag{9}$$

which is based on the approximate Eq. (7). The paramagnetic component χ_{aa}^{p} is much more difficult to tackle. We shall follow Tillieu's formula obtained by the variation approach [33]:

$$\chi_{cc}^{p} = K(\langle a^2\rangle - \langle b^2\rangle)^2/(\langle a^2\rangle + \langle b^2\rangle) \tag{10}$$

Let us consider for simplicity a linear molecule and assume that the symmetry axis coincides with the a-axis. Employing Eq. (6) one obtains:

$$\chi_{cc}^{p} = K(\mathrm{Am}(a^2) + \varepsilon')^2/(\mathrm{Am}(a^2) + \varepsilon) \tag{11}$$

where $Am(a^2)$ denotes the atomic electron density contribution in the monopole approximation assuming that all atoms are neutral $Am(a^2) = \sum_A Z_A R_{Aa}^2$ and small entities ε' and ε read:

$$\varepsilon' = \sum_A \sum_{\mu}^{A} n_\mu \langle \phi_\mu | a_A^2 - b_A^2 | \phi_\mu \rangle \tag{12}$$

and

$$\varepsilon = \sum_A \sum_{\mu}^{A} n_\mu \langle \phi_\mu | a_A^2 + b_A^2 | \phi_\mu \rangle \tag{13}$$

where n_μ are atomic orbital populations and a_A measures electron coordinate to the nucleus A. Spherical symmetry of atoms tacitly supposed in the IAM model implies that ε' vanishes and ε assumes the form:

$$\varepsilon = 2 \sum_A \sum_{\mu}^{A} n_\mu \langle \phi_\mu | a_A^2 | \phi_\mu \rangle \tag{14}$$

Since the atomic term $Am(a^2)$ is much larger than ε, one can develop $(Am(a^2) + \varepsilon)^{-1}$ in a Taylor series and terminate expansion after the first term linear in ε. Then one obtains:

$$\chi_{cc}^{p} = K[Am(a^2) - \varepsilon] \tag{15}$$

Taking into account that $\chi_{cc}^{d} = -K[Am(a^2) + \varepsilon]$ the total magnetic susceptibility is approximately given by the simple formula:

$$\chi_{cc} = \chi_{cc}^{d} + \chi_{cc}^{p} = -2K\varepsilon \tag{16}$$

or in some more detail:

$$\chi_{cc} = -4K \sum_A \left[\sum_{\mu}^{A} n_\mu \langle \phi_\mu | a_A^2 | \phi_\mu \rangle \right] \tag{17}$$

This highly approximate formula reflects several qualitative but salient properties of the magnetic susceptibility. In the first place, elements of the χ-tensor are origin-independent as it should be. Secondly, the negative sign shows that $|\chi^p| < |\chi^d|$ which is generally the case [33]. Finally, magnetic susceptibility of molecules is really, albeit approximately, given by a sum of atomic terms. Let us in accordance with the IAM model assume that

$$\sum_{\mu}^{A} n_\mu \langle \phi_\mu | a_A^2 | \phi_\mu \rangle = 1/3 \langle 0 | r_A^2 | 0 \rangle .$$

Then the formula in Eq. (17) takes the form:

$$\chi_{av} = C \sum_A \chi_A(FA) \tag{18}$$

Table 6. Comparison between Pascal constants and scaled free-atom diamagnetic susceptibilities derived by the IAM approach (in 10^{-6} cm^3 mole^{-1})

Atom	Pascal Increment[a]	Scaled χ (FA)[b]	Pople[c]
H	−2	−1.8	−2.4
He	−2.02[d]	−1.5	−1.6
B	−7.3	−9.7	
C	−7.36	−8.4	−2.7
N	−9.0	−7.4	−1.5
O	−7.95[e]; −6.4[f]	−6.9	−1.6
F	−6.3	−6.2	−3.0
Ne	−6.96[d]	−5.7	−5.6

[a] Taken from Refs. [31, 32]; [b] based on ab initio atomic χ (FA) values [34]; [c] Ref. [35]; [d] measured values; [e] for carbonyl group for both oxygens in acids and esters; [f] in aldehydes and ketones

which shows that Pacal's atomic empirical constant is essentially proportional to the free-atom magnetic susceptibility. Here is tacitly taken into account that atomic paramagnetic term vanishes because of the spherical symmetry. Pascal of course introduced structural corrections λ_S (8) in order to allow for various bonding effects. The qualitative derivation of Pascal additivity rules presented above gives the value of 5/3 for the C constant which is too high. Empirical adjustement in conjunction with atomic Hartree-Fock magnetic susceptibilities [34] yields C = 0.8 for H, He and the first row atoms. Comparison with empirical Pascal constants is presented in Table 6. Agreement is satisfactory in view of the highly approximate nature of the IAM model and the single adjustable weighting factor C. It should be mentioned, however, that IAM values are closer to Pascal constants than the results of Pople's approximate GIAO—MO model [35]. This approach gave too low absolute values which were in addition very unsystematic (Table 6).

2.2 The Effect of Molecular Vibrations

The ground state molecular geometries based on the idea of the Born-Oppenheimer potential surface offer a crude static picture of molecules. Nuclei of atoms execute vibrational and rotational motions thus affecting all molecular properties. These effects are relatively small but significant [36, 37]. It is therefore of some interest to examine how σ^d and χ^d properties vary with the change of geometry. Both entities in the IAM model explicitly depend on interatomic distances and atomic coordinates, respectively. Hence the influence of structural changes is readily estimated. Variation of second moments in H—Cl, Li—H and H—H ... Ar vdW complex following increase in interatomic distances is illustrated by Table 7. It is supposed that the symmetry axis lies in the z-axis. Hence $\langle x^2 \rangle$ corresponds to one of the two perpendicular components and it is constant in the simple IAM picture. The ab initio results show that $\langle x^2 \rangle$ second moments vary extremly slow with the change of the interatomic distance. Increase in $\langle z^2 \rangle$ and/or $\langle r^2 \rangle$ second moments upon pulling atoms apart are well described by the IAM model (Table 6) although the agreement with rigorous calculations is by no means quantitative. However, good

Table 7. Dependance of second moments on intermolecular distance(s)[a]

Molecular System	R (Å)	IAM		Ab initio	
		$\langle x^2 \rangle$	$\langle z^2 \rangle$	$\langle x^2 \rangle$	$\langle z^2 \rangle$
H—H ... Ar[b]	2.275	2.9	12.8	2.78	12.90
	2.805	2.9	17.8	2.79	17.67
	3.545	2.9	26.5	2.79	26.30
	3.863	2.9	30.8	2.79	30.64
	4.392	2.9	38.9	2.79	38.71
	4.921	2.9	48.1	2.79	47.84
	5.450	2.9	58.2	2.79	58.01
H—Cl[c]	1.058	2.7	3.8	2.85	3.33
	1.186	2.7	4.1	2.88	3.50
	1.294	2.7	4.4	2.90	3.69
	1.376	2.7	4.6	2.92	3.88
	1.588	2.7	5.2	2.95	4.32
	1.905	2.7	6.3	2.98	5.09
	2.223	2.7	7.6	2.99	6.00
Li—H[d]		$\langle r^2 \rangle$		$\langle r^2 \rangle$	
	0.847	12.5		11.01	
	1.111	17.1		16.17	
	1.376	23.0		22.59	
	1.508	26.5		26.30	
	1.640	30.2		30.35	
	1.905	38.7		39.53	
	2.302	53.7		55.98	
	2.566	65.3		68.73	

[a] In A^2 units;
[b] Ab initio results of Arroyo ST, Garcia JE, Del Valle FJO, Requena A (1986) J. Mol. Structure (Theochem) 136: 99;
[c] Origin at Cl — ab initio results of Petke JD, Whitten JL (1972) J. Chem. Phys. 56: 830;
[d] Origin at Li — ab initio results of McLean AD (1963) J. Chem. Phys. 39: 2653

Table 8. Geometric dependence of σ^d (in ppm) and χ^d (in 10^{-6} cm^3 mol^{-6}) in some small molecules as estimated by the IAM model and ab initio calculations

Molecular distance R (Å)	Property	IAM	Ab initio	Property	IAM	Ab initio
CO						
1.004	σ^d(C)	335.8	335.9	χ^d(C)	−40.0	−41.65[a]
	σ^d(O)	451.1	451.3	χ^d(O)	−34.1	−36.14
1.128	σ^d(C)	327.6	326.5	χ^d(C)	−45.6	−48.26
	σ^d(O)	445.0	444.9	χ^d(O)	−38.6	−40.70
1.252	σ^d(C)	321.0	318.9	χ^d(C)	−52.4	−55.77
	σ^d(O)	440.0	439.8	χ^d(O)	−43.6	−45.56
1.376	σ^d(C)	315.6	312.7	χ^d(C)	−59.8	−64.00
	σ^d(O)	434.0	435.9	χ^d(O)	−49.1	−50.75
HF						
0.864	σ^d(F)	481.9	482.2	χ^d(F)	−12.3	−10.76[b]

Table 8 (continued)

Molecular distance R (Å)	Property	IAM	Ab initio	Property	IAM	Ab initio
0,917	$\sigma^d(F)$	481.2	481.6	$\chi^d(F)$	−12.6	−10.98
0.970	$\sigma^d(F)$	480.7	481.0	$\chi^d(F)$	−12.8	−11.19
SiF_4						
1.606	$\sigma^d(Si)$	1084.5	1081.4[c]			
1.556	$\sigma^d(Si)$	1091.3	1088.8			
1.506	$\sigma^d(Si)$	1098.5	1096.7			

[a] R. M. Stevens and M. Karplus, J. Chem. Phys., 49 (1968) 1094. [b] R. M. Stevens and W. N. Lipscomb, J. Chem. Phys., 41 (1964) 184. [c] J. A. Tossell and P. Lazzeretti, J. Chem. Phys., 84 (1986) 369

performance of Eqs. (3) and (6) in a qualitative sense is substantiated by additional evidence (Table 8), where CO, HF and SiF_4 molecules are cosidered. In spite of the fact that atoms in question exhibit widely different electronegativities the IAM gives reasonable estimates of σ^d and χ^d diamagnetic properties. This is gratifying because the IAM model offers a simple and transparent interpretation of these two molecular properties which is apparently qualitatively correct. The range of applicability of this model, which is not expected to hold in e.g. alkali halides, will be discussed later (vide infra).

If a comparison of the calculated molecular properties with experiment is desired, then one has to take their average over molecular vibrations and rotations. In diatomics this is easily achieved by making use of Buckingham's formula [38, 39]:

$$P_{v,J} = P_e + \frac{B_e}{\omega_e}\left[3\left(1 + \frac{\alpha_e\omega_e}{6B_e^2}\right)\left(\frac{dP}{d\Delta}\right)_e + \left(\frac{d^2P}{d\Delta^2}\right)_e\right](v + 1/2) + \frac{4B_e^2}{\omega_e^2}\left(\frac{dP}{d\Delta}\right)_e J(J + 1) \tag{19}$$

where P_e is a property at the equilibrium distance, v and J are vibrational and rotational quantum numbers and the spectroscopic constants have their usual meaning. Stretching and squeezing of the bond distance is denoted by $\Delta = (R - R_e)/R_e$. We shall consider here some typical examples. Good-quality MCSCF results are available for HF molecule [39]. Their nuclear motion corrections to the average diamagnetic susceptibility are given by the formula (in 10^{-6} cm^3 mol^{-1}):

$$(\chi^d_{av})_v = (\chi^d_{av})_e - 0.145(v + 1/2) - 0.0004J(J + 1) \tag{20}$$

where $(\chi^d_{av})_e$ refers to the equilibrium value with the gauge origin at the centre of mass of the molecule. Derivatives of the χ^d_{av} are easily obtained if the IAM formula, Eq. (6) is employed. By substituting the corresponding spectroscopic constants [40] one obtains:

$$(\chi^d_{av})_v = (\chi^d_{av})_e - 0.172(v + 1/2) - 0.0005J(J + 1) \tag{21}$$

which is in susprisingly good agreement with the ab initio Eq. (20). Both formulas indicate that rotational influence is small and can be safely neglected. In order to explore the capability of the IAM model in reproducing vibrational effects in homonuclear diatomics N_2 and H_2 systems will be examined. The ab initio equations for the parallel and perpendicular components of the diamagnetic susceptibility in N_2 read as follows [41]:

$$(\chi^d_{\parallel})_v = (\chi^d_{\parallel})_e - 0.050(v + 1/2) \tag{22}$$

and

$$(\chi^d_{\perp})_v = (\chi^d_{\perp})_e - 0.269(v + 1/2) \tag{23}$$

The IAM model yields

$$(\chi^d_{\perp})_v = (\chi^d_{\perp})_e - 0.252(v + 1/2) \tag{24}$$

which is in good accordance with the preceding ab initio Eq. (23). On the other hand the IAM model predicts that $(\chi^d_{\parallel})_v = (\chi^d_{\parallel})_e$ i.e. it is constant and independent of stretching vibrations. This is not true but the ab initio coefficient (—0.05) is substantially smaller for the parallel component as expected from the IAM picture. This is concomitant with our earlier discussion of second moments and their variation with interatomic distance (Table 6).

Vibrational dependence of the diamagnetic shielding of nitrogen in N_2 molecule is given by ab initio calculations [41]:

$$(\sigma^d_{av}(N))_v = 384.4 - 0.468(v + 1/2) \tag{25}$$

Its IAM counterpart takes the form:

$$(\sigma^d_{av}(N))_v = 385.8 - 0.033(v + 1/2) \tag{26}$$

Here, the sign of the vibrational correction is predicted correctly but the coefficient is smaller by an order of magnitude.

Finally, we consider the $\langle r^2 \rangle$ second moment in H_2 molecule where a highly accurate wavefunction is known [42]. The IAM model gives the following approximate formula:

$$(\langle r^2 \rangle_{cm})_v = (\langle r^2 \rangle_{cm})_e + 0.0805R_e^2(v + 1/2) \tag{27}$$

We shall compare results for the first four vibrational states (v = 1–4) with the ab initio values of Wolniewicz [42]. They read correspondingly (in Å^2): 0.047(0.037), 0.095(0.076), 0.142(0.119) and 0.189(0.164) where the ab initio data are given in parentheses. It should be pointed out that vibrational corrections are calculated relative to the zero point vibration contribution. It appears that the IAM model yields reasonable values.

As a corollary we can say that the IAM model gives a good qualitative description of changes in σ^d and χ^d properties induced by vibration of the nuclei. More than that cannot be expected because vibrational effects are small and the IAM model accounts only for gross molecular features.

2.3 The IAM Model as a Treshold in Qualitative Appraisal of Approximate Wavefunctions

Description of the electronic structure in larger molecular systems is necessarily approximate. Semiempirical or simplified ab initio methods should be efficient and feasible and capable of reproducing the main features of the electronic charge distribution. This is obviously not an easy task. Equally difficult is a choice of suitable criteria for quantitative apprasial of the calculated approximate wavefunctions. The most important criterion is the total energy. However, the latter is not a very stringent test because the total energy is at minimum. Thus the variation in wavefunctions of the first order affects the energy only in the second order. Substantially better probes are provided by some one- and two-electron properties [43]. They "measure" electronic charge density in various segments of the molecular volume depending on the functional form of the corresponding operator. Consequently, testing of approximate methods and wavefunctions requires calculations of a large number of molecular properties. We shall concentrate on the properties which are one-electron in nature because they as a rule do not depend strongly on electron correlation [44], the dipole moment being a notable exception [45]. The point is that we would like to find molecular properties which would provide a test for the lower threshold of quality thus enabling detection of approximations in electronic structure calculations which are apparently too crude and inappropriate. The methods to be examined are obviously simple (if not oversimplified) and do not involve treatments for correlation of electrons. Common sense tells us that properties related to the lower treshold in quality should be sought at the IAM level. For if some properties are well reproduced by the independent neutral and spherical atoms placed at the equilibrium positions and less satisfactorily calculated by some approximate (semiempirical) methods, then something is seriously wrong with the latter. Obvious candidates for this type of the riddle properties are diamagnetic shielding and susceptibility. As we shall see shortly, they indeed represent necessary but not sufficient conditions which approximate methods should satisfy [46].

Surprisingly good results in estimating $\langle r^{-1} \rangle$ and $\langle r^2 \rangle$ properties by the IAM model indicate that the corresponding operators are extremely insensitive to finer details of the electronic density distributions. It is therefore not unreasonable to assume that if a particular theoretical procedure yields e.g. σ^d_{av} values at great variance with the Hartree-Fock method, then its foundations should be carefully reexamined. To support this conclusion we present results of the calculation of free-atom diamagnetic shieldings by the statistical Thomas-Fermi-Dirac [47] approach. This method is known to be inadequate for a satisfactory description of atoms and particularly of molecules. The TFD results [48] are compared with ab initio data [34, 49] in Table 9. It appears that TFD estimates of σ^d_{av} values are

Table 9. Comparison of the Hartree-Fock and Thomas-Fermi-Dirac diamagnetic shieldings of free atoms (in ppm)

Z	Atom	HF	TFD	Z	Atom	HF	TFD
2	He	60	89.4	26	Fe	2053	2210
3	Li	102	143.1	27	Co	2166	2321
4	Be	150	205.6	28	Ni	2282	2433
5	B	202	272.1	29	Cu	2405	2546
6	C	261	341.6	30	Zn	2522	2661
7	N	326	413.8	31	Ga	2639	2776
8	O	395	488.8	32	Ge	2757	2892
9	F	471	566.5	33	As	2877	3010
10	Ne	552	646.9	34	Se	2998	3128
11	Na	629	729.7	35	Br	3121	3247
12	Mg	709	815.0	36	Kr	3246	3367
13	Al	790	902.7	37	Rb	3370	3489
14	Si	874	992.6	40	Zr	3740	3857
15	P	961	1085	45	Rh	4400	4488
16	S	1050	1179	50	Sn	5090	5140
17	Cl	1143	1275	55	Cs	5780	5810
18	Ar	1238	1372	60	Nd	6510	6499
19	K	1329	1472	65	Tb	7280	7207
20	Ca	1423	1573	70	Yb	8090	7933
21	Sc	1521	1676	80	Hg	9730	9441
22	Ti	1623	1780	85	At	10560	10223
23	V	1727	1885	90	Th	11400	11025
24	Cr	1837	1992	95	Am	12280	11847
25	Mn	1942	2101	100	Fm	13180	12688

clearly unsatisfactory as anticipated. This means that TFD theory does not satisfy the σ^d test and at the same time that σ^d criterion is a useful means for detecting intrinsic drawbacks of the theoretical approach in question.

In theoretical schemes based on the LCAO approximation σ^d criterion will give some information about the adequacy of the basis set employed, because the choice of the basis functions actually parametrizes the molecular hamiltonian. This plausible assumption, which will be substantiated later (vide infra), is based on an observation that single center terms in the total molecular energy expression, are dominant contributions. On the other hand the nuclear-electron attraction $\langle r^{-1} \rangle$ is apparently important in monocentric energy terms. In order to illustrate the use of σ^d and $\langle r^2 \rangle$ criteria in judging the quality of the basis set we shall consider Gaussian (GTO) and Hermite-Gaussian (HG) expansion of hydrogenic orbitals. For this purpose some mathematical preliminaries are necessary.

HG-functions were introduced by us to circumvent some difficulties in the calculation of molecular integrals over Cartesian Gaussians [50]. They have some advantageous properties particularly for atomic orbitals with higher angular quantum numbers and should be convenient in calculations performed in momentum space. Basic integrals and the matrix elements for many molecular properties can be expressed in closed forms [51–53]. HG-functions are defined as:

$$f_{n_1n_2n_3}(a_A, \vec{r}_A) = a_A^{n/2} H_{n_1}(a_A^{1/2} x_A)\, H_{n_2}(a_A^{1/2} y_A)\, H_{n_3}(a_A^{1/2} z_A) \exp(-a_A r_A^2) \quad (28)$$

where coordinates $x_A = x - A_x$ determine the electron position relative to a nucleus A, $n = n_1 + n_2 + n_3$ and H_{n_i} denotes Hermite polynomials. An alternative and more useful way of expressing HG-functions reads:

$$f_{n_1 n_2 n_3}(a_A, \vec{r}_A) = \partial_{A_x}\partial_{A_y}\partial_{A_z} \exp(-a_A r_A^2) \tag{29}$$

This form, involving derivatives over nuclear coordinates, immediately suggests a great simplification in the calculation of molecular integrals and indicates that the geometry optimization and computation of force constants can be naturally incorporated in the general algorithm.

The hydrogenic functions were chosen as the test orbitals in comparing HG and GTO basis sets, because they are exact solutions of the Schrödinger equation. They can be expressed in general but real form:

$$\Psi_{KLM} = x^K y^L z^M R(r) \tag{30}$$

It is convenient to develop R(r) in terms of spherical Gaussians multiplied by even powers of r:

$$R(r) = \sum_{i=1}^{N} \left(\sum_{j=0}^{T_i} k_{ji} r^{2j} \right) \exp(-a_i r^2) \tag{31}$$

where the limiting indices N and T_i depend on the tolerance in error. An equivalent formula based on HG functions reads:

$$R(r) = \sum_{i=1}^{N} \sum_{k=0}^{T_i} c_{ki} k! \sum_{k_1=0}^{k} \sum_{k_2=0}^{k-k_1} [k_1!\, k_2!(k - k_1 - k_2)!]^{-1} \times f_{2k_1, 2k_2, 2(k-k_1-k_2)}(a_i, \vec{r}) \tag{32}$$

Multiplying R(r) by $x^K y^L z^M$ one obtains:

$$\Phi_{KLM} = \sum_{i=1}^{N} \sum_{k=0}^{T_i} c_{ki}[k!/k_1!\, k_2!\, k_3!]\, f_{2k_1+K, 2k_2+L, 2k_3+M}(a_i, \vec{r}) \tag{33}$$

where Φ stands for approximate description of the corresponding hydrogenic orbital Ψ and summation over k implies a triple sum over k_1, k_2 and k_3 confined to give a fixed value of $k = k_1 + k_2 + k_3$ ($k = 0, 1, 2, \ldots$ etc). The numerical experiments discussed here encompass a range of values of N = 1–5 for two particular cases: (a) $T_i = 0$ and (b) $T_i = 2$. The former corresponds to GTO basis whereas the latter represents HG expansion of the fourth degree which will be denoted as HG-4. The number of nonlinear parameters is determined by N.

Adjustable linear and nonlinear parameters in the series (33) are obtained by the integral least-square fit criterion minimizing Δ:

$$\Delta = \int (\Phi_{KLM} - \Psi_{KLM})^2 \, dv \tag{34}$$

which is equivalent to the maximum overlap requirement between the approximate representation Φ_{KLM} of the hydrogenic orbital and the exact solution Ψ_{KLM}. One can define the similarity measure SM as:

$$SM = 100[1 - (1/2)\Delta] \quad (\text{in } \%) \tag{35}$$

which means that it is given by the overlap integral between the approximate and exact wavefunction. The quality of the GTO-N and HG4-N expansions of hydrogenic orbitals should be judged by estimates of a number of atomic properties [54, 55]. For the present purpose σ^d, $\langle r^2 \rangle$ and SM criteria will suffice. The choice of their lower thresholds is of course arbitrary. We shall set the limits of tolerance in inaccuracy of σ^d and $\langle r^2 \rangle$ properties of 0.2 ppm and 1×10^{-16} cm^3, respectively. It should be pointed out that an error of 0.2 ppm in diamagnetic shielding corresponds to an inaccuracy in the related electrostatic potential $\langle r^{-1} \rangle$ of 7 kcal/mole. The similarity measure SM should be at least 99%. These requirements are mild but could be slightly shifted in both directions without changing the main conclusions. They serve here only for illustrative purposes. We shall make some general remarks before going into a detailed discussion. The most dificult orbitals to tackle are *s*-orbitals. The complexity of their description increases with an increase in the principal quantum number n. The hydrogenic AOs possessing $l \neq 0$ are susceptible to most accurate representation with fewest GTOs or HGOs if the condition $n - l = 1$ is satisfied. If is interesting to observe that high overlapping sometimes as large as 100% does not insure always good accordance for other properties. This should be kept im mind when the maximum overlap results are discussed. Finally, σ^d (or alternatively $\langle r^{-1} \rangle$) is more important criterion than $\langle r^2 \rangle$) for lower AOs whereas the contrary is true for AOs with higher n quantum numbers. This is plausible in view of the spatial extension of these orbitals.

Perusal of the data presented in Table 10 shows that the single Gaussian (N = 1) of the hydrogenic AOs is obviously inappropriate as evidenced by the above mentioned criteria. This is not unexpected because Gaussian AOs do not satisfy neither the nuclear cusp condition nor exibit a correct asymptotic behaviour. Inadequacy of the simple Gaussian representation is clearly reflected in $\langle r^{-1} \rangle$ properties. The *1s* hydrogenic AO should be expanded by at least four Gaussians apparently due to difficulties in accounting for the nuclear cusp. Orbitals *2s* and *3s* require three GTOs whereas *4s* should be represented by four Gaussians — the forth being necessary for a satisfactory value of the $\langle r^2 \rangle$ second moment. Hydrogenic orbitals *2p* and *3p* are well described by two Gaussians, *4p* by three GTOs and *3d* by a simple Gaussian. Larger number of nonlinear parameters is obviously needed for a good description of inner and outer portions of atomic orbitals. The HG basis set is more flexible because of the additional polynomial part of functions. Hence their expansions are expected to be shorter. This is really the case as evidenced by the corresponding $\langle r^{-1} \rangle$ and $\langle r^2 \rangle$ values.

We give within parentheses the number of HG function in expansion of particular hydrogenic AOs: *1s*(2), *2s*(2), *3s*(3), *4s*(2), *2p*(1), *3p*(2), *4p*(2) and *3d*(1). The second moments for higher AOs are substantionally better reproduced by HG functions. This is logical because r^2 and r^4 terms in the HG expansion (31) improve to some extent a behavior of the gaussians at large distances. This brief analysis illustrates

Table 10. Comparison of the $\langle r^{-1} \rangle$ and $\langle r^2 \rangle$ properties and similarity measure SM obtained by GTO and HG4 basis sets

N		$\langle r^{-1} \rangle$		$\langle r^2 \rangle$		SM	
		GTO	HG4	GTO	HG4	GTO	HG4
1s	1	8.30643(−1)	9.27907(−1)	2.76804	2.93778	95.7	99.3
	2	9.59098(−1)	9.92985(−1)	2.97212	2.99810	99.7	100.0
	3	9.89205(−1)	9.99094(−1)	2.99612	2.99985	100.0	100.0
	4	9.96838(−1)	9.99696(−1)	2.99935	3.00001	100.0	100.0
	5	9.98974(−1)	9.99970(−1)	2.99988	2.99999	100.0	100.0
Exact		1.00000	1.00000	3.00000	3.00000	100.0	100.0
2s	1	1.97711(−1)	1.71833(−1)	4.88585(+1)	4.15260(+1)	83.3	91.2
	2	2.30693(−1)	2.43239(−1)	4.19722(+1)	4.20135(+1)	99.6	100.0
	3	2.47141(−1)	2.49568(−1)	4.18288(+1)	4.19984(+1)	99.9	100.0
	4	2.46426(−1)	2.49704(−1)	4.20052(+1)	4.20001(+1)	100.0	100.0
	5	2.49373(−1)	2.49844(−1)	4.19910(+1)	4.20000(+1)	100.0	100.0
Exact		2.50000(−1)	2.50000(−1)	4.20000(+1)	4.20000(+1)	100.0	100.0
3s	1	8.23251(−2)	8.79217(−2)	2.81797(+2)	2.13309(+2)	59.6	93.8
	2	9.24912(−2)	9.42268(−2)	2.17357(+2)	2.06915(+2)	94.6	98.9
	3	1.04796(−1)	1.09131(−1)	2.07110(+2)	2.07024(+2)	99.9	100.0
	4	1.10017(−1)	1.10956(−1)	2.06919(+2)	2.06999(+2)	100.0	100.0
	5	1.10177(−1)	1.11074(−1)	2.06988(+2)	2.07000(+2)	100.0	100.0
Exact		1.11111(−1)	1.11111(−1)	2.07000(+2)	2.07000(+2)	100.0	100.0
4s	1	4.49957(−2)	4.41294(−2)	9.43321(+2)	6.51222(+2)	45.5	91.6
	2	4.71616(−2)	5.31577(−2)	7.73225(+2)	6.48762(+2)	81.1	99.0
	3	5.39137(−2)	6.17168(−2)	6.62257(+2)	6.47505(+2)	97.8	99.9
	4	5.99133(−2)	6.18161(−2)	6.48055(+2)	6.48022(+2)	100.0	100.0
	5	6.21164(−2)	6.24327(−2)	6.47998(+2)	6.48003(+2)	100.0	100.0
Exact		6.25000(−2)	6.25000(−2)	6.48000(+2)	6.48000(+2)	100.0	100.0
2p	1	2.23133(−1)	2.40930(−1)	2.84145(+1)	2.96233(+1)	95.2	99.3
	2	2.45178(−1)	2.49537(−1)	2.98282(+1)	2.89921(+1)	99.7	100.0
	3	2.49096(−1)	2.49980(−1)	2.99799(+1)	2.99993(+1)	100.0	100.0
	4	2.49813(−1)	2.49981(−1)	2.99973(+1)	2.99999(+1)	100.0	100.0
	5	2.49957(−1)	2.49997(−1)	2.99996(+1)	3.00000(+1)	100.0	100.0
Exact		2.50000(−1)	2.50000(−1)	3.00000(+1)	3.00000(+1)	100.0	100.0
3p	1	8.00221(−2)	8.13433(−2)	2.20926(+2)	1.79937(+2)	78.2	89.2
	2	1.06475(−1)	1.10066(−1)	1.80015(+2)	1.80043(+2)	99.5	100.0
	3	1.10711(−1)	1.11085(−1)	1.79372(+2)	1.79994(+2)	99.9	100.0
	4	1.10609(−1)	1.11091(−1)	1.80015(+2)	1.80000(+2)	100.0	100.0
	5	1.11053(−1)	1.11110(−1)	1.80000(+2)	1.80000(+2)	100.0	100.0
Exact		1.11111(−1)	1.11111(−1)	1.80000(+2)	1.80000(+2)	100.0	100.0
4p	1	4.14005(−2)	4.67052(−2)	8.25384(+2)	6.43032(+2)	60.5	90.5
	2	4.77674(−2)	5.64639(−2)	6.52003(+2)	5.99819(+2)	91.0	98.7
	3	6.04153(−2)	6.21617(−2)	6.00450(+2)	6.00048(+2)	99.8	100.0
	4	6.22617(−2)	6.24854(−2)	5.99751(+2)	5.99999(+2)	100.0	100.0
	5	6.23150(−2)	6.24982(−2)	5.99938(+2)	5.99996(+2)	100.0	100.0
Exact		6.25000(−2)	6.25000(−2)	6.00000(+2)	6.00000(+2)	100.0	100.0
3d	1	1.02380(−1)	1.08556(−1)	1.20941(+2)	1.24886(+2)	95.0	99.3
	2	1.09790(−1)	1.11030(−1)	1.25486(+2)	1.25981(+2)	99.7	100.0
	3	1.10910(−1)	1.11110(−1)	1.25947(+2)	1.26000(+2)	100.0	100.0
	4	1.11080(−1)	1.11111(−1)	1.25994(+2)	1.26000(+2)	100.0	100.0
	5	1.11110(−1)	1.11111(−1)	1.25999(+2)	1.26000(+2)	100.0	100.0
Exact		1.11111(−1)	1.11111(−1)	1.26000(+2)	1.26000(+2)	100.0	100.0

rather nicely a fact that $\langle r^{-1}\rangle$ (or σ^d) and $\langle r^2\rangle$ (or χ^d) are useful in determining indices N and T_i in Eq. (31), or in other words, in choosing a satisfactory basis set.

Summarizing results of the discussion so far it is justified to say that the diamagnetic shielding of the nuclei and the Langevin's diamagnetic susceptibility are properties par excellence in determining minimal criteria for acceptable molecular wavefunctions (parametrized by the choice of the basis set and the adopted theoretical scheme). Since these two properties are extremely insensitive they provide necessary but not sufficient conditions which good atomic and molecular wavefunctions should satisfy.

We shall apply qualitative σ^d and χ^d test in appraisal of some semiempirical methods of current interest. The examined schemes encompass two large categories: (a) methods which are based on some variants of the neglect of the differential diatomic overlap (NDDO) approximation and (b) procedures based on some effective hamiltonians fully appreciating overlap of atomic orbitals. Typical representatives of the first class methods are MINDO/3 [56] and MNDO [57]. The EHT [58], SCC—MO (alias IEHT) [59] and iterative maximum overlap (IMO) [60] methods belong to the second class of approximate schemes. Results of the IMO method will be considered in the second volume of this series [61]. In addition to semiempirical methods mentioned above, approach of Ray and Parr [62] is discussed here. They employed a simple two parameter formula of Gadre and Parr [63] for estimating the total molecular energy.

$$E_t = -0.51130(Z_A^{2.40073} + Z_B^{2.40073}) \tag{36}$$

which was extended in obvious way to encompass polyatomic molecules. Incidentally, Eq. (36) is a slight generalization of the expression for the total binding energy of Z electrons in a neutral atom [64]:

$$E_A = -Z_A^{2.4}\ \mathrm{Rydb} \tag{37}$$

Hence the Eq. of Gadre and Parr falls within the framework of the MAM model where atoms are indeed only slightly modified. Finally, it should be mentioned that the MINDO/3 and MNDO methods were used in their original versions. In SCC—MO calculations the basis set of Clementi-Raimondi AOs [65] was utilized.

Since molecules involving hydrogen and first row atoms ($n = 2$) are studied it directly follows that σ^d is a much more important criterion than the Langevin's susceptibility. Hence we shall discuss diamagnetic shieldings estimated by semiempirical methods and compare them with the ab initio results of Snyder and Basch [66]. The latter are close to the double zeta (DZ) quality which is sufficient for our purpose. The use of Snyder-Basch calculations ensures uniformity of results for a large set of gauge molecules.

The semiempirical σ^d values are obtained by the formula (38) which allows for intramolecular charge transfer:

$$\sigma^d_{av}(A) = (e^2/3mc^2)\left[\sum_{\mu}^{A} (\xi_{A\mu} Q^A_{\mu}/n_{A\mu}) + \sum_{B}{}' Q_B/R_{AB}\right] \tag{38}$$

where Q_μ^A and Q_A are gross orbital and gross atomic electron populations, respectively. The principal quantum number of the corresponding AO is denoted by $n_{A\mu}$ whereas $\xi_{A\mu}$ is the screening constant. In the development of Eq. (38) it was tacitly assumed that the monopole approximation for the calculation of the 1/r expectation values holds to a good accuracy. A careful analysis of the DZ ab initio and semiempirical results has conclusively shown that this supposition is justified [67]. Finally, it should be pointed out that the inner shell electrons were treated as highly localized and nonpolarizable cores possessing maximal electron occupancy permitted by the Pauli principle. Since Eq. (38) involves charge drift between atoms of different electronegativities, the underlying physical model is that of modified atoms in molecules to be discussed later (Sect. 3). It is noteworthy that Ray and Parr formula for calculating σ^d shieldings involves the nuclear contribution, a derivative of the total molecular energy over the nuclear charge and the chemical potential given by the negative of the molecular electronegativity:

$$\sigma^d = (e^2/3mc^2)\left[-(\partial E_e/\partial Z_A)_{Z_{B,Q}} + \sum_{B \neq A} (Z_B/R_{AB}) + \mu\right] \tag{39}$$

where E_e denotes the equilibrium total electronic energy.

Table 11. Comparison of the diamagnetic shielding of the nuclei in some diatomic and linear molecules as computed by several semiempirical methods, independent atom model (IAM) and ab initio DZ approach (in ppm)

Molecule	Gauge origin	MINDO/3[a]	MNDO[a]	SCC—MO	IAM	Ray-Parr[b]	Ab initio DZ
H_2	H	35.8 (35.6)	36.3 (37.8)	33.9	30.4	34.5 [29.9]	32.2
N_2	N	392.9 (392.7)	396.7 (396.1)	383.0	385.4	391.3 [387.9]	384.1
F_2	F	484.7 (483.3)	543.7 (550.6)	525.4	530.7	533.4 [526.6]	529.9
HF	H	102.7 (110.9)	104.1 (112.0)	102.9	110.2	114.1 [108.5]	107.9
	F	435.5 (436.0)	497.8 (498.6)	478.3	481.3	483.9 [478.3]	482.2
CO	C	327.3 (326.7)	330.2 (327.6)	324.6	327.3	335.1 [330.2]	326.1
	O	463.9 (463.6)	468.2 (467.3)	444.9	445.0	451.6 [446.8]	445.1
BF	B	262.8 (266.4)	272.3 (271.9)	265.4	269.0	274.8 [270.5]	267.4
	F	463.1 (464.7)	523.8 (523.7)	507.0	503.2	510.8 [506.5]	507.8
C_2H_2	H	102.7 (101.7)	102.6 (103.0)	101.8	93.6	102.8 [98.6]	99.1
	C	323.5 (323.4)	326.4 (326.6)	319.2	320.6	328.4 [324.2]	321.5
HCN	H	105.8 (102.8)	103.2 (103.3)	102.4	100.3	104.6 [100.1]	99.6
	C	327.5 (327.5)	331.2 (330.8)	324.2	326.4	334.3 [329.7]	326.6
	N	385.7 (386.4)	391.7 (391.3)	377.8	378.5	386.2 [381.7]	378.6
CO_2	C	357.4 (382.6)	391.0 (387.7)	386.0	390.1	398.0 [392.7]	386.8
	O	495.8 (494.4)	499.3 (497.3)	475.4	476.0	482.6 [477.3]	476.2
N(1)N(2)O	N(1)	424.4 (425.7)	427.2 (427.0)	412.9	416.3	424.0 [418.5]	414.4
	N(2)	444.3 (444.8)	454.6 (454.3)	442.7	447.2	455.0 [449.5]	443.1
	O	501.2 (501.0)	503.8 (508.6)	479.8	479.1	485.7 [480.2]	479.6
$C(1)(C(2)O)_2$	C(1)	415.2 (413.9)	419.5 (418.9)	404.9	411.0	—	410.3
	C(2)	408.4 (407.1)	412.8 (411.5)	407.0	412.4	—	408.9
	O	517.1 (516.4)	519.7 (518.6)	496.1	497.5	—	497.1

[a] Entries in parentheses refer to optimized geometries.
[b] Entries in square brackets were obtained by taking into account electronegativity correction

Table 12. Comparison of the diamagnetic shielding of the nuclei in some medium size molecules as obtained by several semiempirical methods, independent atom model (IAM) and ab initio DZ approach (in ppm)[a]

Molecule	Gauge origin	MINDO/3[b]	MNDO[b]	SCC—MO	IAM	Ray-Parr[c]	Ab initio DZ
BH_3	H	71.6(71.7)	72.6(74.2)	69.9	66.2	70.2[66.1]	68.6
	B	218.3(218.4)	228.6(229.4)	222.3	225.6	231.5[227.4]	226.4
CH_4	H	90.4 (89.8)	91.0 (90.2)	88.6	85.0	89.2 [84.8]	87.1
	C	296.5 (296.1)	298.6 (298.2)	293.6	295.1	302.8 [298.4]	296.2
H_2O	H	105.3 (105.9)	106.7 (107.7)	104.7	102.4	106.6 [101.5]	102.1
	O	435.4 (435.5)	439.7 (440.0)	415.9	414.7	421.3 [416.1]	416.8
B_2H_6	H_t				94.6	98.6 [94.5]	97.0
	H_b				110.9	114.9 [110.8]	111.2
	B				265.6	271.4 [267.3]	266.0
C_2H_4	H	113.4 (112.2)	113.7 (113.1)	111.6	108.4	112.4 [108.0]	110.0
	C	330.9 (331.1)	333.9 (333.7)	327.5	329.4	336.8 [332.4]	330.1
N_2H_2	H	125.4 (123.6)	125.1 (125.1)	123.1	120.5	124.8 [119.9]	120.9
	N	400.2 (401.3)	405.7 (437.4)	391.9	393.1	400.8 [395.9]	392.9
H_2CO	H	114.8 (114.4)	115.0 (115.1)	112.0	110.1	114.4 [109.6]	112.3
	C	337.0 (338.0)	341.5 (341.0)	336.7	339.7	347.6 [342.8]	338.8
	O	471.9 (472.1)	476.2 (475.7)	453.1	451.0	457.6 [452.9]	452.7
C_2H_6	H	121.2 (115.4)	121.7 (115.9)	119.3	115.7	120.0 [115.6]	118.1
	C	336.9 (338.2)	339.7 (340.0)	334.2	335.7	343.6 [339.1]	337.2
CH_4	H	90.4 (89.8)	91.0 (90.2)	88.6	85.0	89.2 [84.8]	87.2
	C	296.5 (296.1)	298.6 (298.2)	293.6	295.1	302.8 [298.4]	296.7
N_2H_4	H(1)	132.8 (131.4)	132.8 (133.5)	131.1	128.0	132.3 [127.4]	129.0
	H(2)	134.2 (131.3)	134.4 (131.3)	132.3	129.3	133.6 [128.8]	129.9
	N	404.3 (408.9)	410.9 (415.0)	396.2	397.1	404.9 [400.0]	396.5
H_2O_2	H	144.7 (142.0)	146.0 (144.0)	143.7	142.2	146.5 [141.1]	141.4
	O	468.9 (471.5)	484.2 (491.7)	461.3	461.1	467.7 [462.3]	462.0
$CH(1)H(2)_2$—	H(1)	124.7 (133.5)	125.0 (124.9)	121.6	120.3	124.2 [119.5]	122.1
OH(3)	H(2)	124.8 (132.2)	125.2 (123.7)	122.3	119.4	123.4 [118.7]	121.2
	H(3)	139.9 (145.5)	141.1 (141.5)	139.8	136.5	140.7 [136.0]	136.4
	C	343.0 (345.7)	346.6 (347.4)	341.2	344.1	351.7 [347.0]	344.2
	O	466.6 (477.3)	483.6 (484.6)	460.9	457.9	464.6 [459.9]	460.3
CH_3F	H	125.8 (125.7)	126.1 (124.9)	122.8	121.4	125.4 [120.6]	122.8
	C	345.2 (345.3)	349.2 (350.3)	345.2	347.9	355.3 [350.5]	347.2
	F	480.9 (481.3)	520.8 (544.0)	527.0	525.7	528.3 [523.5]	527.7
$H_2C(1)C(2)O$	H	130.5 (129.2)	131.0 (129.7)	128.1	126.2	130.1 [125.6]	126.6
	C(1)	355.1 (354.7)	357.1 (356.3)	346.7	351.8	359.1 [354.6]	352.1
	C(2)	374.6 (373.9)	379.8 (378.0)	373.4	378.0	385.3 [380.7]	375.8
	O	492.2 (491.8)	495.1 (493.9)	472.0	472.5	479.0 [474.5]	473.2
$CH_2N(1)N(2)$	H	133.3 (131.6)	133.2 (131.8)	131.0	128.5	132.5 [127.9]	129.5
	C	359.2 (360.8)	360.5 (360.2)	351.0	355.3	362.7 [358.0]	355.2
	N(1)	436.3 (437.5)	446.4 (445.5)	432.3	436.8	443.8 [439.2]	433.9
	N(2)	422.0 (422.7)	425.3 (424.0)	411.1	414.1	421.1 [416.5]	412.5
$H_2N(1)CN(2)$	H	147.3 (140.3)	146.7 (141.0)	145.0	143.7	147.7 [143.0]	143.0
	C	376.4 (376.5)	380.4 (380.1)	372.4	376.1	383.4 [378.7]	375.2
	N(1)	418.1 (417.5)	427.6 (426.6)	409.7	414.9	421.9 [417.2]	414.1
	N(2)	415.0 (415.3)	419.1 (419.5)	404.8	406.3	413.4 [408.7]	406.3
BH_3CO	H	121.5 (122.3)	123.5 (124.1)	119.5	116.2	120.2 [115.8]	118.6
	B	283.1 (285.7)	296.8 (298.3)	286.7	290.5	296.4 [292.0]	292.1
	C	367.1 (368.5)	369.8 (368.6)	365.9	371.1	378.5 [374.1]	368.2
	O	477.1 (489.7)	491.3 (490.3)	468.9	471.4	478.0 [473.6]	470.6

Table 12 (continued)

Molecule	Gauge origin	MINDO/3[b]	MNDO[b]	SCC—MO	IAM	Ray-Parr[c]	Ab initio DZ
c-CH_2N_2	H	139.2 (137.6)	138.5 (137.2)	136.0	134.0	138.0 [133.3]	135.0
	C	368.5 (370.3)	371.1 (370.4)	362.3	367.2	374.5 [369.8]	366.4
	N	432.8 (434.1)	437.5 (437.0)	423.3	426.3	433.2 [428.6]	424.7
O(1)O(2)O(1)	O(1)	508.4 (508.3)	511.5 (518.5)	484.7	488.5	495.0 [488.8]	488.0
	O(2)	521.5 (523.1)	526.7 (533.7)	506.3	512.9	519.3 [513.1]	509.2
CF_2	C	385.7	391.5 (390.8)	385.3	391.3	398.6 [393.0]	387.6
	F	510.3	571.3 (570.0)	552.2	555.5	558.1 [552.5]	555.5
FNO	N	447.3 (453.6)	433.5 (461.8)	441.1	448.4	455.3 [449.3]	443.3
	O	508.5 (504.4)	512.2 (514.6)	487.0	492.1	498.6 [492.6]	490.8
	F	506.3 (507.7)	567.8 (573.4)	547.1	548.8	551.4 [545.4]	550.2
H(1)CO(1)	H(1)	150.4 (150.6)	150.0 (147.4)	148.1	146.7	150.9 [145.9]	147.4
O(2)H(2)	H(2)	163.2 (159.5)	164.4 (163.3)	162.6	161.2	165.5 [160.4]	159.8
	C	388.7 (389.0)	394.3 (392.2)	389.5	393.1	400.6 [395.6]	390.7
	O(1)	493.2 (491.5)	510.2 (509.5)	486.0	483.9	490.6 [485.5]	485.7
	O(2)	491.8 (491.4)	508.5 (509.1)	484.3	484.6	491.3 [486.7]	484.9
c-C_3H_6	H	143.8 (142.2)	144.1 (142.6)	140.6	138.5	142.7 [138.4]	140.7
	C	371.0 (371.1)	374.4 (373.2)	365.2	369.7	377.2 [372.9]	370.7
c-$C_2H(1)_2$	H(1)	143.6 (143.6)	143.4 (143.5)	141.6	140.6	145.6 [141.2]	140.1
$H(2)_2NH(3)$	H(2)	143.5 (143.3)	143.5 (143.8)	140.9	138.3	144.7 [140.2]	139.4
	H(3)	157.7 (153.7)	158.3 (156.1)	156.1	152.5	157.6 [153.1]	154.2
	C	374.9 (375.1)	378.1 (376.4)	370.2	373.6	381.2 [376.7]	373.5
	N	432.2 (436.1)	438.9 (404.4)	423.9	427.3	435.7 [431.3]	426.0
BF_3	B	387.4 (385.9)	398.9 (395.2)	392.2	398.3	404.0 [398.6]	394.2
	F	535.8 (535.0)	597.7 (595.7)	579.4	582.9	585.5 [580.1]	583.3
F_2CO	C	444.5 (444.2)	452.0 (449.3)	446.1	453.7	461.0 [455.3]	448.1
	O	539.2 (535.1)	543.6 (540.3)	516.5	520.4	526.9 [521.2]	520.8
	F	539.8 (545.9)	601.3 (600.6)	582.8	586.7	589.2 [583.5]	586.3
N_2F_2	N	479.7	488.8 (493.5)	473.5	480.7	487.7 [481.8]	476.9
	F	531.2	590.0 (593.8)	572.2	575.1	577.6 [571.7]	574.9
CHF_3	H	199.0 (199.9)	199.0 (192.0)	195.6	196.9	—	196.5
	C	450.4 (452.5)	458.1 (454.4)	452.1	460.4	—	454.6
	F	549.3 (550.5)	611.8 (610.7)	593.1	596.0	—	595.5
CHOF	H	151.5 (156.0)	151.2 (147.9)	148.7	148.6	—	148.1
	C	390.3 (390.8)	396.5 (394.6)	390.2	396.7	—	392.7
	O	505.4 (503.4)	510.0 (509.2)	483.3	485.7	—	486.0
	F	506.4 (505.7)	568.3 (568.5)	550.0	552.0	—	551.8
C(2)	H	128.8 (127.1)	128.8 (127.8)	127.1	124.0	—	125.1
$(C(1)H_2)_2$	C(1)	350.4 (350.0)	352.6 (345.8)	354.1	351.5	—	348.7
	C(2)	366.5 (366.2)	371.1 (371.0)	361.6	365.2	—	364.8

[a] In azirine *cis*- and *trans*-protons are denoted by H(1) and H(2), respectively
[b] Entries in parentheses refer to optimized geometries.
[c] Entries given in square brackets were obtained by taking into account electronegativity correction

Chemical potential μ obviously plays an important role in determining σ^d values. If it is neglected the results are in extremely poor agreement with ab initio data which is substantiated by calculated diamagnetic shieldings for some linear molecules presented in Table 11. Inclusion of the chemical potential considerably improves the quality of results which are given within the square brackets.

It has been frequently argued that MINDO/3 and MNDO results should be used only at the optimized geometries. We employed both experimental and calculated structural parameters. The resulting σ^d values are about equally unsatisfactory.

In order to give our analysis a statistical relevance a large set of molecules exhibiting widely different bonding properties is examined. The results are given in Table 12. A careful scrutiny of the data shows that the IAM model is the closest to ab initio values as evidenced by the average absolute error $\Delta_a = 1.5$ ppm. The SCC—MO procedure comes next with $\Delta_a = 1.9$ ppm. A relatively good performance of the SCC—MO method is not suprising because it proved to be the most successful of all semiempirical schemes in calculating one-electron properties [67–72].

The Ray-Parr formula with ommission of the chemical potential yields a relatively high average absolute error ($\Delta_a = 5.8$ ppm). Full appreciation of the chemical potential leads to a decrease of the average absolute error by 57% ($\Delta_a = 2.5$ ppm). The MINDO/3 and MNDO seriously fail, yielding Δ_a errors as large as 10.0 and 9.1 ppm, respectively. The MNDO method is somewhat better than the MINDO/3 scheme as expected. The largest discrepancy of the MINDO/3 approach is found for fluorine.

The next atom which is not satisfactorily described in the MINDO/3 scheme is oxygen. The MNDO procedure does not perform well for fluorine, nitrogen and oxygen, the latter atom being the worst case. However, clustering of the errors around the average value is very dense indicating that MNDO describes the studied systems in a more uniform way. This is not the case of the MINDO/3 approach. Its errors are not systematic in nature but scattered instead exibiting sometimes erratic behavior.

The important question arises what is origin of these considerable discrepancies. The answer is not unexpectedly that MINDO/3 and MINDO methods do not employ proper basis sets, or to be more specific, the screening constants ξ of AOs are not adequate. This conjecture is substantiated by the following analysis. Let us consider first simple homodiatomic molecules N_2 and F_2. In these systems the redistribution of electron population takes place only between *2s* and *2p* orbitals since interatomic charge transfer is zero. The screening constants ξ in the MINDO/3 and MNDO methods are optimized by reproducing some observables (ΔH_f, geometry) together with a large number of other parameters. Their numerical values are at appreciable variance with those of more common Clementi-Raimondi AOs [65]. We shall, therefore, use Clementi-Raimondi ξ constants, which are quite successful within the SCC—MO framework, and retain MINDO/3 and MNDO orbital populations. This simple approach will shed some light on the role of the screening parameters. Improvement of the results appears to be quite dramatic. In N_2 molecule the MINDO/3 and MNDO $\sigma^d_{av}(N)$ values are both not 380.6 ppm. Further increase in accuracy is obtained by the use of Ransil's best atom and best limited ξ [73]. The latter distinguish between the $p\sigma$ and $p\Pi$ orbitals which is of importance in linear molecules. The MINDO/3 (MNDO) results based on the best atoms are 381.8 (381.8) ppm, whilst the best limited ξ yield 385.2 (385.0) ppm, which is now in very good agreement with ab initio value (384.1 ppm). Similar improvement is found in F_2 molecule. Ransil's best atom and Clementi-Raimondi ξ values give both ≈ 525 ppm for $\sigma^d_{av}(F)$ for MINDO/3 and MNDO methods. This is in fair agreement with the DZ—SCF value of 529.2 ppm although room for improvement still remains. The original MINDO/3 (MNDO) values of $\approx 485 (\approx 544)$ ppm were clearly unacceptable. Let us focus attention to some heterodiatomics where the charge transfer is highly pronounced. Molecules BF and CO will serve the purpose well. In the former molecule MINDO/3 (MNDO) methods with Clementi-Raimondi ξ yield $\sigma^d_{av}(F) = 507.1$ (504.8) ppm. The corresponding Ransil's best atom results are 507.4 (505.1) ppm which is in fine accordance with DZ—SCF value of 507.8 ppm. Improvement is dramatic indeed. Boron shielding σ^d_{av} is predicted to be 265.5 (265.9) and 266.6 (266.9) ppm for Clementi-Raimondi and Ransil's best atom ξ, respectively, the MNDO results being cited always in parentheses. Degree of agreement with DZ—SCF value of 267.4 ppm is remarkable. Clementi-Raimondi and Ransil's best atom ξ give $\sigma^d_{av}(O)$ values of 443.3 (442.1) and 444.0 (442.8), respectively, offering thus a substantial improvement in description of the recalcitrant oxygen atom (Table 11). Hence it follows that the diamagnetic shielding of nuclei is useful in diagnostics of basis set imperfections — *quod erat demonstrandum*. This is concomitant with our previous analysis of GTO and HGO expansions of hydrogenic orbitals.

A point of considerable interest is observation that σ^d shielding strongly depends on ξ's of the atom in question, but not on the orbital populations. This conclusion is illustrated by a comparison of orbital populations in BF which by MINDO/3 (MNDO) methods read: $Q^B_{2s} = 1.98$ (1.89), $Q^B_{2px} = 0.11$ (0.60), $Q^B_{2pz} = 0.31$ (0.18), $Q^F_{2s} = 1.94$ (1.82), $Q^F_{2px} = 1.97$ (1.70), $Q^F_{2py} = Q_{2pz} = 1.69$ (1.82). In spite of considerable differences in orbital populations both methods give practically the same diamagnetic shieldings for constituent B and F atoms, respectively (vide supra). Apparently, σ^d is sensitive to the form of AOs used as a basis set, and relatively insensitive to their populations.

The second moments were also examined by the IAM model and MINDO/3, MNDO, SCC—MO and ab initio DZ methods [46]. Apart a few "pathological" molecules all three semiempirical methods and the IAM model reproduce DZ results in a satisfactory manner. The average absolute errors of MINDO/3 and MNDO methods is 0.4×10^{-16} cm^2, if the experimental geometries are used. The errors are practically doubled if MINDO/3 ($\Delta_a = 0.8$) and MNDO ($\Delta_a = 0.7$) optimized geometries are employed, because prediction of structural parameters is moderately good as it is well known. It is interesting to mention that the second moments do not depend on the coupling between s and p AOs (hybridization) in the first row atoms in contrast to dipole moments. The same holds for the coupling of s and d AOs in atoms of the second and higher rows of the system of elements. The fact that the second moments obviously do not strongly depend on the choice of AOs can be easily rationalized by a fact that the single center average value of the second moment operator is proportional to the inverse square of the orbital screening constant. If we denote an error in the atomic screening parameter by $\Delta\xi$ then the Taylor expansion gives

$$1/(\xi + \Delta\xi)^2 = (1/\xi^2) - 2(\Delta\xi/\xi^3) + \dots \quad (40)$$

Hence the linear term in $\Delta\xi$ is divided by ξ^3. On the other hand, σ^d shielding is proportional to ξ meaning that inaccuracy in screening straightforwardly affects results to significant extent.

The errors in σ^d by MINDO/3 and MNDO methods are huge being on average roughly ≈ 10 ppm per atom. These errors translated into energy, i.e. $\langle r^{-1} \rangle$ expectation values, are as large as 354 kcal mol^{-1}. The question arises how these two semiempirical methods can work at all with so large inaccuracy per atom. The point is that both schemes do not minimise the total molecular energy, but heats of formation instead. In the ΔH_f expression these errors are largely cancelled out because of the difference of molecular and free-atom terms.

However, semiempirical methods should not give only heats of formation but other molecular properties too. It is not the aim of this chapter to propose a new semiempirical scheme, but we would nevertheless like to mention that optimization of ξ screening constants in MINDO/3 and MNDO methods performed on an equal footing with other adjustable parameters is not justified. On the contrary, ξ values shold be determined by using σ^d criterion and perhaps some other molecular properties [46].

Quantitative appraisal of approximate treatments for calculating electron density distributions in molecules by the σ^d test is further illustrated by examining LCAO-Xα results in some diatomics [74] and SCF-Xα SW/MT prediction in H_2O molecule [75]. These results are compared with IAM estimates of diamagnetic shieldings and accurate ab initio values obtained by large basis sets in Table 13. It appears that the mentioned varitions of the X_α method are obviously unsatisfactory. The

Table 13. Average diamagnetic shielding of the nuclei in some diatomics and small molecules as estimated by IAM, LCAO-X_α and ab initio approaches (in ppm)

Molecule	Gauge origin	IAM	LCAO-X_α [a]	Ab initio
Li_2	Li	112.5	129.5	112.4[b]
N_2	N	385.8	361.2	384.5[b]
F_2	F	530.5	522.0	529.5[b]
CO	O	444.9	423.9	444.8[c]
	C	327.5	306.0	326.5[c]
HF	F	481.2	466.8	481.6[b]
	H	109.6	100.0	110.7[b]
BF	F	508.2	496.4	507.8[d]
	B	269.0	258.4	268.4[d]
BH	B	209.6	198.8	209.6[d]
	H	55.7	51.9	58.1[d]
LiH	Li	107.9	102.3	107.9[e]
	H	35.5	24.9	39.5[e]
H_2O	O	414.7	425.8[f]	416.1[g]
	H	102.8	101.5[f]	103[g]

[a] Bieger W, Seifert G, Eschrig H, Grossman G (1985) Chem. Phys. Lett. 115: 275;
[b] Karplus M, Kolker HJ (1963) J. Chem. Phys. 38: 1263 obtained by using wavefunctions of Ransil BJ (1960) Rev. Mod. Phys, 32: 245;
[c] Stevens RM, Karplus M (1968) J. Chem. Phys. 49: 1094;
[d] Hegstrom RA, Lipscomb WN (1966) J. Chem. Phys. 45: 2378; (1968) 48: 809 based on wavefunctions of Huo WM (1965) J. Chem. Phys. 43: 624;
[e] Stevens RM, Pitzer RM, Lipscomb WN (1963) J. Chem. Phys. 38: 550;
[f] SCF-X_αSW/MT Calculation, Woodruff SB, Wolfsberg M (1978) Chem. Phys. Lett. 56: 125;
[g] Neumann D, Moskowitz JW (1968) J. Chem. Phys. 49: 2056

IAM model is in good accordance with rigorous results as usual. Let's mention in passing that the use of the IAM and the modified atom model (MAM) in determining a choice of atomic domains in the X_α formalism will be briefly discussed in Sect. 4.

2.4 Miscellaneous Applications of the IAM Model

Atomic force constants introduced by King [76, 77] have a remarkable property of being practically independent of bonding multiplicity and other details of the electronic density distribution in molecules. The sum of squares of normal frequencies are determined to a very good approximation as a sum of atomic force constants multiplied by the corresponding reciprocel masses m_α

$$\sum_i v_i^2 = (N/4\Pi^2c^2) \sum_A m_{\alpha A} F_A \tag{41}$$

Here $F_A = \nabla_A^2 U$ is the Laplacian of the Born-Oppenheimer potential function U of a molecule. It appears that characteristic atomic force constants F_A are transferable within a wide variety of compounds. Furthermore, F_C, F_N, and F_O constants are approximately equal. Hence, atomic force constants belong to a class of properties which can be well described by the IAM model. They proved useful in rationalizing primary isotopic effects [78].

Results of Spackman and Maslen deserve close scrutiny [79]. They utilized the IAM model in calculating a number of relevant molecular properties. The wave function describing independent neutral atoms is given by a product

$$\Psi_{IA} = \prod_i \hat{A}_i \Phi_i \tag{42}$$

of properly antisymmetrized free-atom wavefunctions Φ_i. Since it is assumed that atoms are completely independent Ψ_{IA} is not antisymmetric to exchange of electrons between two different atoms. Therefore the wavefunction Ψ_{IA} violates the Pauli exclusion principle. Nevertheless, it offers a lot of useful information. The corresponding density distribution $\varrho_{IA} = \sum_i \varrho_i$, where ϱ_i denotes spherically averaged atomic electron density centred at nuclei, was used for calculation of atomic bond radii employing the criterion of minimum in ϱ_{IA} along the internuclear distance. The obtained results display similar trends as the atomic radii of Slater [80] and the covalent radii of Pauling [6].

Applying the formula of Bonham and Fink [81] involving atomic scattering factors Spackman and Maslen [79] calculated classical electrostatic energy in molecules within the scope of the IAM model. It was found that it is always negative an roughly proportional to the molecular binding energy. Extending their approach to solids they have been able to show that the IAM model yields simple rationalization of their electrostatic, Madelung and cohesive energies [82]. Finally, it was found that the correlation energies in a large number of diatomics and polyatomic molecules are closely related to the classical electrostatic interactions between neutral spherical atoms [83]. More specifically, the correlation energy is a linear function of electrostatic energy E_{es}, 25% of E_{es} being its lower bound. The root-mean-square dispersion from the least-squares fit line is 1.04 eV. Atomic additivity of molecular correlation energy within such a simple model is remarkable indeed.

Electrostatic potential of bare nuclei is apparently a property which can be ascribed to independent atoms. It is very similar to the molecular electron density distributions [84, 85] as intuitively expected. Although some finer details of the electron charge distributions are missing [86], general features are well reproduced by the nuclear potential.

Spackman and Maslen argue that the primary source of binding in molecules is simply the classical electrostatic interaction between spherical atoms [79]. This assertion should not be adopted too literally in view of the energetic importance of electron charge transfer [87, 88], but if it is taken *cum grano salis*, then it represents an appropriate concluding statement about the IAM model.

Natura enim simplex est
I. Newton

3 Modified Atom Model (MAM)

As emphasized earlier, atoms are not scrambled in molecules, but modified in a way that it is always useful to compare their properties in chemical environments with those of free atoms. Atomic orbitals of modified atoms are somewhat distorted, polarized and their nonlinear screening parameters are either diminished or enlarged corresponding to their expansion or contraction, respectively, following formation of chemical bonds. We shall not give a formal theoretical presentation of the MAM model here. Instead, it will be discussed at the qualitative and conceptual level in keeping up with a general character of this chapter. In particular, modified atoms will be described by descriptors like hybridization, atomic multipoles etc. regardless how they are derived. The MAM model will be used for interpretation of theoretically determined or observed properties, but in some simple cases it will serve for predictive purposes. In describing the salient features of the MAM model it is useful to dissect formation of molecules into four stages:

(1) Neutral and spherical free atoms are placed at equilibrium positions.

(2) Atoms are reoriented to insure favourable screening of the bonded nuclear charges allowing for efficient overlapping of singly occupied AOs. Concomitantly, atomic orbitals are shrunk or extended undergoing radial distortion.

(3) AOs are further chemically adapted by hybridization in order to conform the local symmetry and the coordination number and increase beneficial screening of nuclear charges.

(4) Intramolecular charge transfer is invoked to cancel differences in electronegativity.

Although all the stages in the chemical bond formation given above are intermingled implying that their classification is arbitrary and that optimal bonding parameters can be obtained only by an iterative procedure, this systematization is nevertheless very useful because some features are more important than the others when specific molecular properties or family of compounds are considered. Stage (1) corresponds to the IAM picture which suffices for discussion of some properties at a qualitative level (vide supra). Stage (4) may be skipped in hydrocarbons where ionicity is less pronounced. On the other hand hybridization can be neglected in alkali-halides but ionic model is absolutely necessary for approximate description of their behavior. Similarly, electron charge distribution may be well reproduced by atomic multipoles, but a good description of ESCA chemical shifts requires only reliable atomic monopoles (vide infra). It follows that by focusing attention to some of the characteristics (1)—(4) and abandoning others one obtains a variety of related simple models which in turn all fit the MAM concept. The salient feature of the latter is that molecular properties are given in an additive way, i.e. by a summation of monoatomic terms. The strength of the MAM model is its computational ease and interpretational simplicity. It enables reduction of complex problems related to structure and properties of large molecules to the problem of description of constituent modified atoms and their bonding features.

3.1 The Effect of Atomic Distortion

One can distinguish angular and radial distortion of atomic charge density caused by the first order effect of nearest neighbours. The former leads to familiar hybrid atomic orbitals (HAOs). Full account of the hybridization would not be in place here, because it fits better the second volume of this series [61] dealing with covalent bonding. We shall summarize, however some basic ideas and mention that the hybridization concept is well founded and has a physical meaning of local symmetry adapted orbitals which maximize electron density in bonding directions and at the same time lead to its depletion in nonbonded spatial segments. One should also mention that sceptics and critics of the hybridization model often argue that hybridization "does not exist" because the Hartree-Fock (HF) function is invariant to all orthogonal transformations of the basis set. This is however a rather amusing argument, because if the particular theoretical scheme is invariant to orthogonal mixing of basis AOs then it can say nothing about hybridization. It has been emphasized by us that HF method based on free-atom AOs provides a satisfactory model for chemistry because of this very fact that s, p_x, p_y, p_z ... atomic orbitals are equivalent to hybrids. The latter represent actually properly chemically adapted local atomic orbitals. Although HF method can not "see" hybrids directly, information about hybridization is stored in the first-order-density matrix. On the other hand arguments in favour of hybridization can be put forward *a priori* by symmetry consideration [89] and *a posteriori* [61] by its intimate relations with large number of molecular local and global properties. Hence, hybridization "exists" no more and no less than the orbital concept itself. As to the basis set virtues of hybrid AOs they are well suited for various semiempirical procedures [90–93]. The point of considerable interest is certainly a finding that VB method, which (unlike the SCF—MO theories) is not invariant to orthogonal transformations, yields hybrid AOs without any preconceptions [94]. Another support for the hybridization idea comes from the shape of the Fermi holes which remarkably resemble HAOs [95, 96]. We have provided an extensive evidence that hybridization is a phenomenon. Although it takes places in some atoms less then in others (first row atoms exhibit typical hybridization features fluorine being an exception it is one of the corner stones of the chemical theory of the covalent bonding. Hybridization is indispensable in particular in organic chemistry [61]. Last but not least hybridization is the underlying principle which can rationalize many interrelations between various observables established by experiments. It gives enlightening insight into intimate connection between spatial and local bond properties.

Finally, a word about radial deformations in atomic orbitals is necessary. Radial adjustement in direction of covalent bonding is of significant importance in determining molecular stability [97]. It has been pointed out many times that hybrid AOs have a distinct advantage that for each bonding direction the nonlinear screening parameter ξ can be separately optimised. In MOs formalism with pure free-atom AOs basis this is very difficult to achieve except in linear molecules. However, even then the $p\sigma$ orbital coinciding with the symmetry axis can not assume two distinct ξ for two localized σ-bonds emanating in opposite directions from a common atom. It is noteworthy that the radial distortion of hybrid orbitals is conveniently obtained by employing Coulson-Fischer prescription [98, 99] of constructing semi-localised orbitals. They increase not only mutual overlap within the bonded couple of the semilocalized orbital but take into account also the ionic character to some extent.

In conclusion, one can assert that hybridization with accompanying radial readjustement of AOs provides a transparent model of covalent bonding which is difficult to surpass in amount of useful chemical information. To put it in another way, one can say paraphrasing Mulliken that a little hybridization has far reaching chemical consequences.

3.2 The Effect of Intramolecular Charge Transfer

The idea that atoms carry some electric charges in molecules goes back to Berzelius and Faraday, who in his book "Experimental Researches in Electricity" published in 1833 writes explicitly: "The atoms of matter are in some way endowed or associated with electrical powers, to which they owe their most striking qualities, and amongst them their mutual chemical affinity". Modern quantum chemistry has confirmed his prophetical words. Indeed, intraatomic redistribution of orbital electronic populations and interatomic charge transfer are pivotal ingredients of the bonding process. In what follows we shall concentrate first on energetic effects of the charge reorganization. The change in electronic density distribution of atoms in molecular environment could be divided to isotropic and anisotropic parts. The former is described by atomic monopole whereas the latter is given by higher multipole moments. The monopole term give rise to the concept of atomic charge which faithfully serves chemistry for a number of decades. It should be pointed out that apportioning of the total electron density $\varrho(\vec{r})$ to atomic contributions strongly depends on the quality of the basis set and a prescription of partitioning of the mixed charge. It is therefore rejected by quantum fundamentalists and purists. This attitude is not useful and we shall adopt an optimistic point of view necessary in constructing models, and assume that there is a *bona fide* partitioning which yields sensible atomic charges. An in depth discussion of this topic was given in the previous chapter [88]. For present practical purposes we shall employ the simplest possible population analysis of Mulliken [104], which divides the mixed charge on the democratic 50:50 basis.

One of the early applications of the point atomic charges was a development of the crystal fild theory [100]. Additionaly, they were frequently associated with chemical reactivity and provide a useful index in charge controlled reactions where the HOMO-LUMO energy gap between reactants is large [101]. The role of atomic monopoles in determining enthalpies of formations of organic compounds were discussed by Benson [102]. In this connection it should be pointed out that the electrostatic potentials exerted at the nuclei are closely related to the total molecular SCF energy [103]. The formula

$$E_t = \sum_A k_A Z_A V_A \tag{43}$$

recovers most of the total SCF energy. In the monopole approximation [105, 106] the potential V_A takes a form:

$$V_A = -\sum_{\mu}^{A} (\xi_\mu Q_\mu^A / n_{A\mu}) + \sum_B{}' (Z_B - Q_B)/R_{AB} \tag{44}$$

where ξ_μ, Q_μ^A and $n_{A\mu}$ denote orbital screening constant, orbital population and the principal quantum number, respectively. Equation (44) is approximate because it involves Mulliken partitioning. Nevertheless, the SCC—MO monopoles and orbital populations reproduced DZ ab initio SCF energies of Snyder and Basch [66] with a standard deviation of 0.1 au [106]. The same quality of results was obtained by using atomic monopoles extracted from the Snyder-Basch DZ wavefunctions.

The role of electrostatics will be illustrated by just two additional examples. The first is provided by energy in alkali-halides. The most common potential describing interaction between an alkali-halide pair of atoms is that suggested by Rittner (in au) [107]

$$\begin{aligned} V_{MX}(R) = {} & -R^{-1} - [(\alpha_M + \alpha_X)/2R^4] - [2\alpha_M\alpha_X/R^7] - \\ & - [C_{MX}/R^6] + A_{MX} \exp(-R/\varrho_{MX}) \end{aligned} \tag{45}$$

R stands for the interatomic distance whilst α_M and α_X are polarizabilities of alkali and halide free atoms, respectively. The first term represents the Coulomb attraction, the second and third arise from the polarization of the ionic charge clouds which leads to induced atomic dipoles. Van der Waals attraction is given by $-C_{MX}/R^6$ whereas the last term describes repulsion due to penetration of two closed shells of electrons. The Coulomb term is obviously a dominating contribution since others decrease either exponentially or by high inverse powers of the internuclear distances R. Additionally, the polarization and dispersion stabilization will approximately cancel out the repulsion of full electron shells. Assuming that one electron is transfered from an alkali to a halide atom one should obtain reasonable estimates of MX dissociation energies to the ionic limit by taking into account only R^{-1} term. This is indeed the case, as shown im Table 14. The $M^{+1}X^{-1}$ model in the monopole form accounts for main features in variation of the D_e (ion) dissociation energies in alkali halides. For instance, the energy decreases along the series MX for a fixed M and X=F, CL, Br and I. Alternatively, if the halogen atom X is kept fixed, the bond strength decreases along the MX family (M=Li, Na, K, Rb and Cs).

Table 14. Comparison of the dissociation energies in alkali-halides obtained by ionic $M^{+1}X^{-1}$ MAM model and the experimental values

Molecule	Dissociation energies kcal mol^{-1}			
	Re/Å	$M^{+1}X^{-1}$ model	$M^{+q}X^{-q}$ model	Exptl.
Li—F	1.564	212.3	81.0	184.1
Li—Cl	2.0207	164.3	43.0	153.3
Li—Br	2.170	153.0	34.8	147.8
Li—I	2.392	138.8	21.1	138.7
Na—F	1.926	172.4	63.1	153.9
Na—Cl	2.361	140.6	31.6	132.6
Na—Br	2.502	132.7	28.2	127.7
Na—I	2.7115	122.5	17.0	120.3
K—F	2.1715	152.9	62.0	139.2
K—Cl	2.667	124.5	32.8	118.0
K—Br	2.821	117.7	29.5	113.6
K—I	3.048	108.9	18.9	106.1
Rb—F	2.2703	146.2	60.5	133.6
Rb—Cl	2.787	119.1	32.5	113.4
Rb—Br	2.945	112.7	29.3	109.0
Rb—I	3.177	104.5	19.0	101.9
Cs—F	2.345	149.6	60.4	130.5
Cs—Cl	2.906	114.2	32.6	112.3
Cs—Br	3.0722	108.1	30.9	108.6
Cs—I	3.3152	100.2	19.4	101.1

It is interesting to note that the monopole model involving partial +q and −q charges ($M^{+q}X^{-q}$) estimated by electronegativity ϵ values of Little and Jones [108]

$$q_M = 1 - 2\varepsilon_M/(\varepsilon_M + \varepsilon_X) \quad \text{and} \quad q_X = 1 - 2\varepsilon_X/(\varepsilon_M + \varepsilon_X) \tag{46}$$

is not satisfactory. The absolute values of dissociation energy are almost an order of magnitude too small although the trend of changes is correct. Hence, the dissociation energies require the use of the ionic $M^{+1}X^{-1}$ model for alkali halides just like it is a case for the second moments (vide infra).

The second illuminative example which illustrates the role of atomic monopoles in determining molecular energetics is given by ESCA shifts [109]. Binding energies of the localized inner core electrons exhibit strong dependence on the chemical environment as mentioned earlier, thus providing a sensitive probe of the electronic structure of molecules. Binding energy shifts parallel the changes in electrostatic potentials exerted at the nucleus in question. Extensive calculations of ESCA shifts of B, C, N, O, F, Si, S and Ge atoms by employing SCC-AMEP (self consistent charge electrostatic potential expressed by atomic monopoles) model have conclusively shown that the basic features of the ESCA spectre are well accounted for by atomic charges [110]. It should be stressed that the MINDO/3 and MNDO schemes are less successful in this respect [111] whereas EHT and CNDO/2 are notoriously unreliable. A good performance of the semiempirical SCC—MO method within the AMEP approximation is of general importance because the ESCA technique gives the most direct insight into the distribution of atomic monopoles if the effect of reorganization of electron density upon ionization is taken into account [88].

All these results strongly indicate that the nuclear-electron Coulomb potential within a molecule plays a decisive role in molecular bonding which is compatible with findings of Spackman and Maslen obtained by IAM model. Significance of the extramolecular electrostatic potential will be briefly discussed later, because introduction of higher atomic multipoles is then necessary.

In modelling of the electronic structure of molecules determination of the limits of validity of the approximate description is of utmost importance. In considering the second moments and diamagnetic susceptibility, it is expected that the IAM model will fail in highly ionic species e.g. alkali-halides. It is plausible to assume

Table 15. Comparison between the molecular second moments[a] of the electronic charge distributions and diamagnetic susceptibilities[b] and the corresponding ab initio values for some alkali halides

Compound	Second moments	Neutral	Ionic	Ab initio	Diamagnetic susceptibilities	Neutral	Ionic	Ab initio
LiCl	$\langle z^2\rangle_{Cl}$ =	15.8	10.58	11.828	$\chi^d_\perp$(Cl) =	−81.9	−57.65	−65.02
	$\langle z^2\rangle_{Li}$ =	73.1	76.39	75.263	$\chi^d_\perp$(Li) =	−325.0	−335.53	−333.61
	$\langle x^2\rangle$ =	3.5	2.7	3.373	$\chi^d_\parallel$ =	−29.7	−22.91	−28.62
NaCl	$\langle z^2\rangle_{Cl}$ =	66.3	59.2	60.582	$\chi^d_\perp$(Cl) =	−302.5	−265.99	−273.77
	$\langle z^2\rangle_{Na}$ =	99.7	103.75	103.422	$\chi^d_\perp$(Na) =	−444.2	−454.99	−455.72
	$\langle x^2\rangle$ =	5.0	3.5	3.95	$\chi^d_\parallel$ =	−42.4	−29.7	−35.51
NaF	$\langle z^2\rangle_{F}$ =	44.0	38.86	38.99[e]	$\chi^d_\perp$(F) =	−201.5	−173.34	−173.98
	$\langle z^2\rangle_{Na}$ =	36.7	38.86	38.28	$\chi^d_\perp$(Na) =	−170.5	−173.34	−170.97
	$\langle x^2\rangle$ =	3.5	2.0	2.02	$\chi^d_\parallel$ =	−29.7	−16.97	−17.14
KF	$\langle z^2\rangle_{F}$ =	94.1	88.37	88.87	$\chi^d_\perp$(F) =	−418.3	−389.75	−390.81
	$\langle z^2\rangle_{K}$ =	46.9	50.65	49.58	$\chi^d_\perp$(K) =	−218.1	−229.73	−224.13
	$\langle x^2\rangle$ =	4.5	3.5	3.25	$\chi^d_\parallel$ =	−38.2	−29.7	−27.58
RbF	$\langle z^2\rangle_{F}$ =	203.9	196.70	197.82[g]	$\chi^d_\perp$(F) =	−892.6	−853.15	−857.63
	$\langle z^2\rangle_{Rb}$ =	54.5	57.86	56.851	$\chi^d_\perp$(Rb) =	−258.8	−264.56	−259.72
	$\langle x^2\rangle$ =	6.5	4.5	4.369	$\chi^d_\parallel$ =	−55.2	−38.18	−37.07
NaBr	$\langle z^2\rangle_{Br}$ =	74.8	67.09	68.286	$\chi^d_\perp$(Br) =	−342.8	−303.71	−311.49
	$\langle z^2\rangle_{Na}$ =	225.1	229.82	227.597	$\chi^d_\perp$(Na) =	−980.4	−994.08	−987.35
	$\langle x^2\rangle$ =	6.0	4.5	5.136	$\chi^d_\parallel$ =	−50.9	−38.18	−43.58
LiBr	$\langle z^2\rangle_{Br}$ =	18.4	12.96	14.271	$\chi^d_\perp$(Br) =	−97.2	−70.68	−79.79
	$\langle z^2\rangle_{Li}$ =	166.5	170.32	168.623	$\chi^d_\perp$(Li) =	−725.5	−738.26	−734.61
	$\langle x^2\rangle$ =	4.5	3.7	4.536	$\chi^d_\parallel$ =	−38.2	−31.39	−38.49
KCl	$\langle z^2\rangle_{Cl}$ =	141.0	132.94	133.966	$\chi^d_\perp$(Cl) =	−623.6	−585.20	−590.19
	$\langle z^2\rangle_{K}$ =	126.8	132.94	131.356	$\chi^d_\perp$(K) =	−563.4	−585.20	−579.12
	$\langle x^2\rangle$ =	6.0	5.0	5.125	$\chi^d_\parallel$ =	−50.9	−42.42	−43.71

[a] In 10^{-16} cm^2

[b] In 10^{-6} cm^3 mole^{-1}

that the ionic $M^{+1}X^{-1}$ model will perform better. The second moments then take a form

$$\langle y^2 \rangle = (Z_M - 1)\, Y_M^2 + (Z_X + 1)\, Y_X^2 + k_{M-1} + k_X \tag{47}$$

because one electron is transfered from alkali metal M to halide atom X. Concomitantly the additive constant k_M assume the value appropriate for the preceding row of elements k_{M-1}. Results presented in Table 15 illustrate rather nicely not only good results offered by the ionic $M^{+1}X^{-1}$ picture, but also the interpretative power of models in general. It is puzzling, namely, that the ab initio results of Matcha [112] gave in molecules NaF and KCl practically the same second moments irrespective of the choice of the nucleus as gauge origin. In general, these two values are widely different. (Table 15). A simple rationalization of this finding is given by the MAM $M^{+1}X^{-1}$ model. In diatomic NaF molecule transfer of one electron produces a pseudomolecule Ne_2 formed by two Ne atoms. It is irrelevant which of the two neon atoms are chosen as origin of the coordinate system. Similarly, KCl is well represented by the noble gas pseudo-diatomic-molecule Ar_2. Slight difference between $\langle z^2 \rangle_K$ and $\langle z^2 \rangle_{Cl}$ ab initio values indicates limitations of the adopted model. The IAM approach yields substantionally worse results. We would like to draw attention to good estimates of the perpendicular second moments $\langle x^2 \rangle$ provided by the MAM model. It does not depend on the choice of the gauge origin being determined solely by the additive isotropic constants k_{M-1} and k_X. The choice of the k_{M-1} constant for the metal ion corresponding to the preceding period of atoms thus seems to be justified.

Finally, we shall comment on the use of the MAM model in determining atomic domains in the X_α method. Chesnyi et al. [113] used one-electron properties as a criterion in determining overlap between spherical atomic regions. The optimal overlapping is obtained when the best agreement with Hartree-Fock values of selected one-electron properties is achieved. Some characteristic results for the second moments in diatomics are summarized in Table 16. Perusal of the data shows that the MAM model gives second moments in good accordance with the near-HF values and the optimal X_α—SW results. The extent of intersphere overlapping ranges from 15% to 30% in the latter approach. As it is already mentioned, in the case study of Chesnyi et al. the best overlap was estimated by making use of available HF data of some one-electron properties. In general such a strategy would not make any sense, because then X_α—SW computations are redundant particularly in view of their approximate nature. Concomitantly, one has to resort either to experimental data or perhaps to the MAM or IAM estimates of σ^d and $\langle r^2 \rangle$ properties, which in turn require only a back of the envelope calculation. The gauge dependence of these diamagnetic properties is advantageous because one could impose 4N conditions by placing origin at each nucleus and by taking all diagonal components $\langle X^2 \rangle$, $\langle Y^2 \rangle$ and $\langle Z^2 \rangle$ of the second moment tensor. Of course, the gauge origin could be centered at any critical point within the molecule in addition. We feel that this type of the constrained X_α method would represent a useful alternative to the present schemes. It would certainly satisfy at least the necessary condition for a good description of molecular charge densities.

The expectation values of many properties of molecules can be broken down to

Table 16. Comparison of second moments in some diatomics as obtained from the MAM model, X_α-SW and ab initio methods[a]

Molecule	MAM	X_α-SW	Ab initio[g]
Li—H	$\langle z^2 \rangle$ = 3.9[b]	4.6[d]	4.1
Li—F	$\langle z^2 \rangle_{Li}$ = 25.7[c]	25.2[e]	24.2
Na—F	$\langle z^2 \rangle_{Na}$ = 38.9[c]	38.5[f]	39.0
	$\langle z^2 \rangle_F$ = 38.9	39.3[f]	38.3
Rb—F	$\langle z^2 \rangle_{Rb}$ = 57.8[c]	55.2[d]	56.85
	$\langle z^2 \rangle_F$ = 196.6	190.9[d]	197.8

[a] Second moments refer to a center of mass if not otherwise stated. In 10^{-16} cm^2 units. X_α-SW values are taken from Ref. [113]; [b] Obtained by using Pauling electronegativity concept (formula (46)); [c] Ionic model (formula (47)); [d] Overlapping of 15%; [e] Overlapping of 30%; [f] Overlapping of 20%; [g] Ab initio values are taken as cited in Ref. [113]. They are of the near-HF quality

several contributions, the atomic monopole term being usually a dominant member in the series expansion. However, the other contributions could be far from negligible if asymmetry of atomic charge is of importance. Typical cases of this kind are total molecular dipole and quadrupole moments. It is easy to show that they can be expressed as:

$$\mu_x = |e| \sum_A [(Z_A - Q_A) X_A + \mu_{Ax}] \tag{48}$$

and

$$Q_{xx} = [|e|/2] \sum_A [(Z_A - Q_A)(3X_A^2 - R_A^2) + 2(3X_A\mu_{Ax} - \vec{R}\vec{\mu}) + Q_{xx}^A] \tag{49}$$

where μ_{Ax} and Q_{xx}^A are components of local atomic dipole and quadrupole moments, respectively. The first term in Eqs. (48) and (49) correspond to atomic monopole contributions. We shall briefly comment on the empirical scheme of Flygare et al. [114] for calculating molecular dipoles and quadrupoles which enjoyed considerable popularity. They made the following assumptions:

(a) atomic quadrupoles Q_{xx}^A can be neglected;

(b) each atom has exactly Z_A electrons.

The first supposition is probably justified. The second, however, means that atoms are neutral and that $Z_A - Q_A = 0$. It follows that only the second terms in Eqs. (47) and (48) survive. They both involve the atomic dipole μ_A contribution which reflects intraatomic mixing of orbitals (hybridization) or angular deformation of atomic density but the charge transfer is neglected. This type of approximation gave excellent results in considering bonding properties of hydrocarbons [18, 61]. We have also seen that interatomic transfer of electron density may be safely abandoned in calculating σ^d and χ^d properties as evidenced by the very good results of the IAM model. However, in treating total molecular dipole and quadrupole moments the monopole term is very large because it is proportional to nuclear coordinates X_A or their anisotropy $(3X_A^2 - R_A^2)$ and omitting it is a serious conceptual error. A fact that the empirical scheme of Flygare et al. [114] gives results in good accordance with experiment can be traced down to a large number of adjusted parameters (about forty for the first row atoms) i.e. atomic dipole moments. Hence this scheme is a characteristic example of an over-parametrized model which gives good results for wrong reasons. It follows that a revision of the additivity pattern is well advised.

A more detailed microscopic picture of the electron charge distribution in molecules is provided by polycentric multipole expansion [115–117]. This approach gives a simple description of the gross molecular multipoles and a good representation of extramolecular potentials which is valid at all accessible distances. Molecular electrostatic potentials (MEP) is an extremely useful concept because it offers a useful means in exploring protonation sites in large heterocyclic compounds like e.g. nucleic acid bases, or in determining places of attack of larger cations like Li^+ [118, 119]. By changing sign of the MEP one obtains useful information about the most probable positions of fixation of spherical anions (e.g. Cl^-) [120]. Further, MEP is an important determinant of biological activity [121] and drug-receptor recognition [122] etc. The single center molecular multipole expansion is not so well suited for MEP description because of the convergency problems at distances of chemical interest. It is intuitively clear and more importantly it can be mathematicaly verified that (atomic) polycentric expansion will perform better because atomic monopoles are generally different from zero even in neutral molecule where the total molecular monopole vanishes. Hence very high gross molecular multipoles have to be included which in turn are not easily calculated with a desired accuracy.

Recently, distributed multipole analysis was used to predict geometries of van der Waals complexes [123]. Additional careful theoretical analysis has shown that electrostatic interaction in these systems is the main net effect, because exchange, polarization and charge transfer contributions tend to cancel out [124]. This is gratifying because it offers conceptual and computational advantage over the supramolecule approach. Namely, it is always beneficial to treat large system in terms of its subunits and their interactions. In the distributed (atomic) multipole model this strategy found its full realization, because the charge distribution in monomer units is further reduced to irreducible local multipoles.

To summarize, it is obvious that the idea of modified atoms im molecular environments is a rich concept which at its qualitative level of sophistication combines interpretative power and computational feasibility. It is an important link between the rigorous quantum theory and chemical phenomenology.

Die Begriffe ohne Anschauungen sind leer,
die Anschauungen ohne Begriffe sind blind.
I. Kant

4 Conclusion

Discussing tasks of the quantum theory of molecules, Hirschfelder says: "Theoretical chemistry seeks to explain and predict quantitatively the physical and chemical properties of materials, to relate these macroscopic properties to individual molecules, and to predict the structure and the properties of these individual molecules . . . (it aims) to develop simple models which capture the essence of the phenomena . . ." [125]. The model of modified atoms in molecules is one of the intuitively appealing models which provides a rationale for molecular behavior at least at the qualitative level. It is conceptually clear and pictorialy vivid, offering a useful elementary language and vocabulary in everyday chemistry being at the same time computationaly simple and flexible enough to account for a wide range of properties of molecules. We would like to stress once again its visuality because, as I. Kant put it very nicely, notions without images are empty. Importantly, a good performance of the MAM model indicates that it is viable and that it has some semblance of truth.

Properties considered in previous sections do not exhaust all the properties which exhibit atomic additivity. Molecular polarizability [126] and zero-point vibration energies [127] should be added to the list *inter alia*. They were not discussed because of space limitations. A detailed presentation of the polarized atom model described by the hybridization of AOs will be given in vol. 2 of this series.

It is interesting to mention that the notion of atoms is tacitly assumed in a number of approximate treatments of molecules. The method of molecular mechanics is based on the picture of atoms linked by the elastic springs. Intermolecular interactions are most conveniently described by the pairwise atom-atom potentials. Pseudopotential theory of molecules, crystals and metals is deeply rooted in the concept of atoms imbedded in large systems. They are all successful within their limits of applicability.

The valence bond (VB) and molecular orbital (MO) theory in its LCAO approximation employ the notion of atoms by explicitly using atomic orbitals. Whereas VB approach retains its conceptual simplicity in its quantitative form [94, 128], refined MO theory with its extended basis sets and massive CI has some difficulties in recognizing atoms in very accurate molecular wavefunctions. One way out of this problem is the calculation of zero-flux boundary surfaces of Bader [9–13]. Another argument was put forward by Ruedenberg et al. [129, 130]. Their careful analysis has shown that it is possible to build basis sets which are quantitative in nature and yet they lend themselves to qualitative interpretation. Ruedenberg et al. infer that the concept of the minimal basis set is theoretically sound. Interestingly, the latter can be deduced from free-atom SCF AOs with slight readjustments. These modifications are relatively small but have substantial energetic impact. Needless to say, results of this analysis are in full accordance with the idea of modified atoms and offer additional support to the corresponding model.

To conclude, the MAM is a physically meaningful and theoretically well founded model which posses both explanatory and predictive power. It provides a reconciliation of the rigorous quantum theory and chemical experience.

Acknowledgement: A part of this work has been done at the Organisch-chemisches Institut der Universität Heidelberg and the author would like to thank Alexander von Humboldt-Stiftung for financial support and Professor R. Gleiter for hospitality.

5 References

1. Bošković R (1763) Theoria Philosophiae Naturalis, 2nd ed, Venice
2. Dadić Ž (1987) Ruder Bošković, Školska knjiga, Zagreb and the references cited therein
3. Feynman R, The Feynman Lectures on Physics, vol 1, Addison-Wesley, Reading
4. Lewis GN (1916) J. Am. Chem. Soc. 38: 762
5. Pauling L (1931) J. Am. Chem. Soc. 53: 1367
6. Pauling L (1960) The Nature of the Chemical Bond, 3rd ed, Cornell University Press, Ithaca, New York
7. Monkhorst HJ (1987) In: Maksić ZB (ed) Modelling of Structure and Properties of Molecules, Ellis Horwood, Chichester and the references given therein
8. Sutcliffe B (1990) Chapter 1 of this volume and the references cited therein
9. Bader RFW, Tal Y, Anderson SG, Nguyen-Dang TT (1980) Isr. J. Chem. 19: 8
10. Bader RFW, Nguyen-Dang TT (1981) Adv. Quant Chem. 14: 63
11. Bader RFW (1985) Acc. Chem. Res. 18: 9
12. Bader RFW, Tang TH, Tal Y, Biegler-Koenig FW (1982) J. Am. Chem. Soc. 104: 940, 946
13. Bader RFW, Slee TS, Cremer D, Kraka E (1983) J. Am. Chem. Soc. 105: 506
14. Cremer D, Kraka E, Slee TS, Bader RFW, Lau CDH, Dang TH, MacDougall PJ (1983) J. Am. Chem. Soc. 105: 5069
15. Wiberg K (1990) Chapter of this volume
16. Cremer D, Kraka E (1984) Coat Chem. Acta 57: 1259
17. Cremer D. Kraka E (1987) In: Maksić ZB (ed) Modelling of Structure and Properties of Molecules, Ellis Horwood, Chichester
18. Maksić ZB, Eckert-Maksić M, Rupnik K (1984) Croat. Chem. Acta 57: 1295
19. Maksić ZB (1988) J. Mol. Structure (Theochem) 170: 39
20. Maksić ZB (1989) In: Maruani J (ed) Molecules in Physics, Chemistry and Biology, vol 3, Kluwer Academic, Dordrecht
21. Ramsey NF (1950) Phys. Rev. 78: 699
22. Ramsey NF (1961) Am. Sci. 49: 509
23. Van Vleck JH (1932) Electric and Magnetic Susceptibilities, Oxford University Press, Oxford
24. Kovaček D, Maksić ZB (1989) Croat. Chem. Acta 62: 751
25. Maksić ZB, Bloor JE (1972) Chem. Phys. Lett. 13: 571
26. Blickensderfer RP, Wang JHS, Flygare WH (1969) J. Chem. Phys. 51: 3196; Benson RC, Tigelaar HL, Rock SL, Flygare WH (1970) J. Chem. Phys. 52: 5628
27. de Brouckere G, Broer R (1981) Mol. Phys. 43: 1139
28. Maksić ZB, Mikac N (1978) Chem. Phys. Lett. 56: 363; (1980) Mol. Phys. 40: 455
29. Scapinni F, Guarnieri A (1981) Z. Naturforsch. A36: 1393
30. Straub PA, McLean AD (1974) Theoret. Chem. Acta 32: 227
31. Pascal P (1910) Ann. Chim. Phys. 19: 5
32. Pacault A (1948) Rev. Sci. 86: 38
33. Tillieu J (1957) Ann. Phys. 2: 631
34. Fraga S, Malli G (1968) Many-electron systems, properties and interactions, Saunders, Philadelphia
35. Pople JA (1962) J. Chem. Phys. 37: 53
36. Amos RD (1987) In: Lawley KP (ed) Ab initio methods in quantum chemistry, Part 1, Adv. Chem. Phys. 67: 99
37. Jameson C (1990) In: Maksić ZB (ed) Theoretical models of chemical bonding, vol 3, Springer, Berlin New York Heidelberg
38. Buckingham AD (1962) J. Chem. Phys. 36: 3096
39. Amos RD (1978) Mol. Phys. 35: 1765
40. Huber KP, Herzberg G (1979) Constants of diatomic molecules, Van Nostrand Reinhold, New York
41. Amos RD (1980) Mol. Phys. 39: 1
42. Wolniewicz L (1966) J. Chem. Phys. 45: 515
43. Davis DW (1967) The theory of the electric and magnetic properties of molecules, Wiley, London
44. Brillouin L (1934) Actual. Scient. ind 71: 159; Møller C, Plesset MS (1934) Phys. Rev. 46: 618

45. Grimaldi F (1968) Adv. Chem. Phys. 14: 341; Green S (1974) Adv. Chem. Phys. 25: 179
46. Maksić ZB, Supek S (1988) Theoret. Chim. Acta 74: 275
47. Gombas P (1949) Die Statistische Theorie des Atoms und ihre Anwendungen, Springer, Vienna
48. Bonham RA, Strand TG (1963) J. Chem. Phys. 39: 2200
49. Fraga S, Karwowski J, Saxena KMS (1976) Handbook of atomic data, Physical sciences data 5, Elsevier, Amsterdam
50. Živković T, Maksić ZB (1968) J. Chem. Phys. 49: 3083; Živković T (1968) J. Chem. Phys. 49: 5019
51. Maksić ZB, Graovac A, Primorac M (1979) Croat. Chem. Acta 52: 265
52. Golebiewsky A, Mrozek J (1973) Int. J. Quant. Chem. 7: 623; (1973) ibid. 7: 1021
53. Maksić ZB (1986) Z. Naturforsch. 41a: 921
54. Maksić ZB, Kovačević K, Primorac M (1989) Pure Appl. Chem. 61: 2075
55. Primorac M, Kovačević K, Maksić ZB (1989) J. Mol. Structure (Theochem) 202: 75
56. Bingham RC, Dewar MJS, Lo DH (1975) J. Am. Chem. Soc. 97: 1285
57. Dewar MJS, Thiel W (1977) J. Am. Chem. Soc. 99: 4899
58. Hoffmann R (1963) J. Chem. Phys. 39: 1397; (1964) 40: 2480
59. Van der Voorn PC, Drago RS (1966) J. Am. Chem. Soc. 88: 3255; White WD, Drago RS (1970) J. Chem. Phys. 52: 4717
60. Kovačević K, Maksić ZB (1974) J. Org. Chem. 39: 539; Maksić ZB, Rubčić A (1977) J. Am. Chem. Soc. 99: 4233
61. Maksić ZB (1990) In: Maksić ZB (ed) Theoretical models of Chemical bonding, vol 2, Springer, Berlin New York Heidelberg
62. Ray NK, Parr RG (1980) J. Chem. Phys. 73: 1334
63. Parr RG, Gadre SR (1980) J. Chem. Phys. 72: 3669
64. Jorgensen CK (1989) Top. Curr. Chem. 150: 1 and the references cited therein
65. Clementi E, Raimondi DL (1963) J. Chem. Phys. 38: 2886
66. Snyder LC, Basch H (1972) Molecular wavefunctions and properties, Wiley, New York
67. Maksić ZB, Rupnik K (1983) Z. Naturforsch. A38: 308
68. Bloor JE, Maksić ZB (1971) Mol. Phys. 22: 351; (1973) ibid. 26: 397; Bloor JE, Maksić ZB (1975) In: Proc. Sec. Int. Symp. NQR Spectry, Pisa, P. 1
69. Graovac A, Maksić ZB, Rupnik K, Veseli A (1977) Croat. Chem. Acta 49: 695
70. Maksić ZB, Rupnik K (1983) Theoret. Chim. Acta 62: 397
71. Maksić ZB, Rupnik K (1983) Theoret. Chim. Acta 62: 219
72. Maksić ZB, Rupnik K (1986) J. Mol. Structure 141: 309
73. Ransil BJ (1960) Rev. Mod. Phys. 32: 245
74. Bieger W, Seifert G, Eschrig H, Grossman G (1985) Chem. Phys. Lett. 115: 275
75. Woodruff SB, Wolfsberg M (1978) 56: 125
76. King WT, Zelano AJ (1967) J. Chem. Phys. 47: 3197; Gaughan RR, King WT (1972) J. Chem. Phys. 57: 4530
77. King WT (1972) J. Chem. Phys. 57: 4535
78. King WT (1973) J. Phys. Chem. 77: 2770
79. Spackman MA, Maslen EN (1986) J. Phys. Chem. 90: 2020
80. Slater JC (1965) Quantum theory of molecules and solids, vol 2, McGraw-Hill, New York
81. Bonham RA, Fink M (1974) High energy electron scattering, Van Nostrand-Reinhold, New York
82. Trefry MG, Maslen EN, Spackman MA (1987) J, Phys. C, Solid State Phys. 20: 19
83. Spackman MA, Maslen EN (1986) Chem. Phys. Lett 126: 19
84. Parr RG, Gadre SR, Bartolotti LJ (1979) Proc. Nat. Acad. Sci. U.S.A. 76: 2522
85. Tal Y, Bader RFW and Erkku J (1980) Phys. Rev. A21: 1
86. Politzer P, Zilles BA (1984) In: Maksić ZB (ed) Conceputal quantum chemistry. Models and applications, Croat. Chem. Acta 57: 1055
87. Fliszar S (1983) Charge distributions and chemical effects, Springer, Heidelberg and the references therein
88. Jug K, Maksić ZB (1990) In: Maksić ZB (ed) Theoretical Models of chemical bonding, vol. 3, Chapter, Springer, Berlin Heidelberg New York
89. Maksić ZB (1986) Comp. & Math. with appl. 12B: 697

90. Cook DB, Hollis PC, McWeeny R (1967) Mol. Phys. 13: 553
91. Cook DB (1978) Structures and approximations for electrons in molecules, Ellis Horwood, Chichester
92. Honegger E, Heilbronner E, Schmelzer A (1982) Nouv. J. Chim. 6: 519
93. Eckert-Maksić M, Maksić ZB, Gleiter R (1984) Theoret. Chim. Acta 66: 193
94. Penotti F, Gerratt J, Cooper DL, Raimondi M (1988) J. Mol. Structure (Theochem) 169: 421
95. Luken WL, Beratan DN (1982) Theoret. Chim. Acta 61: 265; Luken WL (1984) Croat. Chem. Acta 57: 1283
96. Luken WL (1990) In: Maksić ZB (ed) Theoretical models of chemical bonding, vol. 2, Springer, Berlin Heidelberg New York
97. Magnusson E (1986) Chem. Phys. Lett. 131: 224; (1988) Austr. J. Chem. 41: 827 and the references cited therein
98. Coulson CA, Fischer I (1949) Philos. Mag. 40: 306
99. McWeeny R, Jorge FE (1988) J. Mol. Structure (Theochem) 169: 459
100. Bethe H (1929) Ann. Phys. 5: 133
101. Klopman G (1983) J. Mol. Structure (Theochem) 103: 121 and the references cited therein
102. Benson SW, Luria M (1975) J. Am. Chem. Soc. 97: 707; Benson SW (1978) Angew. Chem. 90: 868
103. Politzer P, Parr RG (1974) J. Chem. Phys. 61: 4258; Politzer P (1980) Isr. J. Chem. 19: 224
104. Mulliken RS (1955) J. Chem. Phys. 23: 1833
105. Maksić ZB, Rupnik K (1983) Z. Naturforsch. 38a: 308
106. Maksić ZB, Rupnik K (1983) Theoret. Chim. Acta 62: 219
107. Rittner ES (1951) J. Chem. Phys. 19: 1030
108. Little Jr EJ, Jones MM (1960) J. Chem. Ed. 37: 231
109. Siegbahn K, Nordling C, Johansson G, Hedman J, Heden PF, Hamrin K, Gelius U, Bergmark T, Werme LO, Manne R, Baer Y (1969) ESCA applied to free molecules, North Holland, Amsterdam
110. Maksić ZB, Rupnik K (1986) J. Mol. Structure 141: 309
111. Maksić ZB, Supek S (1989) J. Mol. Structure 198: 427
112. Matcha RL (1967) J. Chem. Phys. 47; 4595, 5295; (1968) ibid. 48: 335, 1264; (1970) ibid. 53: 485, 4490
113. Chesnyi AS, Topol IA, Rambidi NG (1980) Chem. Phys. 49: 107
114. Gierke TD, Tigelaar HL, Flygare WH (1972) J. Am. Chem. Soc. 94: 330; Flygare WH (1974) 74: 653
115. Bentley J (1981) In: Politzer P, Truhlar DG (eds) Chemical applications of atomic and molecular electrostatic potentials, Plenum, New York
116. Stone AJ (1981) Chem. Phys. Lett. 83: 233
117. Stone AJ (1990) In: Maksić ZB (ed) Theoretical models of chemical bonding, vol. 3, Springer, Berlin Heidelberg New York
118. Tomasi J (1981) In: Politzer P, Truhlar DG (eds) Chemical applications of atomic and molecular electrostatic potentials, Plenum, New York
119. Bonaccorsi R, Pullman A, Scrocco E, Tomasi J (1972) Theoret. Chim. Acta 24: 51; Bonaccorsi R, Scrocco E, Tomasi J, Pullman A (1972) Theoret. Chim. Acta 24: 51
120. Goldblum A, Pullman B (1978) Theoret. Chim. Acta 47: 345
121. Weinstein H, Osman R, Green JP, Topiol S (1981) In: Politzer P, Truhlar DG (eds) Chemical applications of atomic and molecular electrostatic potentials, Plenum, New York
122. Hadži D, Hodoščєk M, Kocjan D, Šolmajer T, Avbelj F (1984) Croat. Chem. Acta 57: 1065
123. Buckingham AD, Fowler PW (1985) Can. J. Chem. 63: 2018
124. Rendell APL, Bacskay GB, Hush NS (1985) Chem. Phys. Lett. 117: 400
125. Hirschfelder JO (1982) Ber. Bunsenges. Phys. Chem. 86: 349
126. Rhee CH, Metzger RM, Wiygul FM (1982) J. Chem. Phys. 77: 899
127. Schulman JM, Disch RL (1985) Chem. Phys. Lett. 113: 291
128. McWeeny R, Jorge FE (1988) J. Mol. Structure (Theochem) 169: 459
129. Ruedenberg K, Schmidt MW, Gilbert MM, Elbert ST (1982) Chem. Phys. 71: 41, 65
130. Ruedenberg K, Schmidt MW, Gilbert MM (1982) Chem. Phys. 71: 51

Bei Fragen zur Produktsicherheit wenden Sie sich bitte an:
If you have any questions regarding product safety,
please contact:

Walter de Gruyter GmbH
Genthiner Straße 13
10785 Berlin
productsafety@degruyterbrill.com